We Are Our Brains

We Are Our Brains

A NEUROBIOGRAPHY OF THE BRAIN, FROM THE WOMB TO ALZHEIMER'S

D. F. Swaab

Translated by Jane Hedley-Prôle

SPIEGEL & GRAU
NEW YORK

Published in the United States by Spiegel & Grau, an imprint of
Random House, a division of Random House LLC,
a Penguin Random House Company, New York.

The publisher gratefully acknowledges the support of the
Dutch Foundation for Literature.

Library of Congress Cataloging-in-Publication Data

Swaab, D. F. (Dick Frans)
We are our brains: a neurobiography of the brain, from the womb to Alzheimer's /
D. F. Swaab ; translated by Jane Hedley-Prôle.
pages cm
Includes index.
ISBN 978-0-8129-9296-0
eBook ISBN 978-0-679-64437-8
1. Brain. 2. Brain—Research. 3. Neurosciences. I. Title.
QP376.S858 2014
612.8'2—dc23 2013020412

Printed in the United States of America on acid-free paper

www.spiegelandgrau.com

2 4 6 8 9 7 5 3 1

First U.S. Edition

Book design by Caroline Cunningham

To all the scientists who stimulated my brain so intensely, and to Patty, Myrthe, Roderick, and Dorien, who formed my enriched environment at home

Many of the views which have been advanced are highly speculative, and some no doubt will prove erroneous; but I have in every case given the reasons which have led me to one view rather than to another. . . . False facts are highly injurious to the progress of science, for they often endure long; but false views, if supported by some evidence, do little harm, for every one takes a salutary pleasure in proving their falseness.

Charles Darwin, *The Descent of Man*

Contents

Illustrations

Preface: Questions About the Brain to a Supposed Expert

> I know full well that the reader has no great desire to know all this, but I have the desire to tell them of it.
>
> Jean-Jacques Rousseau

Perhaps the two greatest scientific questions of this century are "How did the universe come into being?" and "How does the brain work?" Through a combination of my surroundings and chance, I became fascinated by the second question.

I grew up in a household where I overheard such enthralling conversations about every aspect of medicine that it became impossible to escape going into the profession. My father was a gynecologist who devoted his working life to many aspects of reproduction that were then highly controversial, like male infertility, artificial insemination, and the contraceptive pill. He received a stream of visits from friends whom I only later realized were pioneers in their fields. As a small child I got my first lessons in endocrinology from Dries Querido, who later set up Rotterdam's medical faculty. When I noticed our family dog cock his leg against a tree as we took it for a walk, Professor Querido explained that his behavior was caused by the effect of sex hormones on the brain. Coen van Emde Boas, the first Dutch professor of sexology, used to drop by in the evenings with his wife for a drink. His stories were gripping, particularly for a small boy. I recall an anecdote he told about a patient with whom he was having trouble communicating. Finally the man came out with what had been bothering him: He had heard that Van Emde Boas was a homosexual! Van Emde Boas put an arm around his shoulders and said, "But my dear, surely you don't believe that?" We laughed uproariously when he described the look on the patient's face. It was a household in which there was no question you couldn't ask, and during the weekends my father let me look at his medical books and peer through his microscope at plant cells and unicellular creatures fished out of local ditches.

When I was in secondary school, my father took me with him on a lecture tour of the country. I will never forget the hostile response he got from very religious members of the audience when he lectured on the contraceptive pill, which was going to be tested in the Netherlands for the first time. Despite the insults hurled at him, he went on arguing his case, remaining outwardly calm, while I sat sweating next to him, in an agony of embarrassment. In retrospect it was good training for the extremely heated reactions that my own research would later spark. An occasional visitor to our house at that time was Gregory Pincus, the developer of the contraceptive pill. I got my first sight of a laboratory when I was taken along with him on a visit to Organon, the pharmaceutical factory where the pill was produced.

With such a background it seemed to me self-evident that I would study medicine. At meals I would enthusiastically discuss medicine with my father so directly and in such detail that my mother would regularly beg us to stop, even though, having worked as an operating theater nurse and at the front during the war between Russia and Finland in 1939, she was hardly squeamish.

As my medical studies progressed, I was taken aback to discover that I was no longer expected merely to ask questions but also to provide answers. Suddenly everyone around me saw me as an expert on all kinds of health problems and expected free consultations. At one point I was so fed up with hearing about my aunt's nagging pains that I silenced an entire family birthday party with a loud, "How interesting, Aunt Jopie! Take your clothes off, and let's have a look at it!" It worked like a charm; she never bothered me again. The other relatives, though, proved harder to deter, and they continued to seek my advice.

During my studies I wanted to find out more about the experimental work that underpins medical knowledge. I also wanted to pay my own way through school, an idea to which my parents were very much opposed. There were two places in Amsterdam that would take you on as a part-time researcher once you got a basic medical degree after three years of study: the Department of Pharmacology and the Netherlands Institute for Brain Research. The first vacancy to come up just happened to be at the latter. So much for career planning. At the Institute for Brain Research I chose the logical area of focus, given my background: the new field of neuroendocrinology, involving research into hormone production by brain cells and the effect of hormones on the brain. During my interview with the institute's director, Hans Ariëns Kappers, I mentioned that my main interest lay in neuroendocrinology, so the in-house expert, Hans Jongkind, was called in. He revealed my considerable ignorance on the subject with a series of probing questions. But to my surprise, Hans Kappers then said, "Well, we'll give you a trial," and I got the job.

For my PhD I carried out experiments to determine the functions of hormone-producing neurons. I did this research while continuing to study medicine, which kept me busy around the clock. It was with some difficulty that in 1970, while working as an intern in a surgical department, I obtained the afternoon off to defend my PhD thesis. After qualifying as a general physician in 1972, I decided to continue with brain research. In 1975 I became deputy director of the Netherlands Institute for Brain Research (see chapter 15) and in 1978 its director. In 1979 I also took on the chair of neurobiology at the University of Amsterdam's medical faculty. Despite holding these administrative posts for thirty years, hands-on research remained my main focus. After all, that was why I chose the profession. In my research group I've profited immensely from the knowledge of a host of talented and critical students, PhD students, postdocs, and staff members from over twenty countries, whom I still encounter in brain research centers and clinics all over the world; it's a learning process that continues to this day. In turn, all of those academics owe much to the excellent technicians who have developed and perfected new research skills.

Meanwhile, I was getting more and more requests for information, including about issues outside of my own field. As a doctor, even if you're engaged in research rather than general practice, people will come to you with pressing questions. There's no aspect of life that brain disorders don't touch, so my advice was solicited on the most serious issues. One Sunday morning, for instance, the son of an acquaintance knocked on my door with a few scans under his arm. He said, "I've just been told that I only have three months to live—how can that be?" When I looked at the scans, it seemed incredible that he'd even been able to visit me and ask this question: The front of his brain was one giant tumor, and indeed he did die shortly afterward. At such times all I can do is listen, explain the results of tests, and help desperate people to find their way through the medical jungle. The only accurate judges of my capacities were my children, who, when they were feverish and saw me appear at their bedside with a worried air and a stethoscope, would always insist that they wanted a "real" doctor.

When I set up the Netherlands Brain Bank in 1985 (see chapter 19) and became known for examining the brains of dead people, I was again surprised to find myself the focus of a great many questions on every subject relating to the final stage of life: euthanasia, assisted suicide, and the donation of brains and bodies to science; in short, everything connected to life and death (see chapter 19). My research became intertwined with personal and social issues relating to my field. I spoke with the courageous mothers of schizophrenic children who had committed suicide; they responded by setting up a national support network. At international conferences on Prader-Willi syndrome, I also discovered how often the relatives know more than the researchers do. At those conferences, the parents brought their morbidly obese children from every corner of the globe, taught researchers a great deal about the condition, and encouraged us greatly in our attempts to find out why sufferers eat themselves into an early grave. Interacting directly with researchers is an approach that more patient groups should emulate.

My research group was also involved in setting up the first study on Alzheimer's disease in the Netherlands, at a time when the epidemic proportions of the disease were merely conjecture. Our observation that some brain cells could withstand the aging process and Alzheimer's unscathed while others were destroyed was important in guiding our research of therapeutic strategies for the disorder (see chapter 18). Demographic aging means that many of us now have to see loved ones deteriorate during the last stage of life as a result of dementia. Most of us also experience the huge pressure that psychiatric disorders impose on the lives of patients, relatives, and carers. The questions about these conditions that I am confronted with as a brain researcher go so deep that I can't avoid them.

The general public, blissfully unaware of our daily battles with the technical problems of research, assumes that we know everything about the brain. People want answers to the big questions about memory, consciousness, learning and emotion, free will, and near-death experiences. As a researcher, if you don't fend off those types of queries, you'll get sucked into them sooner or later and discover that they are riveting.

Discussions about the brain always reveal how firmly convinced people are of "facts" whose origins are a mystery to me. Take the myth that we use only 10 percent of our brains. You might well be forgiven for thinking this in the case of certain people, but I haven't the faintest idea what prompted this crazy theory. The same goes for the claim that millions of our brain cells die off every day. But lack of expertise can also be refreshing. When I give lectures, I often get asked intriguing questions by members of the audience. Sometimes even children raise the most thought-provoking issues. One Dutch girl of Japanese origin wanted to write a school assignment on the difference between European and Asian brains—differences that do exist but are rarely acknowledged.

On top of general inquiries, I also had to contend over the years with an avalanche of questions and a heated public response to my own research on the human brain, prompting the need for explanation and debate on issues like gender-based differences in the brain, sexual orientation, transsexuality, brain development, and brain disorders like depression and eating disorders (see chapters 1, 2, 3, and 5).

In the forty-five years that I've been active in this field, brain research has developed from the preserve of a few mavericks into a global discipline that has rapidly brought a host of new insights, thanks to the efforts of many tens of thousands of researchers representing many different disciplines and employing highly diverse techniques. The general neurophobia of the old days has been transformed into an overwhelming fascination with everything connected to the brain, partly thanks to excellent science journalism. Unable to escape the questions that the public was posing, I found myself constantly compelled to step aside from my own line of research to think about every conceivable aspect of the brain and how it could all be explained to a general readership. In this way I developed my own views on features of the brain and our emergence as humans, on the way in which we develop and age, on the origins of brain disorders, and on life and death. Over the course of time my own little answers to and ideas about the big brain questions took shape. They are set out in this book.

The question I am most frequently asked is whether I can explain how the brain works. That's a conundrum that has yet to be fully solved, and this book can of course provide only a partial answer. It shows how our brains differentiate into male and female brains, what goes on in the adolescent mind, how the brain preserves the individual and the species, how we age, how we suffer from dementia and die, as well as how the brain evolved, how memory works, and how moral behavior developed. But the book also shows how things can go wrong. It looks at disorders of consciousness, brain damage caused by boxing, and diseases of the brain like addiction, autism, and schizophrenia, as well as the latest medical advances and possibilities for recovery. Finally, it looks at the relationship between the brain and religion, the soul, the mind, and free will.

The various sections of this book can be read separately. In such a short space that deals with so many different subjects, it's impossible to arrive at deep scientific conclusions. These thoughts are intended

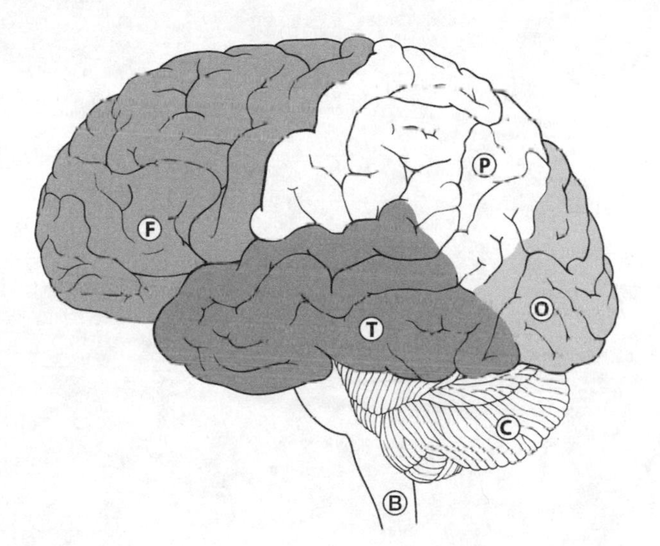

FIGURE 1. The brain seen from the side, facing left, with the parts of the cerebral cortex labeled. F is the frontal lobe (planning, initiative, speech, motor system), which contains the primary motor cortex (fig. 22). P is the parietal lobe, which contains the primary sensory cortex (fig. 22) and integrates sensory information (sight, touch, navigation). This part of the brain is also used for reasoning and calculation, and stores information on the significance of numbers as well as an inner body map. O is the occipital lobe (visual cortex). T is the temporal lobe (memory, hearing, language; fig. 21). At the base is C, the cerebellum (automatic movements and coordination), and B, the brain stem (regulates breathing, heartbeat, temperature, waking and sleeping).

to prompt further debate about why we are as we are, how our brains develop and function, and how they can malfunction. I hope that this book can provide a general readership with answers to many frequently asked questions about the brain. I also hope it will give students and young brain researchers an introduction to neuroculture, encouraging them to cross the borders of their research and engage in dialogue with the general public. That this is necessary is self-evident, not just in view of the social impact of brain research but also because of the support that we, for our part, ask from society for our research.

We Are Our Brains

Introduction

WE ARE OUR BRAINS

> It should be widely known that the brain, and the brain alone, is the source of our pleasures, joys, laughter, and amusement, as well as our sorrow, pain, grief, and tears. It is especially the organ we use to think and learn, see and hear, to distinguish the ugly from the beautiful, the bad from the good, and the pleasant from the unpleasant. The brain is also the seat of madness and delirium, of the fears and terrors which assail us, often at night, but sometimes even during the day, of insomnia, sleepwalking, elusive thoughts, forgetfulness, and eccentricities.
>
> Hippocrates

Everything we think, do, and refrain from doing is determined by the brain. The construction of this fantastic machine determines our potential, our limitations, and our characters; *we are our brains*. Brain research is no longer confined to looking for the cause of brain disorders; it also seeks to establish why we are as we are. It is a quest to find ourselves.

The brain is built from nerve cells called neurons. Weighing around three pounds, the brain contains 100 billion neurons (fifteen times the number of people on earth). And the neurons are outnumbered ten to one by glial cells. It was formerly thought that they were only there to hold neurons together (*glia* comes from the Greek word for "glue"). But recent studies show that these cells, of which humans possess more than any other organism, are crucial to the transfer of chemical messages and therefore to all brain processes, including the formation of long-term memory. That sheds interesting light on the finding that Einstein's brain contained unusually many glial cells.

The product of the interaction of all these billions of neurons is "mind." Just as kidneys produce urine, the brain produces mind, as Jacob Moleschott (1822–1893) so inimitably put it. But now we know what this process actually entails: electrical activity, the release of chemical messengers, changes in cell contacts, and alterations in the activity of nerve cells (see above and chapter 14). Brain scans are used not only to trace diseases of the brain but also to show which areas light up during different activities, so that we know which parts we use to read, think, calculate, listen to music, have religious experiences, fall in love, or become sexually excited. By observing changing patterns of activity in your own brain, you can train it to function differently. With the aid of a functional scanner, for instance, patients suffering from chronic pain can be coached to control activity in the front of the brain, thereby reducing their pain.

Malfunctions in this efficient information-processing machine cause psychiatric and neurological disorders. Paradoxically, these disorders tell us much about the way in which the brain normally functions. Effective therapies have already been devised for some of these conditions. Parkinson's disease has been treated with L-dopa for a long time now, and combination therapy for AIDS now staves off dementia. Genetic and other risk factors for schizophrenia are being rapidly charted: Under the microscope you can see that brain development in schizophrenia sufferers is impaired before they are even born. Schizophrenia can now be treated with medication.

Until recently, neurologists could do little more than pinpoint the exact location of the brain defect you were stuck with for life. Nowadays, the clots that cause strokes can be broken down, hemorrhages stanched, and stents inserted into clogging arteries. Over 3,500 people have donated their brains to the Netherlands Brain Bank (www.brainbank.nl), leading to new insights in the molecular processes that cause diseases like Alzheimer's, schizophrenia, Parkinson's, multiple sclerosis, and depression. The search for new approaches to medication is also in full swing. But research of this kind will only bear clinical fruit for the next generation of patients.

Stimulation electrodes, implanted at exactly the right spot inside the brain, are proving effective. They were first tried on patients with Parkinson's disease (fig. 23). It's impressive to see how violent tremors suddenly disappear when the patients themselves press the button of the stimulator. Depth electrodes are already being used to treat cluster headaches, muscle spasms, and obsessive-compulsive disorder. They help patients who had previously washed their hands hundreds of times a day to lead a normal life. A depth electrode was even used to revive someone who had spent six years in a minimally conscious state. Attempts are being made to treat obesity and addiction with depth electrodes. As always, it takes a while before not only the effects but also the side effects of a new therapy come to light—as is now happening with deep brain stimulation (see chapter 11).

Magnetic stimulation of the prefrontal cortex (fig. 15) has been successfully used to treat depression, and stimulation of the auditory cortex silences the incredibly annoying tunes that can suddenly start playing in the heads of people with inner ear hearing loss. Transcranial magnetic stimulation (see chapter 10) has proved effective in treating hallucinations provoked by schizophrenia.

Neuroprostheses are getting better and better at replacing our senses. At present, over one hundred thousand people have cochlear implants that enable them to hear surprisingly well. Trials are being concluded with electronic cameras that transmit information to the visual cortex (fig. 22) of blind patients. A tiny square with ninety-six electrodes was implanted in the cerebral cortex of a twenty-five-year-old man who had become completely paralyzed after being stabbed in the neck. Merely by thinking of movements he could use a computer mouse, read his email, and play electronic games. The power of thought has even been used to control a prosthetic arm (see chapter 11).

Attempts are being made to carry out cerebral repairs by transplanting pieces of fetal brain tissue into the brains of patients with Parkinson's disease and Huntington's disease. Gene therapy is already being tested on people with Alzheimer's. Stem cells appear highly suitable for repairing brain tissue, but considerable problems, like the possible growth of tumors, still need to be overcome (see chapter 11).

Disorders of the brain are still very difficult to treat, but the era of defeatism has given way to excitement at new insights and optimism about new methods of treatment in the near future.

METAPHORS FOR THE BRAIN

Throughout the ages people have tried to articulate the brain's function in terms of the latest technological advances. In the fifteenth century, for instance, during the Renaissance, at a time when printing was being developed in Europe, the brain was described as "a book containing everything" and language as "a living alphabet." In the sixteenth century the working of the brain was compared to a "theater in the head," and a parallel was also drawn between the brain and a cabinet of curiosities, or a museum in which you could store and view everything. The philosopher Descartes (1596–1650) regarded the body and the brain as a machine, famously comparing the brain to a church organ. He likened the air pumped into the organ by the bellows to the subtlest and most active particles in the blood, "the animal spirits," which he thought were pushed into the cavities of the brain via a system of blood vessels (now known as the choroid plexus). Hollow nerves then transported the animal spirits to the muscles. The pineal gland played the part of the keyboard. It could direct the animal spirits into "certain pores," just as an organist can direct air into certain pipes by pressing a particular key. As a result Descartes has gone down in history as the founder of mind-body dualism, a school of thought that bears his Latinized name: Cartesianism. The ancient Greeks, however, should be credited as the real inventors of dualism, as they already distinguished between body and spirit.

If you regard the brain as a rational, information-processing, organic machine, then the computer metaphor of our time isn't such a bad one. It's a comparison that's hard to avoid, especially if you consider the impressive figures about our brains' building blocks and their connections. There are 1,000 times 1,000 billion points at which neurons connect with one another—or, as Nobel Prize winner Santiago Ramón y Cajal put it, "hold hands"—through junctions called synapses. The neurons are linked by over sixty thousand miles of nerve fiber. The staggering number of cells (see above) and contacts works so efficiently that a typical brain's energy consumption is equal to that of a fifteen-watt lightbulb. Neuroscientist Michel Hofman has calculated that the total energy bill of a single brain during an entire lifetime of eighty years wouldn't exceed $1,500 at today's rates. You certainly couldn't get a decent computer for that price, nor would it last anywhere near as long. For a mere fifteen hundred dollars you can power a billion neurons for your entire lifetime! And your skull comes fitted with a fantastically efficient machine with parallel circuits that can process images and associations better than any computer yet built. It's always an awe-inspiring moment when you carry out a postmortem and hold a person's brain in your hands. You're conscious that you're holding someone's entire life. Of course, you're also immediately aware of how very "soft" the "hardware" of our brain actually is. This gelatinous mass contains everything that this person thought and experienced, coded and recorded in structural and molecular changes to the synapses.

A better metaphor comes to mind when you visit the underground command center in the heart of London where, starting in 1940, Winston Churchill led his war cabinet and a huge staff night and day in efforts to defeat Adolf Hitler. The war rooms are covered in maps displaying all the information (coded and uncoded) that came in from a vast network of lines of communication spanning the globe. Priority was given to the most up-to-date reports, which were checked, evaluated, processed, and stored by a host of well-coordinated departments. Using the information selected (by the front part of the brain, the prefrontal cortex, fig. 15) a draft plan was drawn up, elaborated, and tested based on assessment of all available data. Constant consultations were carried out with an army of specialists, both internal and external, connected by a direct link with America. After weighing all the opinions and information, a plan was either given the green light or shelved. Plans could be carried out by the army (the motor functions), the navy (hormones), units operating stealthily behind the lines (the autonomic nervous system), or the air force (neurotransmitters, cleverly targeting a single brain structure). Most effective of all, of course, was a coordinated operation involving all branches of the service. Yes, our brains are much more like a complicated command center, equipped with the latest apparatus, than a telephone switchboard or a computer with simple one-on-one connections. The command center is engaged in a lifelong battle, first to be born, then to pass exams, to obtain some means of subsistence, to fight off competition, to survive in a sometimes hostile environment, and ultimately, to die as one would wish. It's protected not by the bombproof concrete of Churchill's underground headquarters but by a skull strong enough to survive some very hard knocks. Churchill himself hated his shelter and would stand on the roof during air raids, following the action. He was happy to take risks, an innate quality of some brains.

One could also think of more peaceful metaphors, like the air traffic control center of a large airport. But if you look back at all the metaphors of previous centuries, you realize that you're simply comparing the most recent technology devised by our brains to the brain itself. Indeed, there appears to be nothing more complex than that fantastic machine.

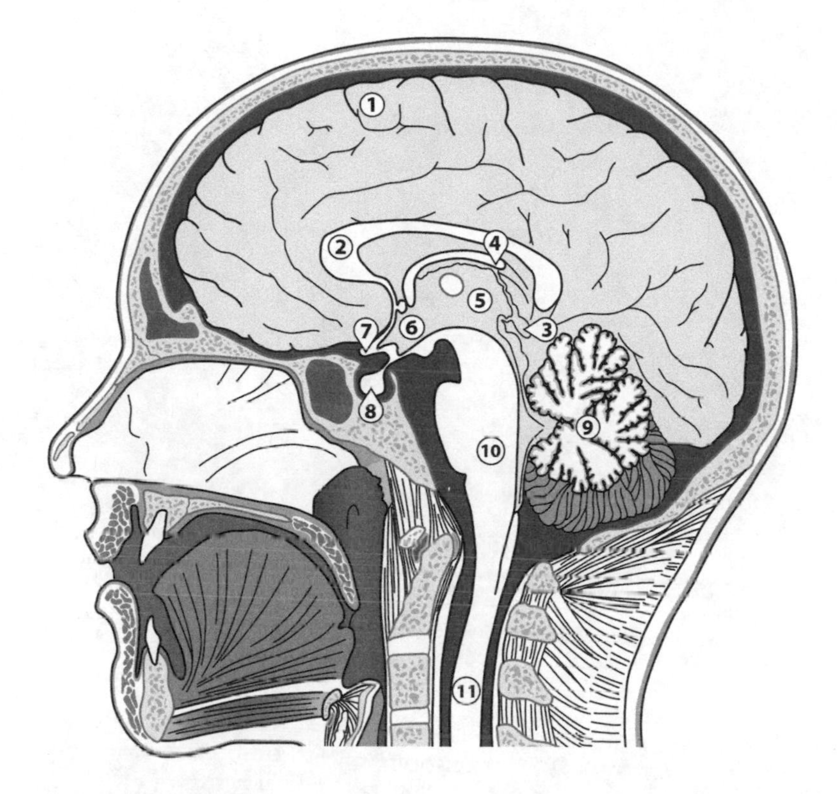

FIGURE 2. Cross section of the brain. (1) The cerebrum, with the convoluted cerebral cortex. (2) The corpus callosum, a bundle of neural fibers connecting the left and right sides of the brain. (3) The pineal gland, which at night secretes the sleep hormone melatonin, a substance that also inhibits puberty in young children. (4) The fornix, which transports memory information from the hippocampus to the mammillary bodies at the base of the hypothalamus (fig. 25), after which it travels on to the thalamus (5) and cortex (1). (5) The thalamus, where sensory information and memory information are sent. (6) The hypothalamus, which is crucial to the survival of the individual and the species. (7) The optic chiasma, where the optic nerves cross. (8) The pituitary gland. (9) The cerebellum. (10) The brain stem. (11) The spinal cord.

1

Development, Birth, and Parental Care

THE SUBTLE INTERACTION BETWEEN MOTHER AND CHILD AT BIRTH

Birth is too important to be left entirely to your mother.

I congratulate my mother on this anniversary of her suffering and thank her for bringing me into the world.

Text message on the occasion of her birthday
from a Chinese daughter to her mother

Somebody once suggested that I went in for brain research because my father was a gynecologist, the theory being that I chose the organ that lay as far as possible from his field of work. This psychoanalytical explanation fails to take account of the research I've done in the function of the brains of both mother and child during childbirth, working with gynecologists like Kees Boer at Amsterdam's Academic Medical Center. The conclusion of Boer's PhD thesis was that smooth childbirth requires good interaction between the brains of mother and child.

Both brains play a role in speeding up labor by secreting a hormone, oxytocin, into the bloodstream that makes the uterus contract. The mother's biological clock imposes a day-night rhythm on the birth process, which explains why most children are born during the quiet phase, at night and in the early hours of the morning. That's also the time when birth progresses fastest and requires the least assistance.

Labor starts when the baby's blood sugar level starts to drop—a sign that the mother can no longer provide the growing child with sufficient nourishment. Michel Hofman has calculated that labor is triggered at a stage when the child accounts for around 15 percent of the mother's metabolism. That point is reached earlier by twins, triplets, and so on, which is why they are born prematurely. While still in the womb, the brain cells in the child's hypothalamus respond to a drop in blood sugar level in the same way that they later respond to a lack of food in adulthood, by stimulating the stress axis. This induces a series of hormonal changes, making the uterus contract (fig. 3). The contractions, stimulated by oxytocin, make the baby's head press against the cervix. This in turn triggers a reflex, via the mother's spinal cord, which causes the release of yet more oxytocin. The baby's head then exerts more pressure, triggering the same reflex. The child can only escape from this chain reaction by being born.

Various psychiatric disorders are associated with a difficult birth. It has long been known that a high percentage of patients with schizophrenia experienced problems at birth, such as delivery by forceps or vacuum pump, low birth weight, premature birth, premature breaking of the waters, or time spent in an incubator. It was once thought that difficult births caused brain damage, leading to schizophrenia. We now know that schizophrenia is an early developmental brain disorder largely caused by genetic factors (see chapter 10). So a difficult birth can be seen as a failure of interaction between the brains of mother and child and thus as the first symptom of schizophrenia, even though the disease doesn't develop fully until puberty. The same applies to autism, another early developmental brain disorder (see chapter 9) that also often goes hand in hand with birth-related problems. Recent studies have shown that girls who suffer from the eating disorders anorexia and bulimia nervosa often had problems at birth, including low birth weight. The more numerous such problems are, the earlier eating disorders manifest themselves in young adults. One wonders whether their hypothalami started out unable to deal well with glucose levels, given that a decrease in them signals the start of labor. So here, too, birth-related problems of this type could be seen as the first symptoms of a malfunction of the hypothalamus, later taking the form of an eating disorder.

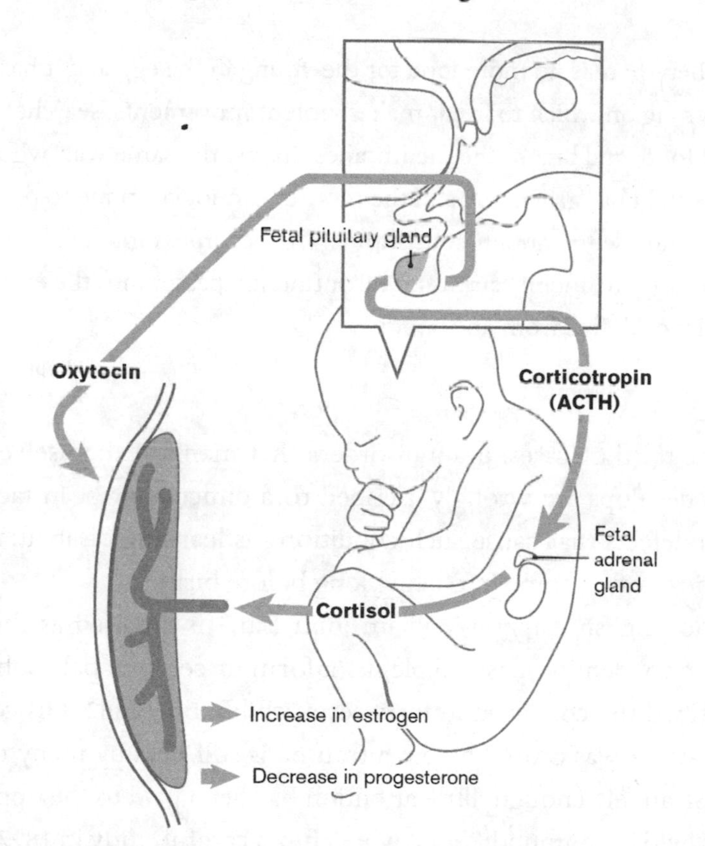

FIGURE 3. When the growing fetus registers that the mother can no longer provide sufficient nourishment, the stress axis in its hypothalamus is activated. Its adrenal gland is stimulated by ACTH to produce cortisol. This reduces the effect of progesterone from the placenta and causes estrogen production to increase. As a result, the uterus becomes more sensitive to oxytocin, activating labor pains and starting the birth process.

A child with a developmental brain disorder can't fulfill its essential role during labor. Extremely delicate interaction is needed between mother and baby for birth to proceed well. Thinking about birth in this way takes a bit of getting used to because it means that a child's say in life starts right from birth.

A DIFFICULT BIRTH AS THE FIRST SYMPTOM OF A DEVELOPMENTAL BRAIN DISORDER

> When there is no more food for the young in the egg and it has nothing on which to live it makes violent movements, searches for food, and breaks the membranes. In just the same way, when the child has grown big and the mother cannot continue to provide him with enough nourishment, he becomes agitated, breaks through the membranes, and incontinently passes into the external world, free from any bonds.
>
> Hippocrates

In one-third of cases, brain disorders that manifest themselves as a child develops are wrongly ascribed to a difficult birth. In fact, the brain defects that cause such conditions as learning disabilities and spasticity often come into being long before birth.

The English surgeon William John Little is credited as the first person to identify spastic diplegia (a form of cerebral palsy), having described the condition in forty-seven children in 1862. His conviction that it was caused by birth trauma is still held by many to this day. Strangely enough, little attention has been paid to the opposing view held by Sigmund Freud, who, after a careful study in 1897, concluded that a difficult birth couldn't cause spasticity but that both the neurological condition and the difficult birth should be seen as the consequence of a developmental disorder of the fetal brain. Problems at birth are often also blamed in the case of children with learning disabilities. Prader-Willi syndrome is a genetic disorder that causes morbid obesity over the course of time (see chapter 5). Many children with this syndrome have a difficult birth and go on to have learning disabilities. These aren't caused by birth-related problems but by the genetic abnormality that was present from conception.

In only 6 percent of children born at due date with spasticity and a mere 1 percent of children with learning disabilities can the disorder be attributed to a lack of oxygen at birth. The vast majority of children with these conditions experience problems long before birth, as is evident from their slow growth and lack of movement in the womb. Spasticity has many different causes, ranging from genetic abnormalities and intrauterine infections to exposure to chemicals, iodine deficiency, and long-term oxygen deficiency in the womb. Conversely, it's striking that serious brain damage often does not result when a normal fetus is suddenly deprived of oxygen at birth, as Freud already noted. To understand why he was right, you need to be aware of the active role played by the fetus in the birth process. Its brain plays a crucial part, both in initiating labor and during its course. The relationship between a difficult labor and impaired brain function is usually the opposite of what is generally assumed. A difficult labor or premature or delayed labor tends to be the *consequence* of a problem in fetal brain development. And that deficiency can in turn be caused by genetic factors, lack of oxygen in the womb, infections, or exposure to medication or addictive substances ingested by the mother, like morphine, cocaine, or nicotine. So efforts to establish the cause of premature or difficult birth are incomplete without examining the child's brain.

That a child's brain plays a very active role in labor is something we established thirty-five years ago, in a study with the gynecologist W. J. Honnebier. We looked at the births of 150 anencephalic infants (children born with most of their brain missing, fig. 4). Babies with this condition are usually born extremely prematurely or very overdue, and labor proceeds much more slowly than normal. That birth takes twice as long (and the birth of the placenta three times as long) is due to the absence of oxytocin in the child's brain. Half of these babies don't survive, which shows how important a well-functioning fetal brain is to the process of labor. Another hormone secreted by the fetal brain, vasopressin, ensures that blood is mainly directed to those organs that are crucial to survival during birth, like the heart, the adrenal gland, the pituitary gland, and the brain. This involves depriving less vital areas like the intestines. Animal studies have revealed the many complex chemical steps that are necessary for the birth process. But it all starts when a child's brain registers that the maternal food supply is becoming inadequate, causing it to give the signal for the onset of labor, just as the Greek doctor and philosopher Hippocrates noted over two thousand years ago.

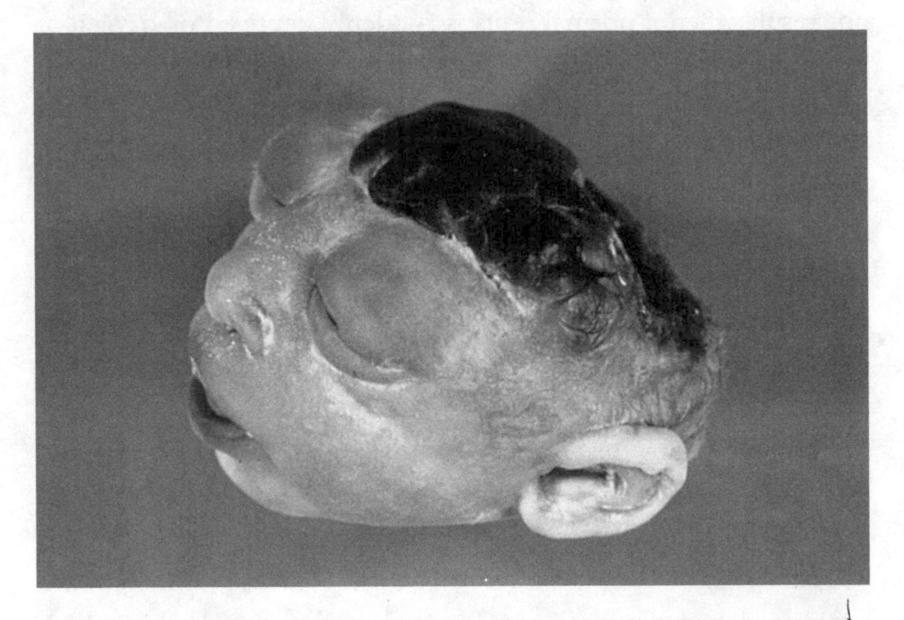

FIGURE 4. An anencephalic newborn, a child born with most of its brain missing. To establish whether the mother's brain or the child's brain initiated the birth process, we studied the birth of 150 anencephalic babies. Their births were usually either extremely premature or extremely overdue, the normal precise timing of birth (around the fortieth week of pregnancy) having been completely disrupted. The birth process itself took twice as long as normal, and the placenta took three times as long to appear. In normal circumstances, a baby's brain determines the moment of birth and speeds up the birth process.

MATERNAL BEHAVIOR

> Behold two horses that appear of the same size and shape: How do you know which is the mother and which the son? Give them hay. The mother will nudge the hay toward her son.
>
> *The Teachings of Buddha*

A woman's brain starts being programmed for maternal behavior right from the onset of pregnancy. Hormones alter the brain in ways that are reinforced after the birth by the presence of the child. The changes in the mother's brain are long-term, perhaps even permanent. Adults sometimes lament that the umbilical cord attaching them to their mothers has never really been cut, while mothers can be burdened by concern about their offspring well after they've grown up. If something happens to their child, some will claim they felt a premonition the previous day—which is true, for the simple reason that they worry about their children on a daily basis.

During pregnancy the pituitary gland secretes the hormone prolactin, which prompts nesting behavior. There's an urge to clean the house and paint the baby's room. Once, when I went to check on my laboratory rats as a PhD student, I was sure that someone had swapped my cages with adult male rats for ones with pregnant rats. Each of the rats had built an enormous nest of sawdust. They were, in fact, my male rats—they had built the nests because the previous day I'd given them a dose of prolactin. Similarly, during his stay in Amsterdam's Wilhelmina Gasthuis hospital, a male patient with a pituitary gland tumor that produced prolactin was never happier than when helping staff clean the bedside lockers.

At the end of pregnancy, both the mother's and the child's brain cells produce oxytocin and release it into the bloodstream. This hormone has many functions. Doctors use oxytocin to induce labor, and some women are given nasal spray containing oxytocin to boost milk release after giving birth. Its role at the end of pregnancy is to stimulate and speed up labor. The mother's brain secretes more oxytocin at night, when the uterus is most sensitive to the hormone, which encourages labor to start while the body is at rest. During labor, extra oxytocin is released when the child's head presses down on the cervix. This signal is passed on to the mother's brain via the spinal cord, triggering the release of more oxytocin to heighten labor. If a woman is given an epidural to offset labor pains, the signal no longer reaches her brain, causing her pituitary gland to release less oxytocin. In such cases women often need an oxytocin drip to restore the strength of their contractions.

After birth, oxytocin ensures that the mother secretes enough milk. When the child sucks at the nipple, it stimulates its mother's brain to release oxytocin, causing milk to be expressed from the mammary gland. After a time, it's sufficient for the child to cry to trigger this reflex. Such a strong dose of oxytocin is produced that milk spurts from the breasts—a potentially embarrassing state of affairs in company. Farmers have known about this reflex for centuries, having seen how milk will spurt from the cow's udders when they come into the cowshed rattling a milk bucket.

Studies are increasingly revealing the importance of oxytocin in many forms of social interaction. One of the new names given to it is the "bonding hormone" because of the tie it creates between mother and child. This process starts in late pregnancy, when oxytocin levels surge, peaking during birth. When delivery is by Caesarean section that peak does not occur, which might explain why the mother's brain reacts less strongly to the child's crying and maternal behavior takes longer to manifest itself. Nursing and playing with the child causes the mother's brain to be calmed by oxytocin, which also stimulates her warm interaction and close bond with the child. A mother who hasn't formed a close attachment to her child doesn't experience a rise in oxytocin levels when she plays with it. Children who grew up in an orphanage have lower levels of oxytocin in their blood than children who grew up in a family. Even three years after being adopted, children who were neglected early in their development don't experience a normal surge of oxytocin during affectionate bodily contact with their carers. In other words, their ability to bond is impaired on a long-term basis, sometimes even permanently. A recent study of women who had been emotionally neglected or physically or sexually abused showed greatly reduced oxytocin levels in their cerebral fluid, causing concern that their problems would be passed on to the next generation. Oxytocin also inhibits the stress axis. When girls between the ages of seven and twelve underwent the stressful experience of giving a talk to a group of strangers, reassuring input from their mothers led to the release of oxytocin. It made no difference whether the child was cuddled or simply reassured by a telephone conversation with the mother.

These findings seem to suggest that it's possible to inhibit that excessive motherly concern that can be so annoying to adult children. Maternal behavior in apes was experimentally blocked with a substance that inhibited oxytocin's effect on the brain. It would seem to be the perfect medication for mothers who simply can't accept the fact that their children can function perfectly well independently. Unfortunately, however, the substance reduces apes' interest not only in their offspring but also in sex.

Thirty years ago, our research group investigated the effect of oxytocin on the brain and behavior. We created oxytocin antibodies, staining them so that they would show up in the brain, and looked for the locations where oxytocin is produced and secreted. We found extensive networks of brain cells and axons containing oxytocin in a number of brain structures (fig. 5). Those fibers made contact with other neurons, transferring oxytocin to them in the form of a chemical messenger. Examination with an electron microscope revealed that transmission locations looked just like the synapses (cell contact points) where other chemical messengers are passed from one neuron to another (fig. 6). These form the basis for oxytocin's behavioral effects.

Different social contexts prompt the release of oxytocin in specific parts of the brain, triggering different types of behavior. It has been identified as the messenger of affection, generosity, tranquility, trust, and attachment. Oxytocin has also been found to suppress fear by affecting the amygdala, the center of fear and aggression. When affectionate social interaction like hugging takes place, not only do oxytocin levels in the blood rise but more oxytocin is also released in the brain. It's also the messenger that tells your brain when you've had enough to eat. Oxytocin affects not only maternal behavior but also adult relationships, influencing our responses to social stress and sexual contact. As a result it has been dubbed the "love hormone." Yet again, this is nothing new to farmers, who have been aware of its effects for centuries, long before anyone had ever heard of the hormone. When an orphan lamb is placed with a foster mother, the farmer stimulates the sheep's vagina and uterus, causing oxytocin to be released and the ewe to bond with the strange lamb.

The brain also produces a closely related substance: vasopressin. Just like oxytocin, vasopressin plays a crucial role in maternal behavior, including maternal aggression prompted by threats to offspring. It's also involved in other aspects of social behavior, like pair bonding. Men with a tiny variation in a DNA building block for the vasopressin receptor (the protein that receives vasopressin's message in the brain) are twice as likely to experience marital difficulties and divorce and twice as likely to be unfaithful. If men are shown an image of a strange man after being given vasopressin, they perceive the facial expression as unfriendly, making them more hostile. In the case of women, exactly the opposite occurs. Vasopressin makes them more likely to approach strangers, because they are better at distinguishing friendly features. It's tempting to imagine how society might be improved by giving men a whiff of oxytocin and women a whiff of vasopressin.

A link has recently been found between autism and malfunctions in the brain's vasopressin and oxytocin systems. People with autism often find it hard to interpret other people's emotions and intentions from their gestures and facial expressions or to feel empathy. Some don't understand what is going on, for instance, when another child cries, or can't detect emotion in a person's voice. Temple Grandin, a professor of animal science with autism, has described her emotional circuitry simply as being disconnected. Indeed, abnormal blood levels of vasopressin and oxytocin have been found in people with autism. They also display small genetic variations in the proteins that are the brain's vasopressin and oxytocin receptors. Conversely, their ability to "read" emotions and predict behavior through facial cues improves when they are given oxytocin. It also helps them to detect emotion in someone's voice and to grasp the emotional significance of intonation. So both oxytocin and vasopressin can play a part in autism, but to regard these two chemical messengers as the "social brain" is a huge oversimplification. Many more messengers and brain structures are involved in social behavior.

There would seem to be considerable scope for capitalizing on our knowledge of the effects of these substances. Psychological experiments with games involving financial payment have shown a link between high oxytocin levels and trust in others, including strangers. That trust remains even if you get cheated a few times. The commercial potential was immediately spotted, and you can now buy "Liquid Trust" online, little bottles of oxytocin that you spray on your clothes to induce confidence in your partner, boss, co-workers, and customers. Since you can only expose people to very tiny amounts in this way, this is at best a placebo (acting via the person who has sprayed the compound on his clothes), if not a downright rip-off.

It's also debatable whether a direct dose of oxytocin in the form of nasal spray can replicate the normal brain processes. After all, the brain produces very precise and limited amounts of oxytocin at a very specific location in response to certain circumstances. Just inhaling the substance may have a completely different effect. And that's actually a general problem when treating disorders of the brain. You can't replace the highly specialized functions of nerve cell systems just by administering their messengers, any more than you can replace a calculator by means of the figures that it produces.

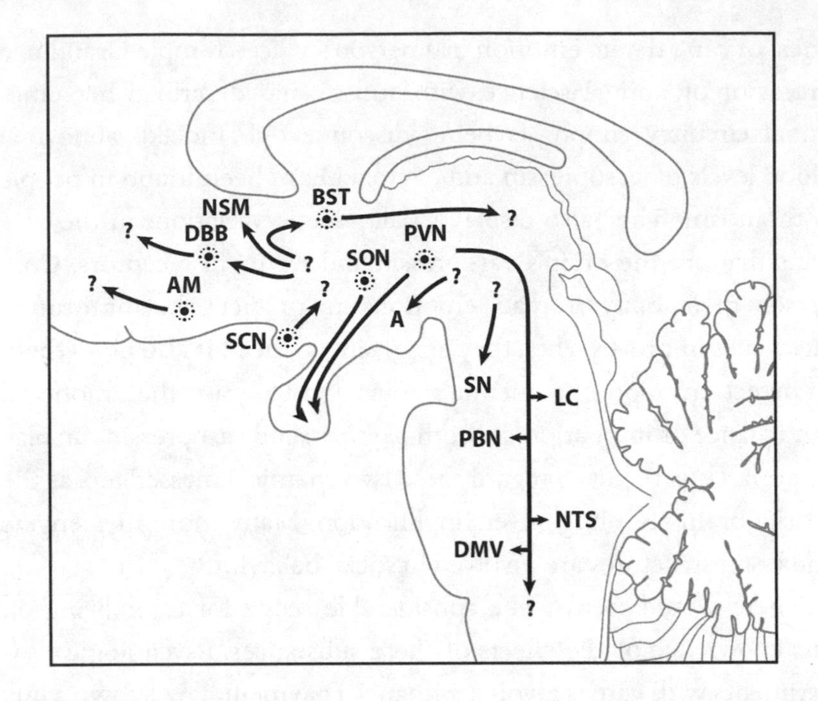

FIGURE 5. The localization of oxytocin and vasopressin in the brain. The two hormones are produced in two regions of the hypothalamus, the paraventricular nucleus (PVN) and the supraoptic nucleus (SON), and released into the bloodstream as neurohormones in the posterior pituitary. Oxytocin causes contractions of the uterus during labor and contractions in the mammary gland during suckling. Vasopressin regulates the body's retention of water by acting on the kidneys. Oxytocin and vasopressin are also transported to many brain areas that we know of (indicated here with abbreviations) and to as yet unknown brain areas, and are released in those locations from synapses as neurotransmitters (chemical messengers).

PATERNAL BEHAVIOR

> A son can never show sufficient gratitude to his parents for their loving kindness, even were he to carry his father on his right shoulder and his mother on his left shoulder for 100 long years.
>
> *The Teachings of Buddha*

We all know cases of mothers who just haven't managed to sever the umbilical cord. Their children might be long grown up, but they are constantly concerned about them and want to know exactly what they are doing, even when they're on the other side of the globe. The tie between mother and child remains a special one. No matter what country he fights for, a wounded soldier on a battlefield will always call for his mother, not his father.

In the case of chimpanzees, females are responsible for teaching cultural skills. So I always thought that a father's role was confined to fertilization, the moment at which less than half the child's DNA has to be delivered, a job that can be done in a few minutes. We fathers could then hide behind the newspaper and leave the child's care and upbringing to the mother. But it turns out that fathers can't get off that lightly. The animal kingdom provides examples of paternal behavior that replicate every aspect of maternal behavior. There is even a male bat that produces milk!

Humans do occupy a special place in the animal world in terms of their focus on the family. The family is the building block of our society, which is not the case with great apes like chimpanzees or bonobos. It's not so much pair formation that's unusual—you see that in gibbons, birds, and voles—but in those species families live isolated from each other. The family-based society is unique to our species. As far back as two million years ago, our ancestors gave birth to offspring that weighed twice as much as those of chimpanzees. Since those heavy, helpless babies couldn't be transported easily, shared child care was crucial to ensure sufficient food for the mother and a chance for her to suckle her offspring. Patriarchy—male dominance within the family—is thought to have developed when our ancestors had to exchange the protection of the jungle for a more vulnerable life in the savanna. In those wide-open spaces it was crucial for the male to protect the female and her child. Incidentally, these human ancestors, who walked on their knuckles, ate fruit, hunted, and used tools, didn't leave the jungle of their own volition, as is so often claimed. The jungle disappeared around them as a result of drastic climate change. Vast tracts were gradually transformed into dry savanna. The male's protection of the female and her child had an evolutionary advantage: Humans were able to reproduce every two to three years, while female chimpanzees, who were solely responsible for their young and therefore had to look after them and feed them for much longer, could reproduce only every six years.

That protective role of the male isn't confined to primates but extends to the entire animal kingdom. A pair of coots have once again built a large nest in the middle of the canal opposite our house. From the moment that the female started to sit on the nest, the male became incredibly aggressive toward any other birds in the vicinity. Not a single egg had yet been laid, but the coot managed to scare off much larger crows and ducks with a great deal of noise and flapping of wings.

A man, too, is prepared for his role as a father during his partner's pregnancy. Hormonal changes take place that affect the brain, making prospective fathers not only behave differently but also feel differently. Even before the child is born, the father's prolactin level increases. That hormone is important for the mother's milk production, but in both women and men it stimulates caring behavior. Conversely, the father-to-be's level of the male sex hormone testosterone declines, reducing aggression toward the child and the urge to procreate—a universal mechanism that affects prospective fathers from New York to Beijing. As a result, even before their child is born, many men feel that something special is happening to them. How those behavioral changes are induced isn't clear, but scents given off by the pregnant partner may play a role. After the birth, prolactin and oxytocin play a role in paternal behavior and bonding between father and child. When playing with their children, an increase in the bonding hormone oxytocin is seen only in fathers who display affectionate and nurturing behavior.

In some animal species, the father has been allotted an extreme role. In the case of the Greater Rhea, an ostrich-like bird, the males incubate the eggs in a nest that they have scraped out themselves, while male seahorses carry their eggs in a pouch until they hatch. Caring paternal behavior comparable to that of humans is seen in a few other species, allowing us to study the changes in the brain that provoke it. Marmoset fathers look after their offspring by carrying them, protecting them, and feeding them. Fatherhood induces changes in the prefrontal cortex. The number of synapses in this area of the brain increases, suggesting a reorganization of the local network. It also becomes more sensitive to vasopressin, the chemical messenger that promotes social behavior and aids fathers in their new tasks.

As children grow up, their fathers may inspire them and affect the course of their lives. This can take many different forms. My grandfather was a doctor, and he succeeded in interesting his son in his profession. My father became a gynecologist, and I knew from the age of six that I would study medicine. My son was uncertain about his choice of studies for a long time, but he knew from an early age that it would not be medicine or biology. He, too, was reacting to his father, albeit in a different way. We later discovered a shared interest in behavioral differences between the sexes and both published findings on this subject (see chapter 21).

Alas, the father's role doesn't confine itself to noble actions like care, protection, and inspiration. Examples of the primitive aggression that males are capable of in the name of fatherhood are common both in the animal kingdom and among humans. Male primates can take over an entire harem of females from another group by chasing away the dominant male. As a rule, they then kill all the young. When a lion takes over a pride of lions, he kills the cubs, despite the lioness's desperate attempts to defend them. This stops her milk production, making her fertile more quickly, ensuring that her young are the offspring of the new dominant male. And human history shows us to be no exception, to judge by what we read in the Bible: "But Moses was angry with the officers of the army, with the captains . . . who had come from the battle [and] said to them: '. . . Now therefore, kill every male among the little ones, and kill every woman who has known a man intimately. But keep alive for yourselves all the young girls who have not known a man intimately' " (Numbers 31:14–18).

Even in this day and age, we haven't yet managed to escape these cruel biological mechanisms. Infanticide and child abuse are more commonly perpetrated by stepfathers than biological fathers, and children of women captured during wars are still regularly killed. Female chimpanzees keep far away from groups of other chimpanzees for years after giving birth—a good strategy for ensuring that their young aren't killed by males who doubt their paternity. The "solution" devised by bonobo females to prevent infanticide is an original one. They mate with all males, so that no single male can ever be certain that he's not a youngster's father. But in the case of humans, mothers must remain vigilant, alert to all the dangers that

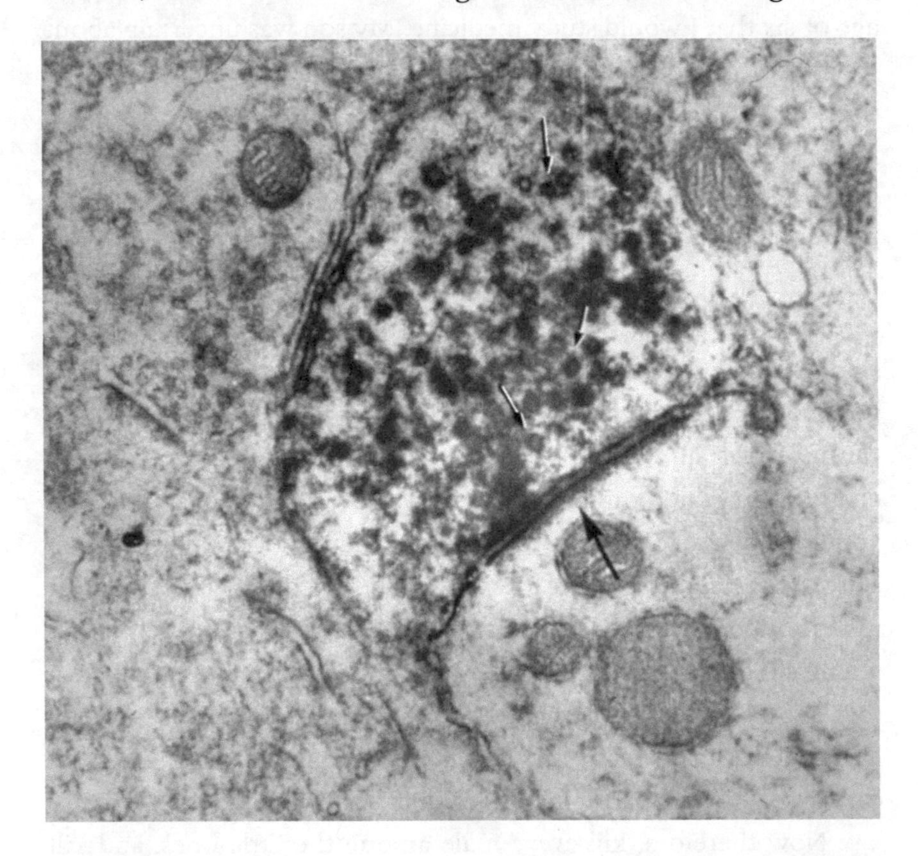

FIGURE 6. A synapse as seen under an electron microscope. Oxytocin and vasopressin appear as black granules. When released in the brain, these substances influence behaviors–for instance, social interaction. From Buijs and Swaab, *Cell Tiss. Res.* 204 (1979): 355–65.

might threaten their children—a state they remain in for the rest of their lives.

THE IMPORTANCE OF A STIMULATING ENVIRONMENT FOR EARLY BRAIN DEVELOPMENT

A good environment is not a luxury, it is a necessity.

R. Wollheim

You come into the world with a brain that your genetic background and your development in the womb have made unique and in which your character, talents, and restrictions have largely been determined (see chapters 3 and 8). For the brain to grow optimally after birth, the developing child needs a safe, stimulating environment that imposes achievable demands on it. Back in 1871, Darwin had already found that the brains of hares and rabbits that grew up confined in boring hutches were 15 to 30 percent smaller than those of their wild counterparts. Conversely, when animals are placed in an "enriched environment," a large enclosure full of objects that are renewed each day and in which they can play with one another, their brains grow and develop more synapses. Children who are seriously neglected during their early development also have smaller brains (fig. 7); their intelligence and linguistic and fine motor control are permanently impaired, and they are impulsive and hyperactive. Their prefrontal cortices can be particularly undersized. Studies have shown that orphans adopted before the age of two go on to develop normal IQs (averaging 100), while children who are not adopted until between the ages of two and six attain average IQs of 80.

The American child psychiatrist Bruce Perry described the case of a grossly neglected six-year-old boy named Justin, who lost his mother and grandmother as a baby and grew up in the care of a dog breeder, who treated him like one of his dogs, keeping him in a cage. He made sure that Justin was fed and changed, but he hardly spoke to him and never cuddled him or played with him. When Justin was later hospitalized, he was unable to speak or walk. He threw his feces at the medical staff. A scan showed his brain to be much too small; it resembled that of someone with Alzheimer's. In the stimulating environment of a foster family, he started to develop, and by the age of eight was able to go to nursery school. It isn't known what lasting damage he has suffered.

In his book *Émile; or, On Education,* published in 1778, the Enlightenment philosopher Jean-Jacques Rousseau (1712–1778) set out his theory of the "noble savage." He believed that children were innately good but were subsequently corrupted by society. However, interaction with one's surroundings is necessary for normal brain development, as is clear from Justin's story and the well-documented tale of the Wild Boy of Aveyron, a feral child discovered in the woods of the region of Languedoc in southern France in 1797. It took three full years before the child was captured by hunters and brought to the local town. At the time he was around ten; having been abandoned at a very young age, he had kept himself alive on a diet of fruit and small animals. A young doctor, Jean Marc Gaspard Itard, took on the task of educating the boy (whom he named Victor), sending lengthy reports on his progress to the French government department for internal affairs. Despite Dr. Itard's best efforts, Victor never developed fully as a person, and the only word he learned to say was *lait* (milk). It makes one wonder about the achievements of those other famous feral children, Romulus and Remus.

The acquisition of our mother tongue also shows how certain brain systems continue to be programmed by the environment after birth. A child's first language isn't determined by its genetic background, only by the surroundings in which it grows up during that critical period of language acquisition. Not only does acquiring language have a very marked effect on the brain, it's also crucial to many other aspects of a child's development. In 1211, the Holy Roman Emperor Frederick II of Germany, Italy, Burgundy, and Sicily tried to establish what language God spoke to Adam and Eve. He believed that children would spontaneously speak it if they were not exposed to other languages and set up a rigorous experiment in which dozens of children were brought up by nurses ordered never to speak to them. However, his hopes were met with disappointment. The children couldn't speak at all, and they all died at a young age. Likewise, one in three orphans brought up in severely understaffed children's homes in World War II died from the consequences of physical and emotional neglect, and those who survived were psychologically scarred. So proper interaction with one's surroundings isn't just a precondition for normal brain development, it's actually crucial to survival.

During the first few years after birth, our surroundings determine the configuration of the brain's language systems. After a certain critical period these systems become fixed. Any attempts to learn a second language are made with, say, Romanian, Uzbek, Dutch, or Italian brains, causing us to speak with an accent. In children between the ages of nine and eleven, the brain areas that process words and visual information still overlap. By adulthood, specialization has taken place, and the two types of information are processed in separate areas. The language environment creates permanent differences in brain structures and functions. Depending on whether your mother tongue is Japanese or a Western language, vowels and animal sounds are processed in the left or right cortex, irrespective of your genetic origins. The frontal cortex is the site of Broca's area, which is crucial for language (fig. 8). When adults learn a second language, another sub-area of this region is involved. But if children are brought up bilingually from an early age, both languages use the same frontal areas. In such cases, the left caudate nucleus (fig. 27) checks which language system is being used.

Our linguistic and cultural environments don't determine only which brain systems are involved in language processing but also how facial expressions are interpreted and how we scan images and their surroundings. Japanese and New Guineans, for instance, find it difficult to distinguish between a face expressing fear and a face expressing surprise. When surveying a scene, Chinese individuals, unlike Americans, don't focus on a single object at a time but look at it in relation to its surroundings. When doing mental arithmetic, the Chinese use different parts of the brain than Western Anglophones. Both use the same Arabic numerals and the lower region of the parietal lobe (fig. 1), but English speakers make more use of language systems to process numbers, while Chinese speakers make more use of visual motor systems. This can be explained by the fact that Chinese grow up learning characters. (The Chinese abacus is no longer so influential in modern China.)

The notion that the environment stimulates brain development was suggested early on by Maria Montessori, who discovered a link between socioeconomic environment and brain development, which she described in her *Handbook* (1913). Socioeconomic status is also an important factor in stimulating the development of children with a

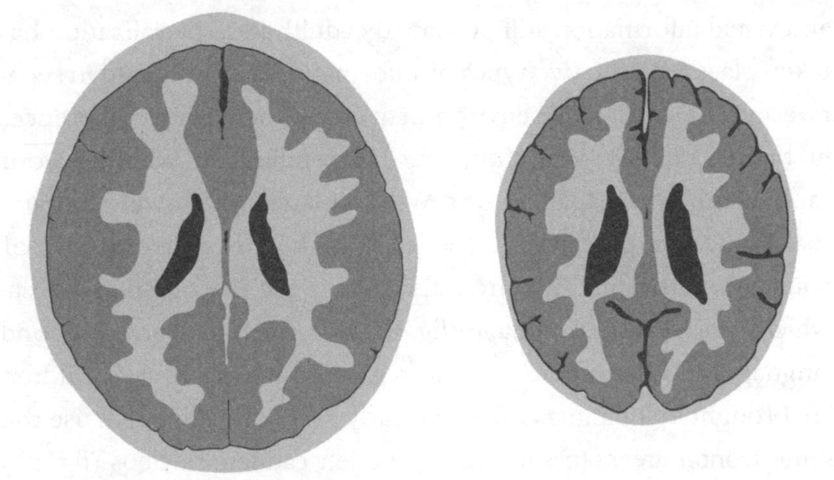

FIGURE 7. On the left, a scan of the brain of a three-year-old child who was brought up normally. On the right, the brain of a three-year-old child who was severely neglected. The neglected child has a much smaller brain, with larger ventricles (the brain cavities, shown in black). There are also much larger spaces between the convolutions of the brain due to shrinking of the cerebral cortex. Based on B. D. Perry, 2002.

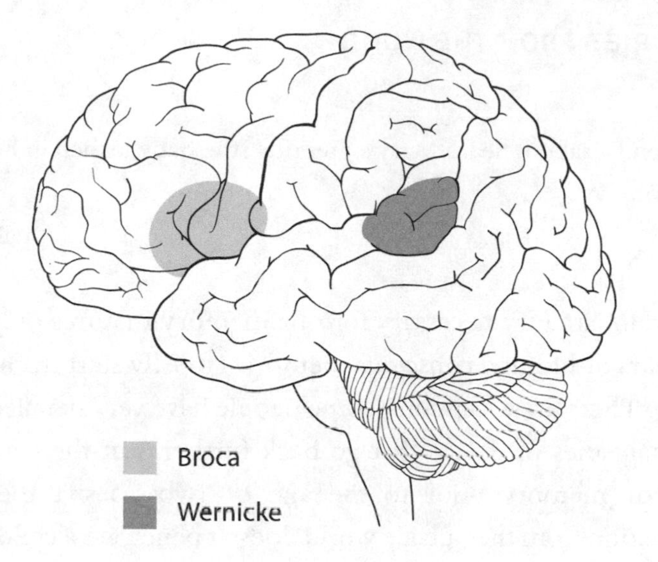

FIGURE 8. Broca's area (frontal, associated with the ability to speak) and Wernicke's area (temporal, associated with the ability to understand language). These centers are also closely involved in processing music and singing. Music and language are very much interrelated.

disadvantage, like being underweight at birth. A highly stimulating "enriched" environment promotes recovery after a developmental brain disorder. Studies have shown that children whose early development is disrupted by malnourishment or emotional neglect can improve radically if placed in a more stimulating environment early on. Children with Down syndrome, too, respond well to intense stimulation in their environment. So children with learning disorders should not be incarcerated in institutions where they receive little stimulation. On the contrary, they have an extra need for stimulus, which will positively affect the rest of their lives.

MEMORIES FROM THE WOMB

> When Elizabeth heard Mary's greeting, the baby leaped in her womb.
>
> Luke 1:41

The brain circuitry necessary for our memory matures only in our first years of life, and conscious memories mostly start from the age of two. There are exceptions; some people have very detailed, verifiable memories of events that go back further. But the general absence of memory prior to the age of two doesn't mean that information from the outside world doesn't penetrate a child's brain. It's a fact that unborn babies respond to external stimuli, but their ability to retain memories from this time hasn't been demonstrated. Are we indeed born as a tabula rasa, a blank slate, as the early Enlightenment philosopher John Locke thought, or with a treasure trove of memories of the best time of our lives, as the painter Salvador Dalí would have us believe?

There's no lack of speculation about the mental baggage we bring with us into the world and the influence that our time in the womb allegedly has on the rest of our lives. "Prenatal universities" have been set up in the United States, at which mothers learn to interact with their unborn children. It's true, your intrauterine history determines your risk of many psychiatric disorders, like schizophrenia and depression. But some therapists go too far when they maintain that traumatic memories from the fetal period are the cause of very specific psychiatric problems later in life. It's been claimed that some headaches in adult life are due to forceps delivery or pain during childbirth. Some blame women's obstetric or gynecological problems on a feeling of being unwanted at birth, because they were girls. Others attribute a penchant for bondage to being entangled in the umbilical cord at birth or a fear of being crushed to a long, difficult passage through the mother's narrow pelvis. Luckily, the same therapists reassure patients that problems like this can easily be solved by regression therapy, the theory being that to identify the cause of your problems is to solve them. A forensic study compared 412 suicide victims who were alcoholics and drug addicts with 2,901 people in a control group. A link was made between events around birth and self-destructive behavior. Suicides by hanging were associated with oxygen deprivation at birth, violent suicides with mechanical birth trauma, and drug addiction with the administration of addictive substances like painkillers during labor. A recent independent Dutch study, however, found no link between opiates administered as painkillers at birth and subsequent addiction. I'm very curious to know the results of future attempts to confirm the other correlations.

Dalí did not need regression analysis or LSD to remember his intrauterine stay in detail, which he recalled as heavenly. "The intrauterine paradise was the color of hell, that is to say, red, orange, yellow and bluish, the color of flames, of fire; above all it was soft, immobile, warm, symmetrical, double, gluey." His most splendid memory was of two fried, phosphorescent eggs. Dalí said he could reproduce a similar image at will by pressing on his closed eyelids ("characteristic of the fetal posture"). Those fried eggs return in many of Dalí's paintings. Indeed, the human fetus does respond to light from the twenty-sixth week of pregnancy. But even if Dalí's mother had lain in the sun in her bikini during her pregnancy, which is highly unlikely, the little Salvador wouldn't have been able to observe much more than a diffuse orange glow. So it would seem that detailed visual memories are the privilege of Surrealists.

However, other types of fetal memory have been demonstrated in a number of species. It's undoubtedly useful for a baby bird to become familiar with the call of its parents while still in the egg. The same applies to humans: The bond between mother and child is first established during pregnancy through the mother's voice. The existence of fetal memory in humans has been shown from three experimental paradigms: habituation, classic conditioning, and exposure learning. Habituation is the simplest form of memory, whereby the reaction to a stimulus declines the more it is encountered. In the human fetus, habituation is present as early as the twenty-second week of pregnancy. Classic conditioning has been demonstrated from the thirtieth week. Vibrations, for instance, have been used as the "conditioned stimulus" (akin to the bell in Pavlov's famous experiment with dogs), while a loud noise has been the "unconditioned stimulus" (akin to the food in Pavlov's experiment). But the level of the nervous system at which this type of learning takes place is debatable. Since an anencephalic fetus (a baby with most of its brain missing, fig. 4) can also be conditioned in this way, such learning may take place at the level of the brain stem or spinal cord. The experiments to determine exposure learning produced a much more interesting finding: When a pregnant woman relaxed every time she heard a particular piece of music, after a while the fetus began to move as soon as the music started. After birth, the same child stopped crying and opened its eyes on hearing the same music. Hearing the mother's voice while in the womb could play a role in the development of language and the bond between mother and child. Newborn babies prefer their mother's voice, particularly if it's distorted in the way that it would have been in the womb. They can also recognize a story repeatedly read aloud by the mother during pregnancy. However, the fetal memory for sounds has its dangers. Newborn babies show a clear response when they hear the theme tune of television soaps obsessively watched by their mothers during pregnancy. They stop crying and listen alertly to the highly familiar tune, and you wonder whether they are doomed to be addicted to such programs when they grow up. The unborn child's great sensitivity to melody might also explain why French babies cry with a rising intonation and German babies with a falling intonation, reflecting the different intonation contours of the two languages. Might this be the first expression of musical ability?

Babies can also remember scent and taste stimuli from the womb. Their mother's smell is instantly recognized after birth, which may be important to successful breast-feeding. Newborn babies normally dislike the smell of garlic, but if a woman eats garlic during pregnancy, her baby will not be averse to its smell. It is interesting to note that culinary differences between the French and the Dutch go back all the way to intrauterine experiences!

In sum, the fetus has a memory of sound, vibration, taste, and smell. So it's possible that we're ruining our children's brains not just by smoking and drinking and by taking medicine and other drugs but also by watching bad television programs. In other words, you'd do well to pick up a good book now and again and read to your unborn child in the hope that the next generation will rediscover literature. And that's not a new idea, by the way, because as far back as A.D. 200–600, the Talmud made mention of prenatal stimulation programs. But memories of the womb aren't detailed and as far as we know disappear within a few weeks, instead of lasting a lifetime, as some therapists and Salvador Dalí would have us believe.

2

Threats to the Fetal Brain in the "Safety" of the Womb

DEVELOPMENTAL BRAIN DISORDERS CAUSED BY ENVIRONMENTAL FACTORS

We pollute our children's amniotic fluid.

Our brains develop with incredible rapidity before birth and in the years immediately after. Moreover, each tiny area of the brain and each cell type within that area develops at a different tempo. During this period of explosive growth, brain cells are extremely susceptible to a number of different factors. First, for the brain to develop normally, the unborn child needs sufficient nourishment. Its thyroid gland also needs to function properly. At this stage, brain development is determined in general by our genetic background and in detail by the activity of our nerve cells. These, in turn, are influenced by the availability of nutrients, chemical messengers from other brain cells (neurotransmitters), growth regulators, and hormones. At that stage, the unborn child's sex hormones regulate the sexual differentiation of the brain. Substances that enter the fetal system via the placenta can derail the delicate process of brain development. These can either come from the environment or be ingested by the expectant mother (for instance, alcohol, nicotine, and other addictive substances and medications).

Sadly, we live in a world in which 200 million children suffer from serious and lasting brain damage due to lack of nourishment. Not only is their mental capacity impaired; they also have an increased risk of schizophrenia, depression, and antisocial behavior. This was shown by a study of children born in the major Dutch cities during the famine ("Hunger Winter") of 1944–1945 (fig. 9). Even in today's affluent society the same problem still occurs when a placenta malfunctions, depriving the fetus of nourishment and causing it to be undersized at birth. Malnutrition in the womb can also occur when a pregnant woman vomits excessively, tries to keep her weight down by dieting, or eats too little because of the Ramadan fast.

Some 200 million people live in regions with an iodine shortage, which affects their children's growth. Such places aren't necessarily remote; they can be found all over the world. Thyroid hormones are necessary for normal brain development but can only function if sufficient iodine is incorporated into the hormone. This happens in the thyroid gland. In mountainous areas, the iodine found naturally in soil can be washed away by rainwater. The resulting shortage of iodine affects the functioning of a child's thyroid hormones, leading to impaired brain and inner ear development. Such children develop huge thyroids that desperately try to store every scrap of iodine ingested. In the worst cases, thyroid hormone deficiency results in cretinism, a condition of severely stunted mental and physical growth. The endocrinologist Dries Querido made it his life's work to find remote places with an iodine shortage. I remember him calling me late one evening to ask if I could get him a sixteen-millimeter film projector for a lecture he had to give the next day in Amsterdam. That was how I became one of the first people to see a film he had shot in the Mulia Valley in New Guinea, then still a Dutch colony—a remote spot that could be reached only by a Cessna airplane—and hear about his expedition's findings. Around 10 percent of the children in that valley were mentally deficient and deaf and had serious neurological disorders. Querido proved that this was caused by iodine deficiency, and he treated the locals with an injection of Lipiodol, an oil containing iodine. Formerly used as a contrast agent in X-ray photos of the lungs, it was found to be potentially damaging to lung tissue. However, it turned out to be extremely effective as a depot injection to treat iodine deficiency. Similarly, the simple remedy of adding iodine to kitchen salt led to the closure of every single institution for deaf-mutes in Switzerland. In the twenty-first century, I myself witnessed developmental disorders resulting from iodine deficiencies in the mountains of Anhui, China. A woman suffering from cretinism—and disfigured by a huge goiter—was sweeping away leaves at a temple. When one of the members of our team, a Chinese professor, asked the woman if she would like to see a doctor, she merely growled and waved her broom threateningly at us.

Heavy metals can also disrupt fetal brain development. The lead added to gasoline to reduce engine knocking entered the atmosphere, causing more children to be born with mental disabilities. The dangers of mercury became apparent only in the 1950s, when cats in the fishing villages around the Bay of Minamata in Japan started acting strangely and dying and fish began to swim in bizarre patterns. The fishermen had been selling their best fish and keeping the worst specimens for the family pantry. As a result of the fish's high organic mercury content—mercury that proved to come from a plastics factory—6 percent of the children in the surrounding villages had suffered serious brain damage before birth. The formation of their brain cells and the growth of brain tissue had been inhibited by the mercury, leading to mental disability. The adults in these villages also developed various forms of paralysis. A monument has now been erected in Minamata's environmental park, dedicated to all the life-forms in the Shiranui Sea that fell victim to this disaster. The park itself was built on twenty-seven tons of mercury-polluted sludge from the Bay of Minamata as well as dozens of sealed containers full of poisoned fish. The Japanese government never gave the victims proper financial compensation.

Disorders of sex development, or intersex, are also often caused by environmental factors during fetal development. They affect up to 2 percent of children, depending on how comprehensively you define such disorders and at what stage of life they are diagnosed. In 10 to 20 percent of cases, no chromosomal cause is found for atypical development of the sex organs, so one may conclude that the disorders are due to environmental chemicals. DDT, PCBs, dioxins, and many other substances present in the environment are now referred to as "endocrinal disrupters" because they can disrupt hormonal regulation of sexual differentiation. As far back as 1940, the pilots of planes spraying DDT were found to have reduced sperm counts. What's more, the effects of these substances on brain development have been demonstrated in many animal species. The possible impact of endocrinal disrupters on the process of sexual differentiation in the fetal brain—and thus on gender identity and sexual orientation (see chapter 3)—is a concern that has arisen only very recently.

DEVELOPMENTAL BRAIN DISORDERS CAUSED BY ADDICTIVE SUBSTANCES AND MEDICATION

Are we harming our children's brains before they are even born?

Title of my inaugural speech, 1980

Fortunately, the most severe developmental disorders that can arise early in pregnancy are rare. Examples of such serious birth defects include spina bifida (the risk of which is increased by taking antiepileptic drugs during pregnancy), anencephaly, which is the absence of a forebrain (often associated with exposure to pesticides), or missing limb parts. The latter type of defect occurred in great numbers during the late 1950s and early 1960s after the appearance of a now-notorious sedative drug called thalidomide that was prescribed for pregnant women. It led to a great number of children being born with teratological abnormalities, usually missing limb sections. The thalidomide disaster made doctors more cautious about prescribing medication during the first three months of pregnancy.

These defects, however, are just the tip of the iceberg of the developmental brain disorders that can be caused by chemical substances during pregnancy, including after the first trimester. Microscopic abnormalities are far more common than classic teratological abnormalities. They occur later in pregnancy only, and the problems they cause manifest themselves much later in life. Children affected in this way appear to be completely healthy at birth, but the defects emerge later, when functional requirements are imposed on their brain systems. The children of pregnant women who smoke, for instance, are much more likely to have learning difficulties as well as behavioral problems in adolescence and reproductive problems in adulthood. These disorders are known as functional disorders or "behavioral-teratological disorders."

Many chemical substances can reach the fetus and threaten its developing brain. Heavy metals in the environment, nicotine, alcohol, cocaine, and other addictive substances, as well as medication taken during pregnancy, can disrupt the rapid development of the brain. Children exposed before birth to the drugs taken by their mothers not only display withdrawal symptoms after birth but can also be left with permanent brain damage. I believe that all substances that affect the adult brain can also influence the development of the fetal brain. I have yet to see a single exception to this rule.

Alcohol

That alcohol can cause birth defects has been common knowledge for a very long time. The Phoenicians of Carthage apparently worried about the effects of alcohol on unborn children, to judge by their law banning the drinking of alcohol on one's wedding day. The English writer Henry Fielding warned of the effects of the British gin epidemic back in 1751, lamenting, "What must become of an infant who is conceived in gin?" It wasn't until 1968 that French scientists established that drinking during pregnancy could impair fetal brain development—rather in the way that Gammas were bred by adding alcohol to the blood surrogate around the developing embryo in Aldous Huxley's *Brave New World* (1932). However, the French publication went unnoticed until it was rediscovered in 1973, with its finding dubbed "fetal alcohol syndrome" in English-language medical journals. To this day, a quarter of pregnant women have the occasional glass of alcohol, even though drinking while pregnant can cause children to be born with undersized brains and severe mental disabilities. It's also responsible for less severe damage, specifically learning and behavioral problems.

In early development, brain cells are created around the brain cavities. They then migrate to the cerebral cortex, where they ripen and sprout tissue to establish contact with other brain cells. This migratory process of fetal brain cells can be so severely disrupted by alcohol that the cells sometimes work their way through the cerebral membranes and end up outside the brain. Alcohol also permanently activates the stress axis of the unborn child's brain, increasing the risk of depression and phobia. In hospitals in the 1960s, alcohol was routinely administered intravenously to women at risk of giving birth prematurely. It inhibited contractions, enabling the baby to remain in the uterus longer. At the time, no one was concerned about alcohol coming into contact with a baby's brain. Whether or not this approach was harmful has never been established.

Smoking

The potential harm that can be caused to an unborn child when its mother smokes during pregnancy is frightening. Smoking is the most common cause of neonatal death. It doubles the risk of sudden infant death syndrome (SIDS). A mother who smokes increases her child's risk of premature birth, low birth weight, impaired brain development, disturbed sleep patterns, poorer school performance, and obesity later in life. Her smoking affects not only her own thyroid function but also that of her child. Her children have a higher risk of ADHD, aggressive behavior, impulsiveness, speech defects, attention problems, and, in the case of boys, impaired testes development and reproductive disorders.

Around 12 percent of women still smoke during pregnancy. Despite the known dangers, very few are able to give up smoking at this stage. (Incidentally, trying to stop by using nicotine patches is also dangerous for the unborn child—animal studies have shown that nicotine has an extremely harmful effect on brain development. In other words, it's not just all of the substances in the smoke but also the nicotine itself that causes developmental brain disorders.) If all pregnant women in the Netherlands were to stop smoking, 30 percent fewer children would be born extremely prematurely, underweight births would decrease by 17 percent, and savings amounting to $33 million could be made in health care. Surely that's an effort worth making for your child?

Aspecific Effects

The functional teratological impact of medication sometimes comes to light by chance. Majid Mirmiran, a PhD student working at our institute in the 1980s, studied the question of whether the high level of REM sleep in fetuses—REM being the phase in which you dream the most—is important for normal brain development. During this stage of sleep, the brain is strongly activated, a pattern that starts already in the womb. Mirmiran carried out an experiment that inhibited REM sleep in rats by giving the rats either chlorimipramine (an antidepressant) or clonidine (a medicine used to combat high blood pressure and migraine). The experiment was conducted on two- to three-week-old rats at a stage at which the rats' brain development was comparable to fetal brain development in the second half of human pregnancy. After a short course of this treatment during their development, the adult animals had less REM and were more fearful.

Moreover, the sex drive in the grown male rats diminished, and they became hyperactive. In other words, a mere two weeks of exposure to these substances during their development caused permanent alterations in the brains and behavior of rats. A subsequent study in Groningen looked at children whose mothers had been prescribed clonidine eight years previously during their pregnancy as a "safe" medication for high blood pressure and migraine. The children proved to have severe sleep disorders; some were even sleepwalkers. One of the problems of functional teratological disorders, in other words, is that doctors must be able to determine, on the basis of animal studies, what disorders they need to look for in humans. What's more, the effects of the substances in question are aspecific. You can't tell from a condition that manifests itself long after birth, such as a sleep disorder, exactly what substance taken during pregnancy caused the brain damage in question. Other examples of aspecific symptoms of functional teratology are learning disorders (caused by alcohol, cocaine, smoking, lead, marijuana, DDT, antiepileptic drugs), depression, phobias and other psychiatric problems (diethylstilbestrol, smoking), transsexuality (phenobarbital, diphantoin), aggression (progestogens, smoking), impaired motor skills, and social and emotional problems.

Additionally, chemical substances are thought to contribute to developmental disorders in which diverse factors play a role, like schizophrenia, autism, SIDS, and ADHD. Depending on her baby's genetic background, a woman who smokes during pregnancy can increase the chances of her child developing ADHD by a factor of nine. The risk of ADHD is also increased when adrenal cortex hormones are administered during pregnancy to promote lung development in babies at risk of being born prematurely. This procedure has been found to impair brain development, potentially causing not only ADHD but also a smaller brain, impaired motor skills, and a lower IQ. These hormones are now administered much more sparingly.

Dilemma

One of the dilemmas confronted by doctors is that patients with schizophrenia, depression, or epilepsy often continue to need treatment during pregnancy, because the mother's condition is potentially harmful to her child. Unfortunately, taking antipsychotics like chlorpromazine during pregnancy has been shown to cause motor disorders in children, and some antiepileptics increase the risk of spina bifida or transsexuality. It's best to treat epilepsy during pregnancy with a single drug (rather than a combination) together with folic acid. Some antiepileptics are more harmful than others: Valproic acid has been shown to impair verbal IQ more than other epilepsy medications. Around 2 percent of pregnant women take antidepressants even when they have only mild depression. Such drugs don't appear to increase the risk of serious birth defects, though the children born to these mothers are somewhat underweight and slightly premature, score somewhat less well on the post-birth Apgar test, and have subtle motor disorders. However, these disadvantages must be weighed against the problems that can result from a mother being stressed and depressed during pregnancy, such as impaired cognitive performance, attention, and language development. When a mother is fearful during pregnancy, she can permanently activate her baby's stress axis, thus increasing the risk of phobia, impulsiveness, ADHD, and depression later in life. If at all possible, it's worth considering treating depression in pregnant women with alternative therapies, like light therapy, transcranial magnetic stimulation, massage, acupuncture, or online therapy. Clearly, doctors treating such patients need to do a lot of careful thinking.

Mechanisms

Brain cells are created with incredible rapidity in the womb and shortly after birth, and this process continues, somewhat more slowly, until around the fourth year of life. Brain maturation goes on much longer; in the case of the prefrontal cortex, it continues right up to the age of twenty-five. Every facet of brain cell development can be disrupted by chemical substances during pregnancy. Disturbances to the migration of brain cells can lead to heterotopias, a condition in which groups of cells making their way to the cerebral cortex end up in the wrong part of the brain. They get trapped in the white matter, the fiber connections, as they journey to the cerebral cortex (fig. 20), a location where they can't function properly. Substances that are regularly taken by pregnant women, such as benzodiazepines, can induce this condition. Drinking during pregnancy also causes malformations and malfunctions of nerve cell fibers. Smoking and drinking during pregnancy alter the receptors for nicotine, and smoking cannabis can alter the dopamine receptors in the fetal brain.

Conclusions

Addictive substances, medication, and environmental substances can permanently disrupt fetal brain development, leading to learning and behavioral disorders in later life. Congenital defects of this kind are known as functional or behavioral-teratological defects.

Tracing the connection between these disorders and the effects of chemical substances is difficult due to the length of time between the child's exposure to such substances in the womb and their effects, which may only be manifested when the child goes to school or—in the case of reproductive problems—perhaps twenty or thirty years later. Moreover, the conditions caused by these substances, like learning and sleep disorders, are so aspecific that they can't be used to identify the substance that caused the brain damage during pregnancy. On top of that, a single substance can produce different symptoms depending on the stage of development at which the child was exposed to it. All of this is complicated by the fact that doctors, especially in the absence of reliable animal studies, don't know what disorders they should be looking for. With women who may require drug treatment during pregnancy, it's essential to discuss potential problems at an early stage so that if a pregnancy is planned, the safest drug or alternative therapy can be prescribed.

THE SHORT-TERM OUTLOOK OF THE UNBORN CHILD

We're programmed in the womb for life after birth. We acquire our feeling of being male or female, our sexual orientation, and our level of aggression while still in the womb (see chapters 3 and 8). Later, our sex hormones activate the brain systems that are programmed before birth, and our sexuality and aggression are manifested. This intrauterine programming is influenced by the hereditary information passed on by our parents. As a result, a significant part of our character is determined from the moment of conception, as is our risk of brain disorders like schizophrenia, autism, depression, and addiction (see chapters 5 and 10). But the information in our DNA is much too limited to program our brains fully in advance. The brain has solved this problem by overproducing cells and synapses. As cells develop, they compete for the best contacts. From these they obtain growth substances that make them more active, enabling them to make more and better connections. Cells that fail to do so die off, and surplus connections are pruned off.

Besides our genetic determination, our developing brains are influenced by all kinds of other factors that affect brain cell activity, like fetal and maternal hormones and nutrients and environmental chemicals passing through the placenta. For instance, sex hormones program us along male or female lines. Our levels of aggression and stress are set before birth for the rest of our lives. Extreme signals from the outside world also lead to fetal brain systems being permanently modified. In this way, the unborn child prepares itself for a hard life outside the womb. In the short term the fetal brain's plasticity promotes survival, but it also makes it more vulnerable to harmful substances like nicotine. In the long term, fetal programming can also contribute to chronic diseases, as a study at Amsterdam's Academic Medical Center shows. Toward the end of the Second World War, the Nazi occupiers of the Netherlands robbed the country of its food supply, leading to the famine of 1944–45. Babies were not only underweight at birth (fig. 9) but also more likely to develop antisocial behavior and obesity in later life. They turned out to prefer fatty foods and to exercise less. They were also more likely to develop high blood pressure, schizophrenia, and depression. The implications are far-reaching because the same mechanisms still come into play when fetuses are malnourished because of placental malfunction, causing babies to be born underweight.

It seems that even before birth, a child registers a shortage of food in its surroundings. What evolutionary advantage might this have? In such cases, all the brain systems that regulate metabolism are programmed in the womb so as to retain every calorie. Later in life, these individuals feel less satiated when they eat. Since they're smaller at birth, they also need less food. So even at this early stage, children adapt their brains and behavior to a life of scarcity outside the womb. Their tendency to antisocial behavior equips them to defend their own interests, giving them an advantage in situations where there isn't enough to go around. Their activated stress axis will also contribute to this survival strategy. But if they are then born into surroundings where there's an abundance of food, this adaptive strategy becomes a handicap. Their inability to feel satiated means they are more likely to be obese and to develop hypertension. They also run a greater risk of addiction. And the fact that their stress axis is constantly switched on heightens their risk of depression and schizophrenia. So the diseases that are more likely to arise after prenatal malnutrition could be regarded as the side effects of an adaptive strategy that improves the fetus's chance of survival in the short term.

Disruption to the sexual differentiation of an unborn child's brain when its mother is severely stressed during pregnancy (see chapter 3) could be regarded in a similar light. When a pregnant woman experiences stress, the brain of a female fetus will become more male and vice versa. This also appears to be an adaptive response. A girl will be able to cope better in later life if she's robust and competitive, while a boy who isn't macho is less likely to get into conflict with alpha males in that stressful environment. This is also an excellent survival strategy in the short term, but in the long term it can impair reproduction and increase the likelihood of developmental disabilities and schizophrenia.

In sum, the fetus appears only to think of survival in the short term, adapting to the difficult circumstances that it anticipates immediately after birth. It's of course wrong to speak of a fetus "thinking." Over a period of millions of years, unborn children have been exposed to threats of this kind. Occasionally, a baby possessed a mutation that enabled it to adapt better to the problems facing it, and

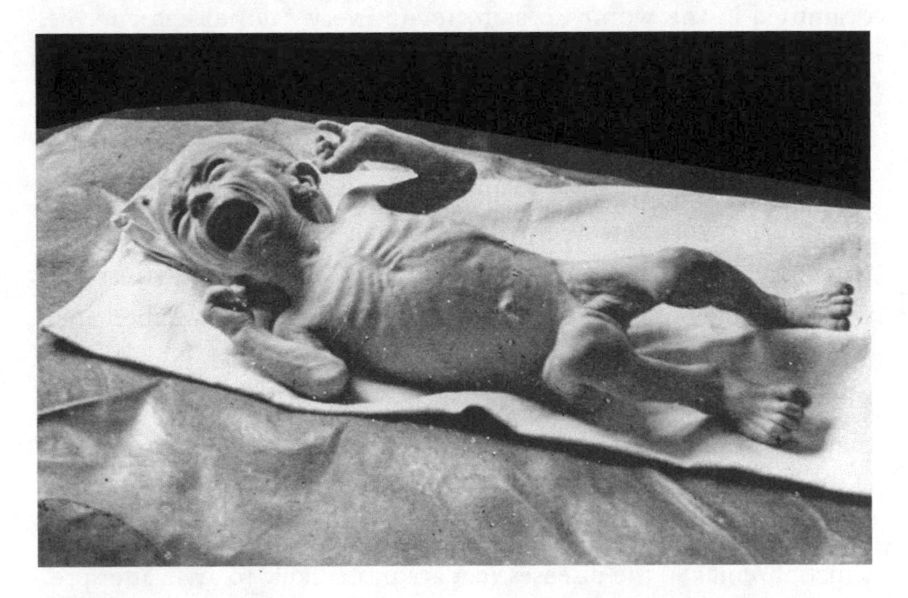

FIGURE 9. A child born in Amsterdam's Wilhelmina Gasthuis hospital during the famine (the "Hunger Winter") of 1944–45. Not only did these babies have low birth weight, in adulthood they were more prone to antisocial behavior and obesity. They showed a preference for fatty foods and were less likely to exercise. They also had greater risk of developing high blood pressure, schizophrenia, depression, and addiction. Photograph: NIOD Institute for War, Genocide and Holocaust Studies.

this favorable mutation then spread through the population. And you can't blame a child for opting for short-term adaptations without taking account of the long-term consequences, because longevity is only a very recent human accomplishment.

Up to now, doctors have been able to treat only the later consequences of fetal programming. Now, armed with knowledge about such programming, doctors can encourage targeted prevention by providing advice on nutrition during pregnancy, for instance.

DOES A FETUS FEEL PAIN?

When George W. Bush was president, impressive footage of a fetus in the womb being touched by a needle and responding with violent movements made the rounds. Pro-life advocates used the video to suggest that fetuses feel pain and will attempt to protect themselves from instruments of abortion. The federal government considered making it mandatory for doctors to inform women that there was "substantial evidence" that an abortion would inflict pain on a fetus. In the case of pregnancies of over twenty-two weeks, it was proposed that anesthesia must be administered to fetuses before an abortion. Doctors who failed to comply would be fined $100,000 and lose their jobs. These proposals met with approval from the pro-life movement, but how much actual evidence is there to show that fetuses truly feel pain?

In a mature state of development, painful stimuli are transported by nerve fibers from the skin via the spinal cord to the center of the brain, the thalamus (fig. 2). From there, the stimuli go to two areas: the primary sensory cortex, where one becomes aware of pain, and the cingulate cortex, the brain's alarm center (fig. 27), which interprets pain and directs the emotional and autonomic responses: emotion, contorted face, stress response, rapid breathing, higher blood pressure, and increased heart rate.

A normal pregnancy lasts for forty weeks. The wiring to conduct pain stimuli from the fetus's cerebral cortex is in place by the twenty-sixth week. Only then can such stimuli travel from the skin to the child's cerebral cortex, but whether they are then consciously received has yet to be established. It seems unlikely that premature babies can consciously feel pain before the twenty-ninth or thirtieth week. The pain sensors in the skin and the nerve pathways that convey pain signals are in place as early as the seventh week, enabling the fetus to respond to touch from a needle. But, contrary to the claims of fanatical pro-lifers, that certainly doesn't constitute proof that the fetus can feel pain. For that to happen, the stimulus must first reach the cerebral cortex, and the cortex must be mature enough to process stimuli meaningfully. Before the cortex is fully mature, the fetal response to pain stimuli is purely based on spinal-cord reflexes. Anencephalic babies, who are born with most of the brain missing, respond in exactly the same way. In their case, the response to pain stimuli is just as violent and generalized—the whole body seems to be involved—as in intact fetuses in the first trimester of pregnancy, precisely because the cerebral cortex hasn't matured and can't keep the spinal-cord reflex in proportion.

Contacts between the thalamus and the cortical plate under the cerebral cortex are established from twelve to sixteen weeks. The cortical plate is a waiting room for fibers that will grow into the cerebral cortex (between twenty-three and thirty weeks). EEG measurements, which determine electrical activity in the brain, and blood circulation in the cerebral cortex of premature babies show a response to pain stimuli from twenty-five to twenty-nine weeks. So at that stage, pain stimuli are arriving in the cerebral cortex. The question is whether the cerebral cortex is already mature enough to receive the pain consciously. Conscious perception is also necessary for pain to be perceived emotionally. EEG measurements in newborns show a difference between the response to touch and the response to the pain of a heel prick only as of thirty-five to thirty-seven weeks.

Nowadays, when treating premature babies in incubators, it's generally thought to be safer to assume that they feel and experience pain. They respond to invasive treatment and the taking of blood through movement and alterations in heart rate, breathing, blood pressure, oxygen pressure, and stress hormone levels. The same applies to operations like circumcision. That doesn't prove, however, that pain is perceived consciously, because these autonomic responses come from regions below the cerebral cortex and therefore might be based on unconscious processes. The same applies to the movements that premature babies make in response to pain stimuli, because these can still be spinal-cord reflexes that don't penetrate through to the cerebral cortex. Not only do anencephalic babies respond to physical stimulus by recoiling, but brain-dead adults in vegetative comas whose cerebral cortex is entirely destroyed also respond in the same way.

So premature babies are seen to respond to pain stimuli in the cerebral cortex as of twenty-five to twenty-nine weeks, but even then we can't be certain that this response is conscious. It's even harder to establish whether a fetus possesses consciousness. The "waking stage" in a fetus's wake and sleep cycles is sometimes regarded as a surrogate for consciousness. But during the late stage of pregnancy, fetuses spend around 95 percent of the time asleep, that is, unconscious, due to the immaturity of their brains and the effects of placental hormones. During the remaining 5 percent of the time, they are "awake," but that period is more like a transitional phase between REM and non-REM sleep than a period of genuine wakefulness or consciousness.

Unpleasant stimuli have been shown to cause changes in cerebral cortex activity in premature babies aged twenty-five to twenty-nine weeks, but there are great differences between a premature baby and a fetus of the same age. Stimuli that prompt a waking response after birth (like a shortage of oxygen) have the completely opposite effect on a fetus: They suppress the waking stage. This allows a fetus to conserve energy in difficult circumstances that it's powerless to escape anyway. Even a potentially "painful" or "annoying" stimulus, like a strong vibration or a loud noise, elicits only a subcortical response in a fetus. Moreover, that a twenty-eight-week-old fetus can "learn" to respond to the stimulus doesn't mean that a conscious memory process is involved. Again, primitive "learning behavior" of this type can also be seen in anencephalic babies. So it's an unconscious form of learning for which the cerebral cortex isn't needed.

As for mandatory anesthesia for fetuses in the case of abortion, one might argue that if it doesn't benefit, it at least can't harm, but abortion under a general anesthetic does increase the risk of complications for the mother. For the same reason, it would be greatly worrying if doctors were obliged to anesthetize fetuses undergoing interventions other than abortion, given that there's no hard evidence that fetuses possess consciousness, whereas there is proof that anesthesia can impair a child's later development.

From all this I conclude that in the case of abortions or interventions in utero up to the twenty-fifth or twenty-sixth week of pregnancy, a general anesthetic is unnecessary for the fetus and may entail extra risks for the mother, that a premature baby should be anesthetized before undergoing painful treatment just in case, and that it should be mandatory to anesthetize boys undergoing circumcision.

SAWING OFF YOUR OWN LEG: BODY INTEGRITY IDENTITY DISORDER, A BIZARRE DEVELOPMENTAL DISORDER

During our early development, not only is our gender identity and our sexual orientation programmed in our brains (see chapter 3) but the functioning of our inner body map is as well. Body Integrity Identity Disorder (BIID) is a bizarre developmental disorder of this latter process. Persons with this syndrome develop a belief early on that part of their body doesn't belong to them, and they become desperate to get rid of it. They don't accept a limb as being part of them, even though it functions perfectly. This leads to an overwhelming desire for amputation. Only when their arm or leg has been amputated—and around 27 percent of these individuals succeed in achieving this—do they feel "complete." Surgeons who comply with these wishes run the risk of being fired for removing a healthy limb. That's curious, because the principle being applied is similar to that used in the case of transsexuals and even circumcision. (Moreover, the latter operation is carried out on baby boys who are incapable of giving informed consent, and it can lead to complications like bleeding, infection, a perforated urethra, narrowing of the urethra, scarring, and malformation.) However, acceptance of the problems of BIID sufferers doesn't seem to be coming any time soon. Psychotherapy or pills don't on the whole change the sufferers' way of thinking. There's a sole recorded case of a BIID patient whose misery was alleviated by antidepressants and cognitive behavioral therapy, but he later declared that, although it was nice to talk to someone, the therapy had done nothing to resolve his BIID issue.

The conviction that a lower leg or arm doesn't belong to them or the desire for paralysis of one or more limbs is something that affects these patients from an early age. A child with BIID cut dolls out of a magazine and then snipped off the leg that he himself didn't want. BIID sufferers can even get excited or jealous when they see amputees or individuals suffering from the paralysis that they so long for. Sometimes it's only then that they realize what they truly desire. They often pretend to be amputees, for instance by strapping their bent lower leg against their thigh, putting on wide trousers that hide the extra girth, folding the trouser leg back, and walking with crutches or sitting in a wheelchair. BIID patients often spend years trying to find a surgeon who will amputate their perfectly healthy, functioning limb. If this fails, which is usually the case, two-thirds of patients who ultimately undergo an amputation damage their unwanted limb to the extent that amputation becomes necessary. Sometimes they endanger their lives by shooting themselves through the knee, freezing their leg, or resorting to a saw. People with BIID are very specific about exactly where a limb should be amputated and have definite ideas as to whether the surgeon got it exactly right, but they are elated once they have had the operation, only regretting that it didn't happen much earlier.

At present, it's impossible to say exactly how the inner body map of such patients became flawed during the brain's development. However, scans show that their frontal and parietal cortices respond differently to a touch on the leg according to whether it's the accepted or the rejected leg. BIID shows similarities with transsexuality (see chapter 3). In both cases, a person knows—typically from an early age—that their body's anatomy doesn't tie in with how they feel. The link with transsexuality is particularly intriguing due to the high percentage of BIID patients (19 percent) who also have a gender identity problem, and the high percentage of homosexual and bisexual BIID patients (38 percent). Since all of these characteristics are programmed early in development, the same probably applies to BIID, though both its cause and location in the brain remain a mystery. There's no reason at all to believe that BIID is caused by memories of a former life in which the person in question was missing a particular limb—as someone who wrote to me believed.

We have the technology to establish what has gone wrong in the body mapping function during the brain's development. But for that to happen, doctors need to lose their fear of being involved in a patient's desire for amputation. They have to stop dismissing them as "simply mad." Researchers, for their part, need to look more closely at these strange variants that can increase our understanding of normal brain development. And, finally, people with BIID have to have the courage to come out of the closet in which many of them currently prefer to stay.

3

Sexual Differentiation of the Brain in the Womb

I am inclined to agree with Francis Galton in believing that education and environment produce only a small effect on the mind of anyone, and that most of our qualities are innate.

The Autobiography of Charles Darwin

My brain? That's my second favorite organ.

Miles Monroe (Woody Allen) in *Sleeper*

A TYPICAL BOY OR GIRL?

Gender identity is sufficiently incompletely differentiated at birth as to permit successful assignment of a genetic male as a girl. Gender identity then differentiates in keeping with the experience of rearing.

J. Money, 1975

Nothing would seem simpler than seeing at birth whether a child is a boy or girl. After all, gender is determined from the moment of conception: Two XX chromosomes will become a girl, an X and a Y chromosome a boy. The boy's Y chromosome starts the process that causes the male hormone testosterone to be produced. The presence or absence of testosterone makes the child develop male or female sex organs between the sixth and twelfth week of pregnancy. The brain differentiates along male or female lines in the second half of pregnancy, due to a male baby producing a peak of testosterone or a female baby not doing so. It's in that period that the feeling of being a man or woman—our gender identity—is fixed in our brains for the rest of our lives.

That our gender identity is determined as early as in the womb has only been discovered fairly recently. Up to the 1980s it was thought that a child was born as a blank slate and that its behavior was then made male or female by social influences. This had huge consequences for the treatment of newborns with indeterminate sex organs in the 1960s and 1970s. It didn't matter what sex you selected for your child, it was thought, as long as the operation took place soon after birth. The child's surroundings would then ensure that its gender identity adapted to its sex organ. Only since then have patient associations revealed how many lives were ruined by assigning a sex on the operating table that didn't match the gender identity imprinted in the brain before birth. The story of John-Joan-John shows how disastrous this approach could be. When a little boy (John) lost his penis at the age of eight months through a botched circumcision, it was decided to turn him into a girl (Joan). His testicles were removed while still an infant. He was dressed in girls' clothes, received psychological counseling from John Money, a sexologist from Philadelphia, and was given estrogen during puberty. Money described the case as a great success: The child was said to have developed normally as a female (see the epigraph to this section). When I remarked during a seminar in the United States that this was the only case I knew showing that a child's gender identity could be changed by its environment after birth, Milton Diamond, a renowned sexuality expert, stood up and said that Money's claim was completely unfounded. Diamond was acquainted with Joan; he knew that Joan had had his sex change reversed as an adult and had married and adopted children. Diamond made these findings public. Sadly, John later lost everything he had on the stock market, suffered an unhappy separation from his wife, and in 2004 committed suicide. This tragic story shows how strongly testosterone programs the brain in the womb. Removing this child's penis and testicles, giving him psychological counseling, and administering estrogen during puberty couldn't change his gender identity.

That testosterone is indeed responsible for causing sex organs and brains to develop along male lines is apparent from androgen insensitivity syndrome (AIS). People with this condition produce testosterone, but their bodies are insensitive to it. As a result, both the external sex organs and the brain are feminized. Even if they are genetically male (XY), they become heterosexual women. Conversely, in the case of girls who have been exposed to a high dose of testosterone in the womb due to congenital adrenal hyperplasia (CAH), the clitoris becomes so enlarged that they are sometimes registered as boys after birth. Almost all of these girls are assigned the female gender. But in 2 percent of cases it later emerges that they did in fact acquire a male gender identity in the womb.

The effect this can have in practice was shown clearly in a Dutch newspaper article by Jannetje Koelewijn (*NRC Handelsblad,* June 23, 2005). The parents of four daughters were overjoyed when their fifth child proved to be a boy. But after a few months the child fell ill, and it turned out to be a girl with CAH. Doctors talked at length with the parents, Turkish Muslims, who refused to consider gender reassignment, partly on religious grounds. So the doctors decided to make the child more like a boy. The clitoris was made larger, to resemble a penis, and the child was given hormones to promote masculine development. Its parents were delighted with the solution. But the brains of girls with CAH mostly develop along female lines. From the above data it seems extremely likely that the "little boy" will later experience gender problems and want to be a girl once more. When he enters puberty, he will also have to be told that he's infertile, that he will need testosterone treatment for the rest of his life, and that his uterus and ovaries will have to be removed. The medical field now agrees that girls with CAH, even those who have become masculinized, should be raised as girls.

In those rare cases where a child's sex is ambiguous and it's uncertain whether its brain has masculinized or feminized, it can be assigned a temporary gender. Far-reaching interventions to turn such children into boys or girls should preferably be postponed until their gender identity has become clear through their behavior, although Katja Wolffenbuttel, a pediatric urologist working in Rotterdam, has shown that even operations can be reversed.

GENDER-BASED DIFFERENCES IN BEHAVIOR

Gender-based differences in the brain and in behavior are also found in areas that don't appear to have a direct connection with reproduction. One of the stereotypical differences in behavior between boys and girls that's often said to be socially conditioned is the way in which they play. Little boys are wilder and more active, preferring to play with cars or to pretend to be soldiers, while girls prefer to play with dolls. Because my observations of animals had left me with strong doubts about the social conditioning theory, when my children (a girl and a boy) were small, more than thirty years ago, my wife and I always would offer them both kinds of toys—but they were both very consistent in making stereotypical choices. Our daughter played only with dolls, while our son was interested only in toy cars. However, two children isn't a big enough sample for proper research. That this difference has a biological basis was subsequently proven by Alexander and Hines, who offered dolls, toy cars, and balls to vervet monkeys. The female monkeys preferred the dolls, whose genitals they sniffed in a display of typical motherly behavior, while the male monkeys were much more interested in playing with toy cars and a ball. So toy preference isn't forced on us by society, it's programmed in our brains in order to prepare us for our roles in later life, namely motherhood in the case of girls and fighting and more technical tasks in the case of boys. The gender difference in the choice of toys by monkeys shows that its underlying mechanism goes back tens of millions of years in our evolutionary history. The peak in testosterone produced by boys in the womb appears to be responsible for this difference. Girls with CAH prefer to play with boys (and with boys' toys) and are unusually boisterous, often getting labeled as tomboys.

There are also clear gender-based differences in the drawings that children make. Not only the subjects but also the colors and compositions of boys' and girls' drawings differ in ways that are influenced by hormones in the womb. Girls prefer to draw human figures, especially girls and women, as well as flowers and butterflies. They use bright colors like red, orange, and yellow. The compositions are peaceful, and the figures often stand on the same line. Boys, by contrast, prefer to draw mechanical objects, guns, conflict scenes, and vehicles like cars, trains, and planes. They often adopt a bird's-eye view perspective and favor dark and cool colors like blue. Drawings done by five- and six-year-old girls with CAH resemble those of little boys, despite their being treated for their condition immediately after birth.

Some gender-based differences in our behavior emerge so early on that they can only have arisen in the womb. As early as the first day after birth, girl babies prefer to look at faces, while boy babies prefer to look at mechanical moving objects. At one year of age, girls already make more eye contact than boys, while girls exposed to too much testosterone in the womb make less eye contact later in childhood. So here, too, testosterone in the womb plays a key role. In daily life, eye contact has a very different significance for women and men. In Western culture, women use eye contact to understand other women better, and they find it satisfying. For Western men, however, the significance of eye contact lies in testing their place in the hierarchy, something that can feel very threatening. That too is pure biology. When you leave the airport in

Aspen, Colorado, you come face-to-face with a warning sign reading, "If you meet a bear, don't make eye contact." The reason is that the bear will immediately attack to show who's boss. My son carried out studies in the United States of the factors that determine success in negotiations. I'd once told him that in my experience, women negotiate differently than men. He wasn't very interested at the time, but when he was in Chicago he suddenly decided to look into my theory. The experiments we subsequently carried out show that gender-based differences in eye contact also affect business transactions. Eye contact between two women during negotiation turns out to lead to a more creative outcome, while eye contact between two men actually prevents them from coming to terms. Men are handicapped by the threatening hierarchical implications of looking into someone's eyes. Feel free to use this practical tip to your advantage.

HETEROSEXUALITY, HOMOSEXUALITY, AND BISEXUALITY

> If a man lies with a male as he lies with a woman, both of them have committed an abomination. They shall surely be put to death.
>
> Leviticus 20:13

> [The] exclusive sexual interest felt by men for women is also a problem that needs elucidating and is not a self-evident fact based upon an attraction that is ultimately chemical in nature.
>
> Sigmund Freud

Alfred Kinsey didn't attract any notice when he published his doctoral thesis on gall wasps. But in 1948, when he produced the report *Sexual Behavior in the Human Male* and then, five years later, *Sexual Behavior in the Human Female,* he became a celebrity. He devised the "Kinsey scale," which went from 0 to 6, 0 signifying exclusively heterosexual and 6 exclusively homosexual. Being bisexual, he himself would have been classified as a "Kinsey 3."

A person's position on the scale is determined in the womb by his or her genetic background and the effects of hormones and other substances on the developing brain. Studies of twins and families show that sexual orientation is 50 percent genetically determined, but the genes in question haven't yet been identified. It is curious that a genetic predisposition for homosexuality should persist in populations over the course of evolution, given that this group reproduces so much less. One explanation for why homosexuality persists is that the involved genes don't just increase the likelihood of homosexuality but also promote fertility in the rest of the family. Heterosexual individuals with the same genes produce a larger than average number of offspring, causing the genes to remain in circulation.

Hormones and other chemical substances importantly affect the development of our sexual orientation. Girls with the adrenal gland disorder CAH who are exposed to high testosterone levels in the womb are more likely to become bisexual or homosexual. Between 1939 and 1960, around two million expectant mothers in the United States and Europe were prescribed the synthetic estrogen known as diethylstilbestrol (DES) in the belief that it would prevent miscarriages. (It didn't, in fact, but doctors like to prescribe things, and patients are always keen to be treated.) DES turned out to increase the likelihood of bisexuality and homosexuality in the daughters of women given the drug. Pre-birth exposure to nicotine or amphetamines also increases the likelihood of lesbian daughters.

The more older brothers a boy has, the greater the chance that he will be homosexual. This is due to a mother's immune response to male substances produced by boy babies in the womb, a response that becomes stronger with each pregnancy. Pregnant women suffering from stress are also more likely to give birth to homosexual children, because their raised levels of the stress hormone cortisol affect the production of fetal sex hormones.

Although it's frequently assumed that development after birth also importantly affects our sexual orientation, there's no proof of this whatsoever. Children brought up by lesbians aren't more likely to be homosexual. Nor is there any evidence at all for the misconception that homosexuality is a "lifestyle choice."

The above-mentioned factors alter the development of the child's brain, particularly the hypothalamus, which is important for sexual orientation. In 1990 Michel Hofman and I found the first brain difference in relation to sexual orientation: The brain's biological clock turned out to be twice as large in homosexual men as in heterosexual men. At the time we were actually looking for something else. I'd previously discovered that Alzheimer's damages the biological clock, which explains why people suffering from this disorder wander around at night and doze during the day (see chapter 18). I did some more studies to see if the same applied to other forms of dementia. In the case of AIDS dementia I found that the biological clock was twice as large as normal. Follow-up studies showed that this wasn't caused by AIDS; it was related to homosexuality. In 1991, Simon LeVay reported a second difference in hypothalamic structure between homosexual and heterosexual men, and in 1992 Allen and Gorski found that the structure on top of the hypothalamus that connects the brain's left and right temporal lobes is larger in homosexual men.

Scans have also revealed functional differences in the hypothalamus with regard to sexual orientation. A study by Ivanka Savic of the Stockholm Brain Institute involved pheromones, the scented sex hormones that are given off in sweat and urine. Pheromones influence sexual behavior unconsciously. A male pheromone stimulates activity in the hypothalamus of heterosexual women and homosexual men but doesn't provoke a response in heterosexual men. It seems that male scents don't turn them on. Lesbian women were found to react differently to pheromones than heterosexual women. Savic also showed that heterosexual women and homosexual men had more extensive functional connections between the amygdala and other brain areas than heterosexual men and homosexual women, proving that brain circuits function differently according to sexual orientation. Functional scanning also showed changes of activity in other brain areas. In the case of heterosexual men and homosexual women, the thalamus and prefrontal cortex responded more strongly to a photograph of a female face, while in the case of homosexual men and heterosexual women these structures responded more strongly to a male face. In other words, sexual orientation is determined by many structural and functional differences in the brain, all of which develop in the womb during the second half of pregnancy. They aren't caused by the behavior of dominant mothers, who are the traditional scapegoats in this context. Just for the record, I made a habit over the years of asking the medical students I taught (250 at a time) which of them did *not* have a dominant mother. No one ever raised their hand.

HOMOSEXUALITY: NO CHOICE

> Xq28—Thanks for the genes, mom!
>
> T shirt referring to research done by Dean Hamer showing that a gene for homosexuality might reside in the q28 marker on the X chromosome

> Homosexuality is God's way of insuring that the truly gifted aren't burdened with children.
>
> Sam Austin, composer and lyricist

Toward the end of George W. Bush's presidency, an "ex-gay movement" that regarded homosexuality as a curable disease gained momentum. Hundreds of clinics and therapists jumped on the bandwagon, and it was claimed (but not proven) that 30 percent of those who went into therapy were cured. At such clinics, you received two weeks of "treatment" for $2,500 or six weeks of treatment for $6,000. The therapists were often homosexual themselves but claimed to have been turned into family men after therapy. A countermovement with the slogan "It's OK to be gay" claimed that the therapies involved conditioning based on stigma and shame as well as discrimination against homosexuals. In 2009 an annihilating report by the American Psychological Association (APA) confirmed that the treatments were causing a rash of suicides among patients. The report concluded that therapy to change homosexuals into heterosexuals didn't work and that the association's 150,000 members should stop offering it to their clients. The report stated that the best such therapy could do was to teach people to ignore their feelings and to suppress homosexual inclinations. It went on to confirm that the therapy could cause depression and even lead to suicide.

All the research indicates that our sexual orientation is programmed in the brain before birth, determining it for the rest of our lives (see earlier in this chapter). Many structural and functional differences have now been found between the brains of homosexual and heterosexual men that must occur early on in development and can no longer be changed by the post-birth environment. Even an upbringing in a British boarding school apparently doesn't make you more likely to be homosexual in adulthood. Initially I thought that "curing" homosexuals was a typical aberration of the Christian community in America, but I was amazed to find that it goes on in the Netherlands too. The Pentecostal Church holds meetings whose prayers can allegedly "cure" you simultaneously of homosexuality and HIV infections, after which you're married off to a woman from the Pentecostal community. It's not just misleading but also potentially life-threatening to make seropositive individuals think that they've been cured in this way and no longer need to take any medication.

The outmoded notion that we're free to choose our sexual orientation and that homosexuality is therefore a wrong choice is still causing a lot of misery. The stories I heard when I gave a lecture to ContrariO, a Christian gay association, showed that homosexuals brought up in the Dutch Reformed Church tradition can still struggle terribly with their sexual orientation. Indeed, until recently, homosexuality was still regarded as a disease by the medical community. Only in 1992 was it removed from the ICD-10 (International Classification of Diseases). Before that time, doctors had striven to "cure" men of their homosexuality.

The idea that our social environment shapes our sexual orientation has led to mass persecution. The Nazi notion, as expressed by Hitler himself, that homosexuality was as infectious as the plague led to the unimaginable in Germany: first voluntary castrations, then compulsory castrations, and finally the systematic murder of homosexuals in concentration camps.

An important argument against the idea that homosexuality is a "lifestyle choice" or caused by environmental factors is the demonstrable impossibility of ridding people of their homosexuality. Every conceivable thing that could be devised has been tried: hormone treatments, castration, and treatments that influence libido rather than sexual orientation. Electroshock therapies have been tried, as well as epileptic insults. Prison sentences have proved equally ineffectual, as seen in the sad case of Oscar Wilde. Testicular transplants have been carried out, leading to a "success story" in which a homosexual man pinched the nurse's bottom after the operation. Psychoanalysis has also been tried, of course, as well as giving homosexuals apomorphine, a drug that induces nausea, in combination with homoerotic images, as a form of aversion therapy. The story goes that this didn't diminish the men's erotic desires; its only effect was to make them start vomiting as soon as the therapist entered the room. Brain operations have also been performed on homosexual prisoners with a view to reducing their sentences if the treatment proved effective. Naturally, the men all said that it was effective.

Since none of these approaches has led to a well-documented change of sexual orientation, there can be little doubt that by adulthood our sexual orientation has been determined and can no longer be influenced. If churches were finally to accept this fact, the lives of many of their young members and clergy would be a great deal happier.

HOMOSEXUALITY IN THE ANIMAL KINGDOM

Homosexual behavior has now been observed in around 1,500 animal species, from insects to mammals. The male penguin couple Roy and Silo in New York's Central Park Zoo are a famous example. They copulated, built a nest together, took care of an egg that a kindly keeper gave them (hatching it out after thirty-four days), and together looked after the baby. If a female rat develops alongside a male rat in the womb, thus being exposed to more testosterone during early development, it will mount other female rats. Two percent of oystercatchers, a monogamous bird species, form a trio of two females and a male, after which all three guard the same nest. A trio of this kind produces more offspring than a conventional pair, because they are better able to look after and protect the nest.

Behavioral scientists have also shown that homosexual behavior in animals is often used to make peace with enemies or obtain the help of others against possible attackers. Primatologist Frans de Waal has found bonobos to be completely bisexual, a perfect 3 in the Kinsey scale. Where possible, bonobos solve problems in the group by sexual means, through both heterosexual and homosexual behavior. De Waal has found that same-sex practices are displayed by other primates, too, like macaque monkeys, in addition to bull elephants (who mount each other), giraffes ("necking"), swans (greeting ceremonies), and whales (mutual caressing). He classifies such behavior as examples of bisexuality rather than homosexuality, since it only manifests itself in certain periods. However, a preference for same-sex copulation has been reported in a bird in the swamps of New Zealand, a female antelope in Uganda, and cows. Lesbian seagulls have been found in Southern California, jointly incubating a double clutch of eggs. These female gulls copulated with each other as a pair. However, this proved to be not spontaneous behavior but a by-product of environmental pollution with DDT, leading to sterility among male seagulls and an excess of females, who formed lesbian couples (see also endocrine disruptors, chapter 2). A few male seagulls must of course have escaped the DDT and had the time of their lives inseminating all of the females at least once, but apparently the ladies had no further need of them. In an albatross colony on a Hawaiian island with an excessive number of females, the females would pair up annually to preen each other, join in ritual mating dances, and guard each other. Together they would hatch out a single egg each year, taking turns to sit on the nest. No male came near them after insemination.

According to Frans de Waal, *exclusive* focus on members of the same sex, as shown among humans, is rare if not absent in the animal kingdom. I don't agree with him. In Montana, Anne Perkins discovered that 10 percent of the rams intended for breeding weren't mounting ewes. They were referred to as "lazy." But out in the meadow they were anything but lazy, as they enthusiastically mounted other males. Some rams even took turns mounting each other. Perkins discovered chemical differences in the hypothalami of these rams that indicated altered interaction between hormones and brain cells. Structural differences were also found in the hypothalami of these homosexual rams just like the ones that we and other researchers described in the case of humans. Of course homosexuality is a natural variation.

TRANSSEXUALITY

> Re: new phalloplasty technique proposal; seeking surgeon. P.S. I am interested in a neophallus uncircumcised in appearance. So I am looking overseas, since a natural uncircumcised penis is more common in Europe than in the U.S.
>
> From a letter to the author from an American female-to-male transsexual

Transsexuals feel that they have been born into a body of the wrong gender and are desperate for a sex change or gender reassignment. This is a gradual process that starts with an individual adopting the social role of the opposite sex and taking hormones, then undergoing a series of extensive operations—which only 0.4 percent later regret. The first person in the Netherlands to respond to the plight of transsexuals was Otto de Vaal, an endocrinologist and pharmacologist who taught at the University of Amsterdam. He treated them for free starting in 1965, feeling that his university pay was sufficient. The gender team of the VU University Medical Center in Amsterdam (VUmc) subsequently took on a pioneering role, headed first by Louis Gooren and now by Peggy Cohen-Kettenis. That's remarkable in itself because the Vrije Universiteit was established as a Calvinist university, and the Bible does say: "A woman shall not wear man's clothing, nor shall a man put on a woman's clothing; for whoever does these things is an abomination to the Lord your God" (Deuteronomy 22:5–6).

Since 1975, 3,500 people have undergone gender reassignment at the VUmc. The first time I learned about transsexuality was as a medical student in the 1960s. Coen van Emde Boas, a professor of sexology, entered the lecture room of the obstetrics and gynecology department with a bearded man. It wasn't exactly the place you would expect a man to be demonstrating anything. But he turned out to be a genetic woman, a female-to-male transsexual. This made a deep impression on me, and set me thinking about the possible underlying mechanism.

Male-to-female transsexuality (MtF) occurs in 1 in 10,000 individuals, and female-to-male transsexuality (FtM) in 1 in 30,000. Gender problems tend to become apparent from an early age. Mothers typically relate how their little boys dressed up in their frocks and shoes, were only interested in girls' toys, and mainly played with girls. But not all children with gender problems want to change sex later. If necessary, puberty can be delayed with the help of hormones to gain extra time in which to decide whether or not to undergo treatment.

All the data indicates that gender problems arise in the womb. Tiny variations in genes associated with the effect of hormones on brain development have been found to increase the likelihood of transsexuality. It can also be increased by abnormal fetal hormone levels or by medication taken during pregnancy that inhibits the breakdown of sex hormones. The differentiation of our sex organs takes place in the first months of pregnancy, while the sexual differentiation of the brain occurs in the second half of pregnancy. Since these two processes take place at different times, the theory is that in the case of transsexuality, they have been influenced independently of one another. If this is the case, one would expect to find female structures in the brains of MtF transsexuals and vice versa in the case of FtM transsexuals. In 1995 we indeed found, in postmortem studies of donor brains, a small structure in which the usual sex difference was reversed. We published our findings in *Nature*. The brain structure in question is the bed nucleus of the stria terminalis (BST), an area that's involved in many aspects of sexual behavior (figs. 10 and 11). The central part of this nucleus, the BSTc, is twice as large in men as in women and contains twice as many neurons. We found MtF transsexuals to have a "female" BSTc. The only FtM transsexual we could study—the material in question being yet rarer than the brains of MtF transsexuals—indeed proved to have a "male" BSTc. We were able to rule out the reversal of the sex difference in transsexuals being caused by altered hormone levels in adulthood, so the reversal must have happened at the developmental stage.

If you publish something truly interesting, the nicest thing you'll probably hear your colleagues say is, "It'll need to be confirmed by an independent research group." And that can take a while, because it took me twenty years to collect the brain material for my study. So I was delighted when in 2008 the group headed by Ivanka Savic in Stockholm published a study involving functional brain scans of living MtF transsexuals. They had not yet been operated on, nor given hormones. As a stimulus they were given male and female pheromones, scents that aren't consciously perceived. In control groups, these were shown to produce different patterns of stimulation in the hypothalamus and other brain areas in men and women. The stimulation pattern for MtF transsexuals fell between that of men and women.

In 2007 V. S. Ramachandran published an interesting hypothesis and provisional research findings on transsexuality. He believes that the neural body map of MtF transsexuals lacks a penis, while that of FtM transsexuals lacks breasts, due to these not being programmed into the map during development. As a result, they don't recognize these organs as their own and want to get rid of them. So everything indicates that the early development of sexual differentiation in the

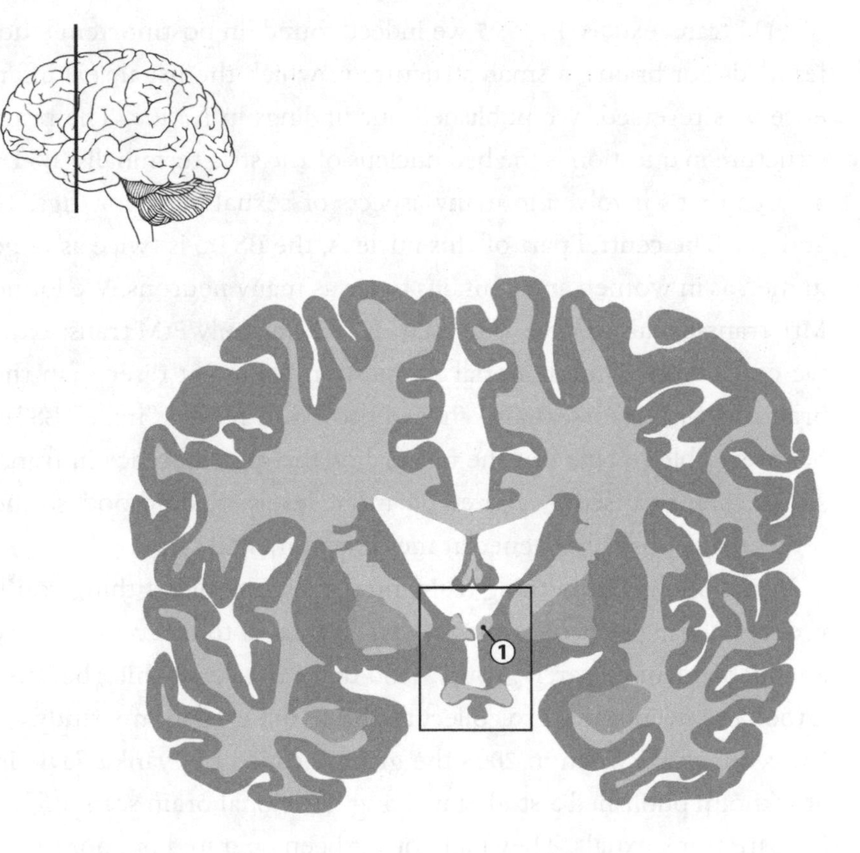

FIGURE 10. Located at the tip of the lateral ventricle (1) is the bed nucleus of the stria terminalis (BST), a region of the brain important for sexual behavior.

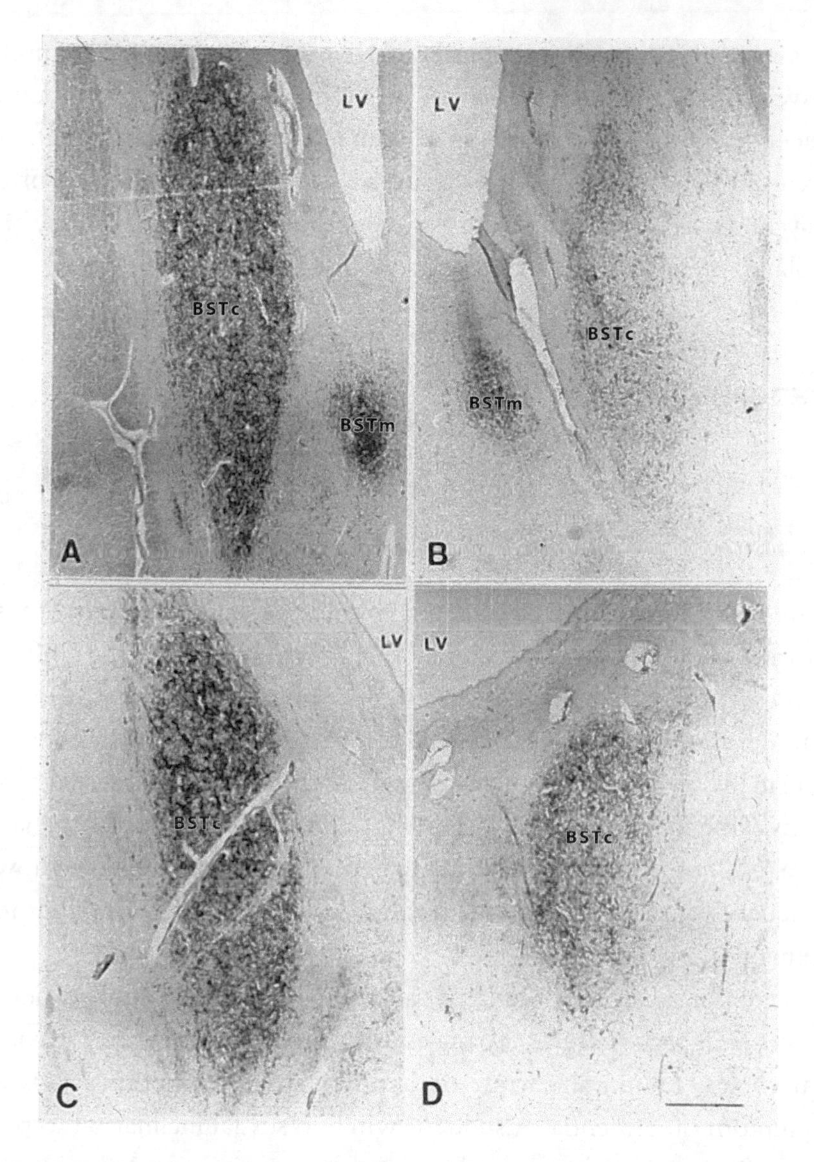

FIGURE 11. The central part of the bed nucleus of the stria terminalis (BSTc) (see fig. 10 for location) is twice as big in men (A, C) and contains twice as many neurons as in women (B). In male-to-female transsexuals we found a female BSTc (D). The only female-to-male transsexual we could study (these brains being rarer than those of MtF transsexuals) indeed proved to have a male BSTc. This reversal of the sex difference in transsexuals corresponds with their gender identity (the feeling of being a man or woman) rather than with their chromosomal sex, or the sex on their birth certificate. LV = lateral ventricle, BSTm = medial section of the BST. J-N Zhou et al., *Nature* 378 (1995): 68–70.

brains of transsexuals is atypical and that they aren't, in fact, simply psychotic, as a Dutch psychiatrist was impertinent enough to claim recently. At the same time it is of course essential, before initiating treatment, to make sure that the desire to change sex isn't part of a psychosis, as it can be an occasional symptom of schizophrenia, bipolar depressions, and serious personality disorders.

PEDOPHILIA

> "May I humbly crave Your Excellency's permission to be castrated?"

The shocking scale of child abuse within the Catholic Church has come to light in recent years. The first cases emerged in the United States, then in Ireland, where, within the bishopric of Dublin alone, hundreds of children were abused between 1976 and 2004. Cases in Germany were subsequently exposed, after which hundreds of victims came forward in the Netherlands. These revelations show that, as a result of the taboo surrounding pedophilia, we have no idea how frequently such abuse actually occurs—not just in the church but in general.

Pedophilia can have different causes. If an adult suddenly experiences pedophilic urges, they may have a brain tumor in the prefrontal cortex, temporal cortex, or hypothalamus. Sometimes it is a symptom of dementia. A sudden switch in sexual inclination to pedophilia has also been caused by operations to cure epilepsy by removing part of the anterior temporal lobe. Such patients can go on to develop Klüver-Bucy syndrome, which involves the loss of sexual inhibition (see chapter 4). In the United States, a man who started to download child pornography after an operation of this kind was recently sentenced to nineteen months of imprisonment! Pedophilia can also be caused by infections of the brain, Parkinson's, multiple sclerosis, and brain trauma.

But a neurological cause for pedophilia is rare. Most pedophiles have always been attracted to children, and the cause can be traced to fetal brain development and early development after birth. Just as gender identity and sexual orientation are determined by genetic background and the interaction between a fetus's sex hormones and its developing brain (see earlier in this chapter), so too pedophilia can apparently be explained by genetic and other factors causing the brain to develop abnormally at an early stage, leading to structural differences. I was once shown a family tree that included three generations of pedophile men. Deviant sexual behavior (like pedophilia) is displayed by a high percentage (18 percent) of first-degree relatives of pedophiles, pointing to a genetic factor. In addition, pedophiles are more likely to have been sexually abused by adults as young children. At the end of 2009, the leader of the Northern Irish party Sinn Féin, Gerry Adams, went public with the painful family secret that his father had abused his own children, while his brother was in turn suspected of having sexually abused his daughter. Whether abuse as a child is a causal factor in the development of pedophilia in adulthood, or whether there's a genetic factor in such families, still needs to be investigated.

Daniel Gajdusek (1923–2008), a man of remarkable talent who studied physics, biology, mathematics, and medicine in the United States, thought that abuse as a child could cause pedophilia. He had himself been abused by an uncle as a child. I once had the dubious honor of chairing a lecture by the hypomanic Gajdusek; my colleagues were amused by my vain attempts to keep him in check. Gajdusek had been researching the cause of mass deaths of young women and children from the disease kuru in villages in the interior of New Guinea in 1957. At the time, it was still a Dutch colony, and he was able to find his way there using Dutch ordnance survey maps that he'd stolen from the Leiden endocrinology department headed by Dries Querido. Gajdusek discovered that the deaths were indirectly caused by cannibalism. Long after they had eaten the brains of conquered enemies, the victims were struck down by a slow-acting virus, one of whose symptoms was dementia. The disease turned out to be caused by prions (infectious agents made of protein), just like mad cow disease. In 1996 Gajdusek was awarded the Nobel Prize for Medicine. However, when he returned from New Guinea and other remote locations it wasn't just with brain tissue for further research; he also brought back fifty-six children, mostly little boys. We always thought this was very odd. He took them into his home and gave them an education but, as an accusation made by a man who had lived with him as a child later revealed, also molested them. He was imprisoned for a year and died in 2008.

There are all kinds of factors in early development that could influence the risk of developing pedophilia. It would seem logical to study them, but the taboo on this condition stands in the way. Who in our society would dare openly admit to being a pedophile and take part in research into the causes of this disorder?

In recent years, the first structural differences have been reported between the brains of pedophiles and those of control groups. A study involving magnetic resonance imaging (MRI) showed that the former have less gray matter (neurons) in various areas of the brain, like the hypothalamus, the bed nucleus of the stria terminalis (whose size also differs in transsexuals; see earlier in this chapter), and the amygdala, which plays a role in sex, fear, and aggressive behavior. It moreover emerged that the smaller the amygdala, the more likely an individual was to commit pedophilic crimes. Exposure to emotional and erotic images of adults sparks less activity in the hypothalamus and prefrontal cortex of pedophile men than in control men, which ties in with the fact that pedophiles are less sexually interested in adults. Convicted pedophiles display greater amygdala activity than control men in response to images of children. Functional scans of the brains of homosexual, heterosexual, and pedophile men shown pictures of men, women, girls, and boys moreover show a clear difference between these groups in terms of brain activity. However, we must bear in mind that research into pedophilia focuses solely on a small, selected group of pedophiles. The majority are able to control their urges, don't commit crimes, and therefore aren't studied.

Sexual abuse damages children and is punished, not only for reasons of atonement but also to prevent further abuse. The latter objective poses a problem, though, because how do you change behavior that has been programmed in the brain at an early stage of development? In the past, every conceivable effort has been made to change homosexual men into heterosexuals (see earlier in this chapter), without any success whatsoever. The same applies to pedophiles. Not so long ago, a court in Utrecht heard the case of a sixty-year-old heterosexual church minister charged with pedosexuality. The prosecution called for a sentence of ten months in prison, but after a great deal of deliberation he was eventually given a community sentence. How things have changed.

There was a time when an obscure mix of arguments bearing on eugenics, punishment, the protection of society, and the repression of homosexuality led to the castration of pedosexuals in the Netherlands. Between 1938 and 1968, at least four hundred sex offenders were "voluntarily" castrated. This practice wasn't laid down by law. These were offenders detained under a hospital order who were given the choice of life imprisonment or castration. They had to submit a standard letter to the minister of justice, the text of which ran, "May I humbly crave Your Excellency's permission to be castrated?" Up to 1950, 80 percent of the castrated men were pedosexuals, a situation complicated by the high legal age of sexual consent (sixteen). In Germany, the hypothalami of pedophiles were surgically lesioned in the hope that this would change their sexual orientation. These brain operations were never scientifically documented.

The incidence of chemical castrations among offenders detained under a hospital order is currently increasing. This involves suppressing the libido with a substance that diminishes the effect of testosterone. It can provide relief at being freed from sexual obsession. However, it's worrying that some of these individuals are being chemically castrated because the authorities would otherwise deny their applications for leave. These substances certainly aren't suitable for every sex offender, and the side effects, including the development of breasts, obesity, and osteoporosis, are serious.

The pedosexual minister from Utrecht can thank his lucky stars that things have changed since the days of formal castration requests. The judge who presided over his case was worried about reoffending, and rightly so. Nevertheless, he thought that the six-week pretrial detention would have a deterrent effect and that the combination of a long conditional sentence and a community order would be more effective than lengthy imprisonment. Whether he was right we'll never know, because the judicial system has no tradition of researching the effectiveness of its punishments. And the medical world, alas, has no tradition of researching the factors in early development that could cause pedophilia. Doing away with the taboo on such research could shed light on these factors and on the best methods of checking pedophile impulses and stopping people from reoffending. This would prevent a great deal of misery for all concerned.

The same applies to female pedophiles. The idea that women can't be guilty of pedosexuality has been found to be a myth. Sexual abuse of children by women is usually perpetrated by mothers on their own offspring. For the most part, the victims are girls with an average age of around six. The mothers tend to be poor and uneducated and often have mental health problems like cognitive impairment, psychoses, or addictions.

An initiative in Canada has shown that it's possible to tackle this issue by quite simple means. There, pedosexuals are helped by a group of volunteers after their detention. The resulting social network has been shown to cut reoffending rates quite considerably. This is much better than the situation in the Netherlands, where in late 2009, a pedophile was first banned from the city of Eindhoven by its mayor, then prohibited from entering a national park in the province of Utrecht. The man now lives in his car and travels from parking lot to parking lot. That's asking for trouble. But the Netherlands is now trying out the Canadian initiative. Another way of preventing child abuse might be to issue smart forms of fake child pornography that don't involve the abuse of real children. Milton Diamond, a renowned sexologist in Hawaii, has found considerable evidence to suggest that this works. However, it will no doubt prove difficult to convince the authorities to consider such an innovative idea.

PUBLIC RESPONSE TO MY RESEARCH INTO SEX DIFFERENCES IN THE BRAIN

> Angry gays got it all wrong.
>
> Dutch gay newspaper

In the 1960s and 1970s the received wisdom was that a child is born as a blank slate and that the development of both its gender identity and sexual orientation are very much determined by social conventions. This notion, one of whose leading proponents was the Philadelphia-based psychologist John Money, had terrible consequences (see the John-Joan-John case earlier in this chapter) but reflected the general thinking at that time that everything could be socially engineered, including whether you felt male or female and were heterosexual or homosexual.

When I gave my first lectures on sex differences in the brain in the 1970s at the medical faculty in Amsterdam, the broadly held views on the importance of social conditioning weren't just being trumpeted by Money and his supporters, they were also espoused by the feminist movement. Its adherents believed that all of the differences between the sexes in terms of behavior, occupation, and interest had been forced on women by a male-dominated society. In those early lectures, female students would sit in the front row of the auditorium, demonstratively knitting and crocheting. They made it abundantly clear that the subject I was discussing and my views on the matter were anathema to them. When the light was switched off so that I could show some slides, they protested vociferously, because they couldn't see their knitting anymore. From then on I turned the lights down and showed slides throughout all classes and lectures. The ladies from the front row sent a delegation to the dean to request a lecturer who would be more sympathetic to women. Apparently none was available, because I never heard any more about it.

Our description of the first sex differences found in human hypothalami in postmortem brain tissue (Swaab and Fliers, *Science* 228 [1985]: 1112–15) provoked a hostile response from feminists. At the time there was widespread denial within the feminist movement of possible biological sex differences in the human brain and behavior. Speaking about our findings in an interview with the Dutch magazine *HP* (January 17, 1987), a woman biologist by the name of Joke 't Hart said, "But if I were to accept that there are differences between the sexes in such fundamental areas as the structure of our brains, I would no longer have a leg to stand on as a feminist." Whatever the case, I never heard any more of her. Many hundreds of sex differences have subsequently been identified between the male and female brains.

After we had reported on the first difference found between the brains of homosexual and heterosexual men (later published in Swaab and Hofman, *Brain Research* 537 [1990]: 141–48; see earlier in this chapter), the unexpected backlash took us by surprise. It all started in December 1988 with an article that appeared in an obscure Dutch publication called *Akademie Nieuws*. Researchers at institutes directed by the Royal Netherlands Academy of Arts and Sciences (KNAW) had been asked what they were working on, so I'd talked about our brain research into sexual orientation and gender. This was picked up by Hans van Maanen, a reporter for the Dutch daily newspaper *Het Parool,* who wrote two articles entitled "Gays' Brains Are Different" and "The Brain Behind Homosexuality," both of which presented an entirely correct picture. But they unleashed an uproar of unbelievable proportions. What exactly caused this overwhelming and emotional response—which was completely off target—is still a mystery to me. The taboo on a biological explanation for our sexual orientation, which was very marked in an age of boundless belief in social engineering, must have played a role. One group of homosexual men made an almost religious pronouncement to the effect that all men are homosexual, but only some of them opt to come out publicly. They called coming out a political choice. I stated that I couldn't see how it was political and that the choice of sexual orientation was made for you in the womb; this only fanned the flames. Many hundreds of articles were published in the space of three weeks. COC Netherlands, a gay rights organization, pronounced itself "amazed by the study." At that time, Rob Tielman, who held the chair of gay studies at the University of Utrecht, was one of my most vocal opponents. He demonized the study by calling it "in extraordinarily poor taste" and claimed, ludicrously, that I should first have asked his permission to carry it out and publish my findings. He later retracted his remarks in an interview in which he said, "My position in the field of gay studies is closest to Swaab's," and "I am among those who are inclined to take the biological component very seriously." But the editor in chief of the *Gay Krant*, Henk Krol, had meanwhile joined the fray, arguing that "a study of this kind underlines the notion of homosexuality as a disease. In turn, this promotes discrimination against gays." Questions were asked in Parliament about my study by Peter Lankhorst (Progressive Radical Party). Those questions landed on my desk via the minister of education and science and the president of the KNAW, and my answers went by the same route in reverse. I received threatening telephone calls day and night as well as a card addressed to "the SS Doctor Mengele-Swaab," which read, "Nazi. Saw your ugly mug on TV. We homosexuals are going to kill you. As an example. Like the leader of Iran did to the Englishman" (fig. 12). At the time I didn't take it seriously and commented that if their assassination skills were as bad as their written Dutch, I wasn't in very much danger. Nowadays I think I would be more concerned. I also got a card that read, "Bet you regret not having been able to work under

Mengele in Auschwitz!" (fig. 13). Committees scrutinized my research, and I was given bodyguards when I lectured at the Academic Medical Center. The Netherlands Institute for Brain Research became a focus for bomb scares (which I didn't take seriously either), our children were teased at school, and a demonstration took place one Sunday morning in front of our house, described in inimitable fashion by the (gay) writer Gerard Reve. It even furnished the title for his essay collection *A Carefree Sunday Morning* (1995). He wrote:

> Only now did it become clear what a serious omission Professor Swaab had been guilty of by failing to ask the homosexual trade union COC permission in advance for his research. The consequences made themselves seen and heard. A large group of motivated individuals appeared in front of Professor's Swaab's home in Amstelveen on Sunday morning, chanting loudly, "Dick, cut up your own d—!" A curious choice of words, given that, although Professor Swaab did carry out a study on sexuality, it involved cutting up brains rather than genitals. But this trade union's adherents don't have brains, only genitals, so in a way it makes sense.

It took three weeks for the storm to die down. Then Ayatollah Khomeini pronounced a fatwa on Salman Rushdie after the publication of *The Satanic Verses,* and suddenly all attention shifted to the British-Indian writer. When the smoke of battle had cleared away and I'd emerged unscathed, the president of the KNAW, David de Wied, gave an interview in the Dutch daily newspaper *De Telegraaf* in which he backed me up and said that an affair like that should never be allowed to happen again. A pity he hadn't done so a few weeks earlier.

But I had some nice responses too, like the cartoon by Peter van Straaten (fig. 14) and personal ads in the prominent Dutch weekly magazine *Vrij Nederland,* like "Nice guy (37, 1.87m, 87kg, fair-haired, blue-eyed) with big hypothalamus is looking for a partner" and "Wanted: BIG suprachiasmatic nucleus, Postbox 654 Wageningen."

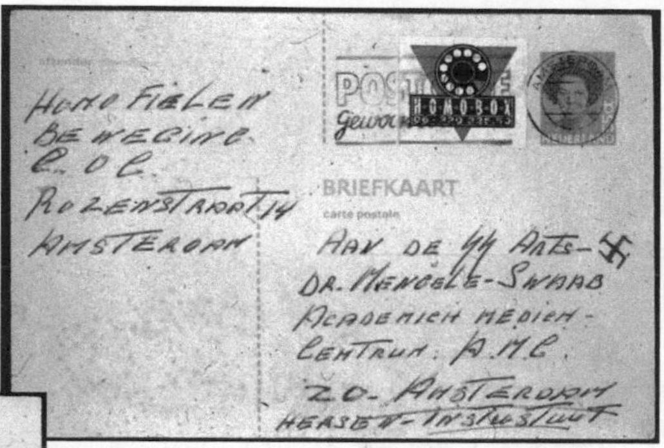
HOMOFIELEN
BEWEGING
C.O.C.
ROZENSTRAAT 14
AMSTERDAM

BRIEFKAART
carte postale

AAN DE ARTS-
DR. MENGELE-SWAAB
ACADEMISCH MEDISCH-
CENTRUM: A.M.C.
Z.O.-AMSTERDAM
HERSEN-INSTITUUT

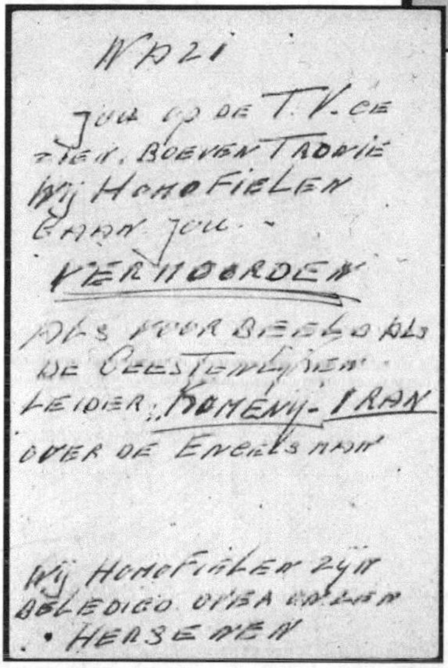
NAZI
JOU OP DE T.V. GE
ZIEN, BOEVENTRONIE
WIJ HOMOFIELEN
GAAN JOU
VERMOORDEN
ALS VOORBEELD ALS
DE GEESTELIJKEN
LEIDER: KOMENY-IRAN
OVER DE ENGELSMAN

WIJ HOMOFIELEN ZIJN
BELEDIGD OVER ONZE
HERSENEN

FIGURE 12. A postcard I received after publishing the first findings of a difference between the brains of homosexual and heterosexual men, in 1989. Said to be sent on behalf of the COC (gay rights) organization, the text reads, "Nazi. Saw your ugly mug on TV. We homosexuals are going to kill you. As an example. Like the leader of Iran [Khomeini] did to the Englishman. We homosexuals are insulted about our brains."

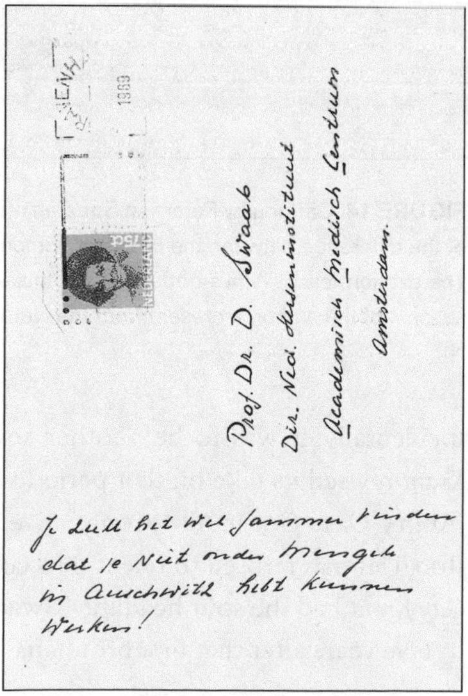
Prof. Dr. D. Swaab
Dir. Ned. Hersenenstituut
Academisch Medisch Centrum
Amsterdam.

Je zult het wel jammer vinden
dat je niet onder Mengele
in Auschwitz hebt kunnen
werken!

FIGURE 13. Another example of correspondence I received after publishing the first findings of the difference between the brains of homosexual and heterosexual men. The text reads, "Bet you regret not having been able to work under Mengele in Auschwitz!"

FIGURE 14. Cartoon by Peter van Straaten after the publication of the first findings of the difference between the brains of homosexual and heterosexual men (1989). The caption read, "Wim's got a big hypothalamus too, eh, Wim?" Original in possession of the author, a present from the Netherlands Institute for Brain Research, NIH.

Incidentally, it was to be another seventeen years before the *Gay Krant* revised its take on that period with an article tellingly headed "Angry Gays Got It All Wrong." Even after all that time, however, Rob Tielman refused to relent. His column in the same issue of the *Gay Krant* had the sour headline "Swaab Headstrong."

Five years after that first brouhaha, the publication of our discovery of a sex-reversed pattern in the brains of transsexuals (Zhou et al., *Nature* 378 [1995]: 68–70; see fig. 11 in this book) met with an entirely positive response. Transsexuals immediately seized on the article in order to get sex changes registered in birth certificates or passports in countries where that had not previously been possible. It was used to the same end at the European Court of Justice and played a role in the drafting of legislation on the issue in Britain.

Nowadays, articles on the differences between male and female or heterosexual and homosexual brains barely cause a ripple (see for instance Swaab, D. F., *Proc. Natl. Acad. Sci. USA* 105 [2008]: 10273–74), and there's a great interest in the topic in popular science publications.

CHECKING THE POPE'S SEX

At the stage when our bodies and brains differentiate along male or female lines, hybrid forms sometimes develop. This can have far-reaching consequences. A controversial example of hybrid sexuality dates back to the Middle Ages, when a woman allegedly became pope, subverting the strict male hierarchy of the Roman Catholic Church. Measures are said to have been taken to prevent such a "disaster" from occurring again.

The story of Pope Joan was chronicled by the Dominican monk Jean de Mailly around 1250, and a film was made of it in 1972. Was it a myth or a cover-up? No one knows for sure. The gist of the legend is as follows. Though born in the German town of Mainz in 833, Joan was of English extraction. Having traversed Europe dressed as a monk, she gained so much respect and authority for her great learning that she succeeded Leo IV as pope in 854, taking the name of Johannes Anglicus (John the Englishman) or John VIII. Three years later, however, she became pregnant, suddenly giving birth during the Easter procession, near the Basilica of San Clemente in Rome. This gave the game away, of course, and she was lynched on the spot. Her successor, Benedict III, is said to have eradicated all trace of her memory. There's no record of Joan in the Vatican's pontifical yearbook.

Although the Catholic Church systematically denies the story, there are indications that there may be some truth in it. In 1276, Pope John XX was said to have changed his name to John XXI in order to account for the female Pope John. What's more, records indicate that the sculpted head of John VIII, "Femina de Anglica," stood alongside the busts of all the other popes in the Cathedral of Siena. In 1600, however, her bust was removed by order of Pope Clement VIII.

And then there's the chair with the hole, referred to in Italian as *La Sedia Gestatoria*—meaning "litter" or "sedan chair." But why would a sedan chair have a hole in the seat? The story goes that, to prevent another female pope from being elected, candidates for the papacy were required to sit on the chair. The youngest cleric present had to stick his hand through the hole, feel the papal candidate's genitals, and then call out loudly, *"Testiculos habet et bene pendentes"* ("He has testicles and they hang well"). The cardinals present would then respond, *"Habe ova nostra Papa"* ("Our father has balls"—as if they were any use to him). The need for such testing suggests that the story of Pope Joan may very well be true. Maria New, a New York pediatric endocrinologist, has put forward the theory that Pope Joan had a form of intersex called congenital adrenal hyperplasia, or CAH (see earlier in this chapter). But this diagnosis is pure speculation.

In her article (1993), New referred to a red marble chair that apparently stood in the Vatican Museum. When I met Dr. New in 2007 at a conference in Rome, however, I asked where exactly the chair was, because I'd been invited to the Vatican the following morning. She told me that she had never gotten to see it. The one depicted in her article had been an identical chair looted by Napoleon, now in the Louvre, to which she had only gained access after negotiating a great deal of red tape.

At the Vatican the next day, on a private tour organized by one of the pope's doctors, a collaborator of mine immediately said to the head of security that I was especially interested in seeing the chair. No problem, our security man responded, though he immediately added that Maria New's theory was nonsense. According to him, the chair was simply an ancient commode. At the time I wondered how he knew about her article. It was highly technical and published in an academic journal intended for specialists—not the kind of reading you associate with security guards.

He gave us a tour through the hushed corridors of the Vatican. We saw the room where the cardinals elect the pope, the "crying room" to which newly appointed popes are ushered to shed a few tears, the containers of white and black smoke, the chamber with the famous balcony where the pope appears (and from which Pope John Paul II called out in something faintly resembling Dutch, "Thanks for the flowers!"), the terrible murals everywhere, and the Swiss guards flanking each important doorway. We were shown the papal gardens, the secret escape route to the stronghold, and so on. Our guide even got out all the pope's gorgeously embroidered mantles for special occasions so that we could admire and feel them. One was pink. "Is that for Saturday evenings?" I asked the cleric, who was lovingly displaying each garment. "No," he answered seriously, "that's for prison visits." The pope's miters were unpacked for us, along with a crucifix that he takes with him on his travels—there was no end to it all.

It was all most impressive and yet not quite what we'd come for, and I reminded our guide of the chair. Yes, yes, it was a bit farther on, he said soothingly. When we had emerged from the silence of the private rooms, had passed the Sistine Chapel with its hordes of tourists, and had gone once again through a series of doors opened and locked behind us by Swiss guards, I mentioned the chair once more.

"Oh, what a pity," said the head of security. "We passed it about fifteen minutes ago. I'm so sorry, I completely forgot about it."

"No problem," I said airily, "we can just go back."

Alas, this was impossible, "for security reasons," the security officer said, as he told us exactly which country had donated each of the many trees and shrubs in the Vatican garden. A complex network of cables hung over our heads, and large antennas betrayed an advanced communications system. The way back to the Middle Ages had been cut off. So that ended my chance to establish the truth of the chair and Pope Joan, but the matter continued to preoccupy me. If the chair didn't exist, why didn't the security officer just say so? Why did he keep stringing us along? Was the chair still being used, or might Pope Benedict XVI have been toying with the idea of reinstating this old custom?

4

Puberty, Love, and Sexual Behavior

THE ADOLESCENT BRAIN

Puberty starts with a kiss.

Dungan et al., 2006

In puberty, the pituitary gland starts to produce sex hormones. These affect the adolescent brain, causing marked and often incredibly annoying behavioral changes. The evolutionary advantage of puberty is clear: Youngsters are being prepared for reproduction. And their annoying behavior, leading to frequent clashes with their families, makes it less likely that reproduction will take place in their own surroundings, thus reducing the risk of inherited defects. The craving for new experiences, the readiness to take great risks, and the impulsive behavior are all part of preparations to leave the nest. Because their prefrontal cortex hasn't yet matured, adolescents can think only in the short term and are unable to take in the negative consequences of risky choices. As a result, they are also more likely to try addictive substances that can permanently damage the developing brain.

A great many chemical changes are needed to initiate puberty. Its onset is triggered by the gene KISS1, which produces the protein kisspeptins in the hypothalamus. The gene is so central to this process that it has been said that "puberty starts with a kiss." The gene was discovered by American researchers in Hershey, Pennsylvania, and named after the most famous local product, the Hershey Chocolate Kiss. People with a mutation in the KISS1 system never enter puberty.

However, puberty is dependent on other systems, too. For instance, women must have sufficient fat reserves to be able to nourish a fetus at times of scarcity. Before puberty, the brain registers whether there's sufficient fatty tissue by monitoring the amount of leptin, a hormone that's produced by fat cells. If the fat reserves are insufficient—because of an eating disorder, for instance, or intense athletic training—leptin levels decline and puberty is delayed, sometimes for good. Similarly, mutations in the leptin gene can impede puberty and also cause extreme obesity. In such cases, the brain registers the absence of leptin—and therefore of fat. The brain then blocks the onset of puberty because pregnancy would be too risky while also sending out a signal to eat copiously to make up fat reserves—unaware that it's merely leptin, not fat, that's lacking.

Melatonin, a hormone produced by the pineal gland, is one of the substances that prevent the onset of puberty in children. Melatonin's inhibiting effect has been known since 1898, when Otto Heubner described a boy of four and a half who had already entered puberty. He turned out to have a brain tumor that had destroyed his pineal gland, which resulted in a lack of melatonin, in turn spurring the onset of puberty. A Dutch girl who started puberty at the age of three and a half was more fortunate. She didn't have a brain tumor and was given inhibitory hormones up to the age of twelve. Then she entered puberty again and is now flourishing at secondary school. Conversely, some people's melatonin level is too high and has to be normalized before puberty can start.

Puberty can also be disrupted by a condition known as Kallmann syndrome. Normally, the brain cells that stimulate the sex hormones develop at the place where a fetus's nose develops. The cells then migrate along the olfactory nerve to the hypothalamus. This process is disrupted in patients with Kallmann syndrome, so that not only do they not enter puberty, they also lack a sense of smell.

So we should stop dismissing the embryonic and adolescent brains as organs in which not much is going on yet. On the contrary, they are both undergoing the most complex and delicate modifications.

ADOLESCENT BEHAVIOR

> Young people today love luxury, they have bad manners, contempt for authority and disrespect for older people. They're too lazy to train; they'd rather sit and chat. . . . They no longer rise when elders enter the room, they contradict their parents, can't hold their tongues in company, gobble their food, and tyrannize their teachers.
>
> Socrates

According to adolescents, it's not their immature brains that are the problem, it's their parents. In fact, they are more or less the same thing: A youngster's parents function as his or her temporary prefrontal cortex (PFC, fig. 15). While the adolescent PFC is still immature, parents have to be responsible for a child's planning, organization, moral framework, and limits. These functions are gradually taken over by the slowly maturing PFC. The problem is that today's youngsters have discovered that their parents aren't in a position to enforce their role as human PFC substitutes.

The PFC plays a central role in regulating other brain areas and is responsible, among other things, for the control of our impulses, complex actions, planning, and organization. It doesn't fully mature until a person is in his or her twenties. According to the neuropsychologist Jelle Jolles, this explains why a recent restructuring of the Dutch secondary school system that focused on independent study is flawed. When your PFC hasn't matured, you're not very good at organizing your work and making independent choices. Functional scans also show clear differences between the brains of adolescents and those of adults. Adults distribute assignments across different brain areas. Adolescent PFCs can sometimes function at an adult level but need to work much harder to do so, as they fail to outsource tasks to other brain areas. As a result, a teenager's PFC reaches the ceiling of its capacity earlier, and distractions can undermine the performance of assigned tasks.

The regulation of day-night rhythms is also influenced by sex hormones. That might explain why it's so hard to get adolescents out of bed in the morning and into it at night. Should we force them to get up early or adapt school hours to their biological clocks?

A lot of drinking goes on in puberty: At the age of fifteen, 52 percent of boys and 46 percent of girls in the Netherlands drink at least five units of alcohol an evening on the weekend, and children of that age regularly end up in a coma in intensive care. The practice of "pre-gaming" before going to parties is now quite common. Alcohol abuse makes the brain shrink, causing permanent damage. In Europe, around fifty-five thousand youngsters die annually of alcohol poisoning or in traffic accidents in which alcohol was involved.

The sudden surge of sex hormones during puberty causes not only sexual awakening but also typical male aggression and risk-seeking behavior. That explains why the incidence of unrestrained, antisocial, aggressive, and delinquent behavior increases during puberty. A survey in the Netherlands showed that one in three children between the ages of ten and seventeen commits a crime, ranging from theft and breaking and entering to vandalism and crimes of violence. After the age of seventeen, young people commit fewer crimes. It seems logical to assume that this curve reflects the gradual development of the PFC, which inhibits impulsiveness and promotes moral behavior. Parents can take comfort in the thought that puberty is a finite process. Teachers, on the other hand, must sometimes despair. As fast as they mold adolescents and launch them into society, a fresh crop of impulsive youngsters enters the school gates. For teachers the process is never over.

THE BRAIN IN LOVE

> LOVE, *n.* A temporary insanity curable by marriage.
>
> Ambrose Bierce (1842–c. 1914)

A great many brain processes are involved in various stages of our love lives, including falling in love, sexual arousal, attachment leading

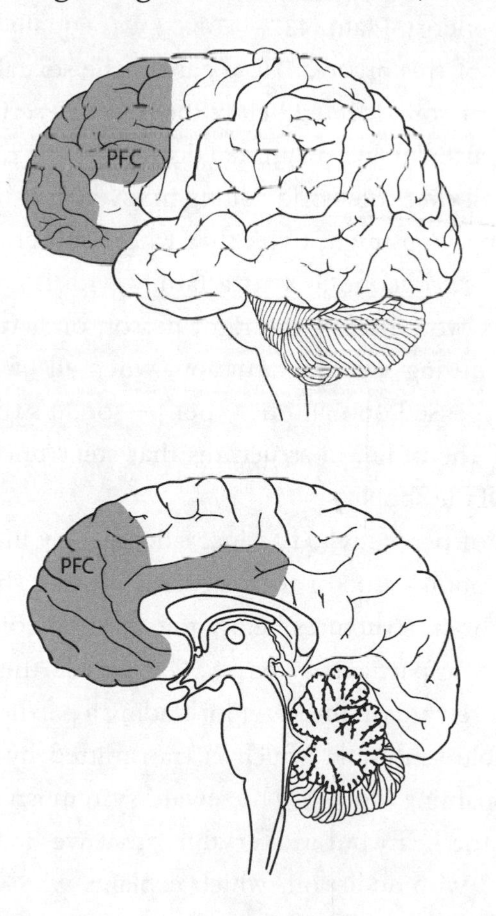

FIGURE 15. The prefrontal cortex (PFC) as seen from the side (external view at top, cross section at bottom).

to long-term partner bonding, and maternal and paternal behavior (see chapter 1). Although it wasn't Mother Nature's "intention," we see on a daily basis that these stages can perfectly well exist independently of one another, and I will therefore look at them separately.

No one who can still remember the suddenness and intensity of falling passionately in love will classify partner choice as a free choice or even a well-considered decision. Love at first sight just happens—it is pure biology—along with all the euphoria and severe physical reactions that ensue, like a beating heart; perspiration and insomnia; emotional dependency; strongly focused attention; an obsessive, possessive, and protective attitude toward the partner; and a feeling of heightened energy. Plato (427–347 B.C.) was equally convinced of the autonomy of this process. He regarded the sexual impulse as a fourth species of soul, located below the navel, describing it as "rebellious and masterful, like an animal disobedient to reason."

For people all over the world, falling in love tends to be the basis for pair forming. You might think that where something as important as choosing someone to start a family with is concerned, our cerebral cortex would select the right person on a fully conscious basis. But no, during severe infatuation, when all of our attention and energy is focused on that one other person, it's the areas down at the base of the brain, in structures that steer unconscious processes, that call the shots.

Brain scans of people who had just fallen deeply in love and who were shown a photograph of their significant other showed activity exclusively in brain structures below the cerebral cortex. Their reward circuitry was particularly active. This part of the brain focuses on obtaining a reward (in this case, for finding a partner) in the form of a pleasurable sensation, which is transmitted by the chemical messenger dopamine (fig. 16). The reward system isn't involved just in matters of the heart but in everything that we find pleasant. It's also associated with addiction, which explains why people experience severe withdrawal symptoms when a love affair ends. Scans show this system to be primarily activated on the right side of the brain, in proportion to the attractiveness of the face in the photo and the intensity of the romantic passion.

People who are in love also have raised levels of the stress hormone cortisol. Being in love is a stressful situation, and the body responds by producing more of this hormone. The level of testosterone (also produced by the activated adrenal gland) increases in women who are in love, while in men, cortisol reduces the testicular production of testosterone.

It's only when love has persisted for a certain length of time that the prefrontal cortex, the front part of the brain involved in planning, deliberation, and assessment, becomes involved. If stable pair formation ensues, the activity in the stress axis dies down and testosterone levels return to normal. The processing of sensory information in the cerebral cortex has of course played a role during that exciting period—we have, after all, seen, smelt, and touched the person we love. But this isn't the same thing as making a conscious

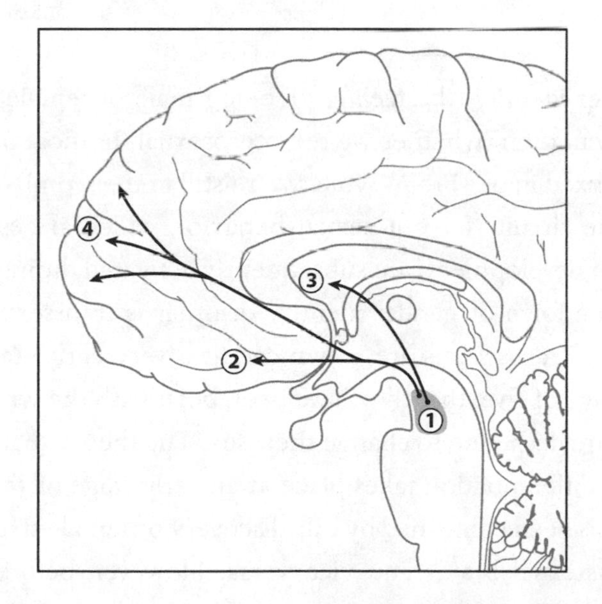

FIGURE 16. The dopaminergic reward system, originating in the cell bodies in the ventral tegmental area (1), whose fibers extend into regions including the ventral striatum (ventral pallidum/nucleus accumbens [2]), the caudate nucleus (3), and the prefrontal cortex (4).

choice for that particular person. Whether they are "Mr. (or Ms.) Right" is determined by our ancient reward circuitry, which thus links reproduction to the "right" partner—or at least the right partner in that moment. Only when the most intense period of infatuation has passed does the cerebral cortex take over. So if your son or daughter suddenly falls for the wrong person, it's no good reproaching them that they should have used their brains. They did, in fact, do so, but those parts of the cerebral cortex (such as the PFC) that could have come to a different decision after a balanced, conscious judgment unfortunately only kick in when it's too late.

DISORDERS OF THE BRAIN AND SEXUALITY

> An intellectual is someone who has found something more interesting than sex.
>
> Edgar Wallace

Our gender identity (the feeling of being male or female) and our sexual orientation (whether we're heterosexual, homosexual, or bisexual) is fixed in our brains while we're still in the womb (see chapter 3). The circuits for our sexual behavior, which are established early on in development, are subsequently activated during puberty. An extreme form of gender identity disorder is transsexuality (see chapter 3). Transsexuals are convinced at a very early stage, often from the age of five, that they have been born with the wrong body, and they are desperate to change their sex. The theory that an atypical sexual differentiation takes place at an early stage of their development has been confirmed by our discovery of female structures in male transsexual brains and vice versa. However, before treating transsexuals, it's important to rule out the possibility that their desire for a sex change isn't part of a psychosis caused by schizophrenia, a bipolar disorder, or a serious personality disorder. In the case of changes in sexual orientation, disorders of the brain also need to be ruled out. A shift from adult heterosexuality to homosexuality or pedophilia is sometimes seen in patients with brain tumors or with brain disorders like damage to the temporal lobe that cause uninhibited sexual behavior.

Since there's a time and place for everything, many areas of the brain are constantly busy inhibiting our sexual urges. This is usually effective for around twenty-three hours a day. Sexual disinhibition, or hypersexuality, is seen in patients who have suffered damage to such inhibitory brain areas. This type of brain damage can also cause paraphilia (atypical sexual arousal, for example in response to inanimate objects), sadomasochism, or pedophilia. Some forms of epilepsy are treated by surgically removing part of the temporal lobe, and operations of this kind occasionally result in a type of hypersexuality known as Kluver-Bucy syndrome. One man who had undergone this operation wanted his wife to have sex with him five or six times a day, in between which he masturbated. The amygdala, an almond-shaped structure located in the front part of the temporal lobe, regulates aggression and inhibits sexual behavior, among other things. Sometimes, in order to treat unmanageable forms of aggression, the amygdala is surgically lesioned, which occasionally also results in patients developing Klüver-Bucy syndrome.

Electrically stimulating the amygdala, on the other hand, has been shown to induce pleasant sexual sensations. In other brain structures, too, this method of stimulation can generate sexual behavior. Patients with an electrode inserted into the septum (fig. 26) were able to induce orgasms in themselves; some even developed a compulsion to masturbate. Patients whose septum has been accidentally perforated by the tip of a plastic tube inserted to drain cerebrospinal fluid to the abdominal cavity (a ventriculoperitoneal shunt) sometimes experience heightened sexual urges, while, conversely, damage to the septum has also been shown to cause impotence. Such cases help shed light on the brain structures that inhibit our sexual impulses and enable us at least to keep up the appearance of being decent members of society.

Orgasms Can Be Seen in the Brain

"Is nothing sacred anymore?"

Sex starts and ends in the brain. Many cerebral systems keep our sexual behavior constantly in check, but when we fall in love, inhibition flies out of the window. Our brain structures spring into action, sending messages via the spinal cord and the autonomic nervous system preparing our sex organs for that one true purpose of existence: the fertilization of an egg. To encourage us to truly commit to that objective, the brain provides orgasm as a reward. Impulses caused by the stimulation of our sex organs travel via the spinal cord to the brain, arriving at the thalamus (fig. 2), the central structure for all erotic sensory information. These then travel on to the ventral tegmental area (fig. 16), with its dopamine-delivering reward circuitry, and the hypothalamus (fig. 18). If all of these impulses lead to orgasm, we're simultaneously rewarded by the release of dopamine into the nucleus accumbens (fig. 16) and of the "love hormone" oxytocin in the hypothalamus (fig. 5), which increases social interaction between partners and promotes the release of opiates in the brain. All of these substances are so addictive that the number of people on earth has now swelled to seven billion.

People differ in all kinds of ways, including the extent to which they are interested in sex. DNA polymorphisms, tiny differences in the gene for the protein that receives dopamine's chemical message (the dopamine D4 receptor), are correlated to the degree of sexual desire, sexual arousal, and sexual activity itself. An overactive dopamine system can also cause problems. People with Parkinson's disease have a dopamine shortage; they are treated with L-dopa, which is transformed into dopamine in the brain. A possible side effect of this treatment is hypersexuality. One of the surgical methods used to treat Parkinson's involves implanting a depth electrode in the brain, in the subthalamic nucleus (fig. 23), in order to reduce tremor. However, stimulus of this kind can occasionally cause patients to develop hypersexuality, with or without mania.

The activity that sex causes in the brain's reward circuitry is visible on scans. In Groningen, the neuroscientist Gert Holstege somehow persuaded people to induce orgasms in their partners while the latter were lying with their heads in a brain scanner. The scans revealed activity in the dopamine-producing system in the ventral tegmental area (fig. 16) that was identical to the response to a heroin injection. That makes sense, because in addition to the dopamine system, the brain's opiate system is also involved in orgasm. Patients who take naloxone, a substance that inhibits the production of opiates in the brain, experience less pleasure from orgasms. Holstege's scans showed that different areas of the brain are activated in men and women during sexual arousal. Activity in the female brain is concentrated in the motor and sensory areas of the cerebral cortex; in the male brain it's mainly in the occipito-temporal cortex (fig. 1) and the claustrum, a thin sheet of gray matter just under the insular cortex (fig. 27). So in contrast to the theory put forward by the Nobel Prize–winning neuroscientist Sir Francis Crick, namely that the claustrum plays a role in our highest brain function, that is, consciousness, it turns out that this structure, at least in men, is in fact occupied with the very down-to-earth activity of sex. In men, sexual stimulation also activates the insular cortex, a brain area that regulates heart rate, breathing, and blood pressure. It's striking that the brains of both sexes turn out to approach the same goal, orgasm, via different routes. The amygdala, one of the structures that inhibits sexual behavior when we're supposed to be concentrating on something else, becomes less active in both men and women during orgasm.

While the scans revealed different routes to orgasm in men and women, they showed largely identical patterns of activity and inhibition during orgasm, with a considerable increase of activity in the cerebellum. It appears that in both sexes the cerebellum regulates muscle contractions during orgasm. The prefrontal and temporal cortices become less active during orgasm, ensuring that sexual behavior is even less inhibited during this stage. So at such moments you're effectively of unsound mind. In the case of men, moreover, an area in the brain stem known as the periaqueductal gray matter is activated during orgasm in exactly the same way as when drug addicts inject themselves with heroin.

Holstege's scanner studies were met with predictable outrage. In an interview, he mentioned that one of his presentations in the United States caused an American colleague to blush and mumble as he walked away, "Is nothing sacred anymore?"

Sexuality and Hormones

> We must remember that all our provisional ideas on psychology will one day be explained on the basis of organic substrates. It seems then probable that there are particular chemical substances and processes that produce the effects of sexuality and permit the perpetuation of individual life.
>
> Sigmund Freud, *On Narcissism*

Hormones are involved in every aspect of sexual behavior. Sexual arousal is influenced by the male sex hormone testosterone. In older men, testosterone levels can decline, lowering their sex drive and causing depression. (Testosterone replacement therapy alleviates both symptoms.)

Testosterone is also responsible for promoting women's sex drive. In their case, testosterone is produced in the adrenal gland and the ovaries. A woman who had a testosterone-producing tumor found that she missed it after it was surgically removed, because one of its side effects had been an exceptionally intense sex life.

The monthly fluctuations in women's hormone levels send signals to their brains indicating their degree of fertility. A study showed that, without being aware of it, American female students dressed more fashionably around the time of ovulation. They were more inclined to wear skirts than pants, wore more jewelry, revealed more skin, and were more sexually active. The conclusion was that they were picking up signals from their hormones that they were at their peak fertility and subconsciously chose clothing and behavior that flaunted it. The unconscious signals that women transmit during their fertile period are also picked up by the outside world. Another study has shown that strippers' tips around the time of ovulation averaged $335 an evening, whereas they "only" netted $195 an evening during the rest of their cycle. Around ovulation, women also show a preference for more masculine faces, voices, and behavior in the males around them. This 2007 study by Geoffrey Miller and Brent Jordan won the Ig Nobel Prize, a parody of the Nobel Prize.

Indeed, the brain's response to erotic images depends not only on gender and age but also on hormone levels. Erotic images provoke greater sexual arousal and more activity in the brains of young men than of young women. In general, in the early stage of sexual stimulation, the hypothalamus and amygdala are more strongly activated in men than women. For women, the degree of activation also depends on the stage of the menstrual cycle. Around ovulation, women react comparatively more strongly to such stimuli than during menstruation. In middle-aged men (ages 46–55) certain brain areas like the thalamus and hypothalamus cease to display activity, a sign that aging reduces arousal in response to erotic stimuli.

Oxytocin is a hormone produced by brain cells in the hypothalamus and released into the bloodstream via the pituitary gland (see chapter 1). It affects muscle fibers in the reproductive organs but also influences behavior when it's released in the brain itself. In men, the release of oxytocin in the hypothalamus also plays an important role in erection. During sexual arousal, high levels of oxytocin are found in the blood of both sexes, and oxytocin is involved in both the male and the female orgasm. In men, oxytocin stimulates contractions of the smooth muscle fibers, facilitating sperm transport. In women, it sends sperm transport one way and egg transport the other way, ensuring that an encounter is inevitable.

Oxytocin's effect on sperm transport means that the female orgasm also affects partner choice, because a partner who can ensure orgasm gains an evolutionary advantage by increasing the likelihood of fertilization. Furthermore, a genetic component has been found in the female orgasm, so there is every reason to regard it as an adaptive mechanism that has developed through natural selection. What's more, the release of high levels of oxytocin, not only in the blood but also in the brain itself, promotes pair forming. The blissful feeling of orgasm is heightened because oxytocin causes opiates to be released from other brain cells. That explains why chronic pain sufferers report a reduction in pain after making love. The peak values of oxytocin in the plasma that occur during sexual arousal inhibit the stress system, which has a relaxing effect. So oxytocin would at first sight appear to be the neurobiological substrate for the credo of the 1960s: "Make love, not war." However, oxytocin doesn't just inhibit aggression within one's own group—it also stimulates aggression toward others. So it certainly isn't an entirely harmless substance.

Neuropsychiatric Disorders and Sexuality

> In the minds of many, the sins of sex overshadow the science of reproduction.
>
> J. Parks

Brain Damage and Disorders of the Brain

Damage to the brain or spinal cord can impair sexual functioning. The type of disorder depends on the site of the lesion, rather than its cause. Damage to the prefrontal cortex can cause apathy and a reduced sex drive or, conversely, a lack of inhibition and an increase in sex drive. In the case of dementia, the loss of inhibition caused by degeneration of the cerebral cortex can lead to abnormal behavior ranging from inappropriate sexual allusions to exhibitionism and sexual offenses. Lesions in the temporal lobe can cause people to develop Klüver-Bucy syndrome, the symptoms of which are hypersexuality and hyperorality (constantly putting things in one's mouth). Considerable loss of inhibition has also been reported in individuals with lesions of the thalamus or the subthalamic nucleus. The majority of MS patients experience sexual dysfunction. This can take any form, depending on the site of the lesions (MS plaques). Two months before her death from MS, one patient developed an extremely rare complication that combined hypersexuality with multiple forms of paraphilia, such as pedophilia, zoophilia (desire for sex with animals), and incest. She had many lesions in her hypothalamus, basal prefrontal cortex, septum, and temporal lobe, so it's impossible to say which lesion was responsible for what behavior. A thirty-four-year-old man became aroused at the sight of sleeping women and at the prospect of painting their nails, especially those of the right hand. He lost control of this paraphilic urge, sedating his wife so that he could indulge in it. She was outraged when she found out, but he proved unable to control this obsession. When he tried to knock her out with pepper spray, she called the police and he was examined by a psychiatrist. A scan showed atrophy in the frontoparietal region, along with severe white matter injury and lesions in the subcortical fiber systems. He had sustained a head injury at the age of ten, after which he had been in a coma for four days. He also proved to have a body image disorder (see chapter 2) with an incomplete mental image of his right hand. Studies have shown that around 50 percent of people who have paraphilia or commit sexual offenses have experienced severe skull trauma causing unconsciousness.

Depression is associated with a loss of libido. That is because the dopamine reward system is inhibited by higher levels of cortisol, the stress hormone, which can deprive patients of all their pleasure in life (a condition known as anhedonia). Depression also causes testosterone levels to decline, further lowering mood. On top of that, antidepressants lower libido and inhibit orgasm. Conversely, patients with bipolar disorder can have a heightened sex drive during manic periods.

Nearly all patients with a disorder of the hypothalamus or the pituitary gland as a result of bleeding, trauma, or infection also experience sexual problems, caused partly by the impairment of the autonomic nervous system and partly by hormonal factors. Diabetics can suffer from sexual dysfunction as a result of damage to nerve fibers. In fact, diabetes is the most common cause of erectile dysfunction in men and painful intercourse in women.

Chronic diseases and certain forms of medication (for instance, for high blood pressure, depression, schizophrenia, and epilepsy) also impair sexual functioning.

Spinal Cord Injuries

A paraplegic whose wife was expecting a baby heard, through the grapevine, reactions like "How could she do that to the poor man, getting pregnant by someone else—as if his life wasn't tough enough already." Indeed, one wouldn't expect someone with paraplegia to be able to perform sexually. Men with a full spinal cord injury (paraplegia or quadriplegia) involving the loss of sensation from the navel downward can't have erections induced by the brain (psychogenic erections), which result from arousal caused by seeing, feeling, and smelling a partner. But they retain the ability to have erections caused by stimulation of the penis (reflex erections), because that reflex travels through the lowest part of the spinal cord, which is still intact. In able-bodied men, psychogenic erections start in the brain and erotic impulses from the sex organs travel up the spinal cord. When you think how much traffic goes up and down the spinal cord during sexual activity, all of it necessary for ejaculation, it's surprising that 38 percent of men with a complete spinal injury can still experience orgasm. There are three explanations for this, none of them obvious. First, some paraplegics find that the skin near the site where the loss of feeling starts becomes so hypersensitive that it turns into a new erogenous zone, stimulation of which can lead to orgasm. This can be in places like the breasts or mouth or even the ears or shoulders. Second, functional brain scans carried out by the American psychol-

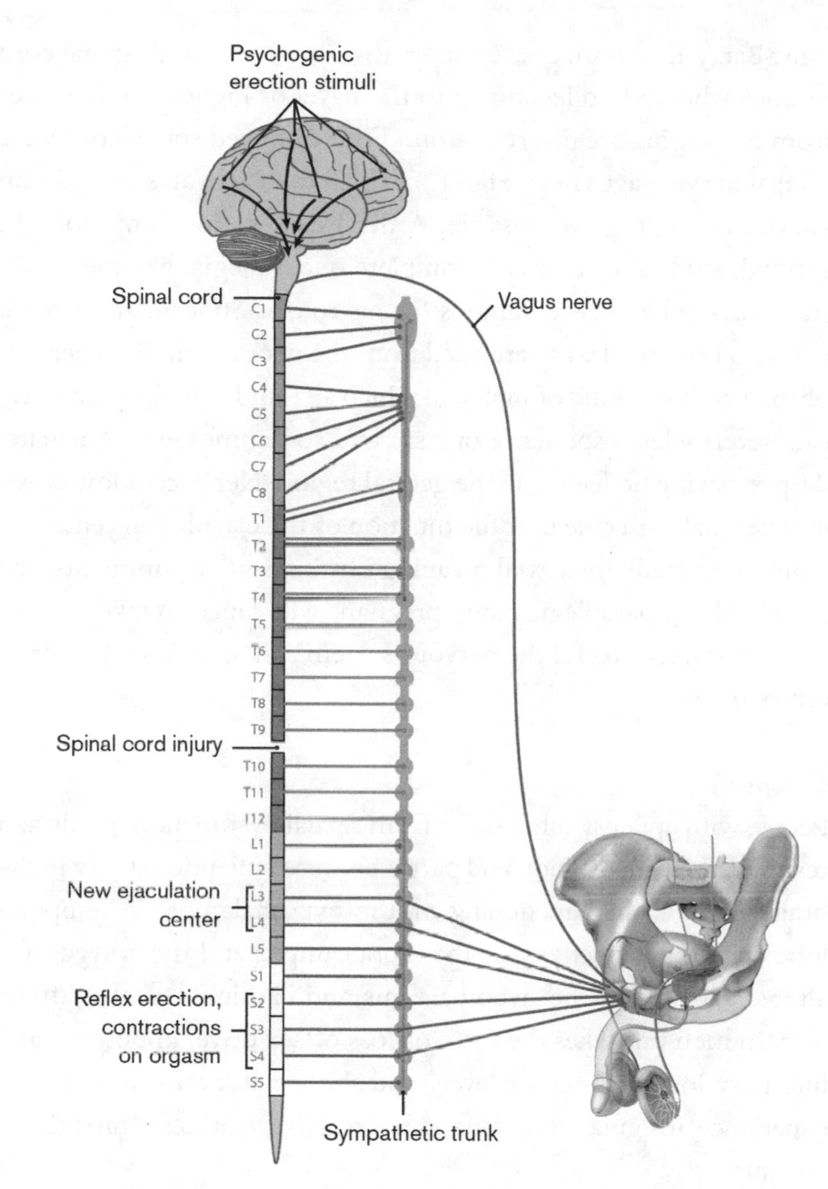

FIGURE 17. Psychogenic erections arise in the brain, and erotic impulses from the sex organs travel up the spinal cord to the brain. Although a great many stimuli travel in both directions along the spinal cord during sexual intercourse, individuals with a full spinal cord injury who have no feeling below the level of the navel can nevertheless experience orgasm. In such cases, stimuli from the sex organs are diverted from the damaged spinal cord and sent to the brain via a cranial nerve (the vagus nerve). Moreover, the function of the ejaculation center can be taken over by nerve cells lower down the spinal cord.

ogist Barry Komisaruk have shown that in women with spinal cord injuries who had no feeling up to the navel or higher, the impulses from the vagina are diverted around the damaged spinal cord via a cranial nerve (vagus nerve, fig. 17), which then activates many brain areas, making orgasm possible. A third explanation came to light through studies of men with complete quadriplegia, like the movie star Christopher Reeve, famous for his role as Superman, who fell from his horse and was paralyzed from the neck down. Fifty percent of men with this kind of injury, in which all four limbs are paralyzed, can nevertheless experience orgasm and sometimes even ejaculate, despite having no feeling in the genital region. Nerve cells low down in the spinal cord take over the function of the ejaculation center. So when they made their well-meaning but insensitive comments, the friends of the paraplegic man's pregnant wife failed to take into account how resourceful the nervous system can be when it comes to reproduction.

Epilepsy

People with epilepsy often suffer from sexual dysfunction, partly as a result of their medication and partly because epileptic activity in the brain disrupts the functioning of the hypothalamus. In temporal lobe epilepsy, the activity of the hippocampus and the amygdala is altered, disrupting the hypothalamus and causing sexual dysfunction. In men this takes the form of loss of sex drive, impotence and infertility, low testosterone levels, and abnormal sperm. Women can experience irregular menstruation, excessive hairiness (hirsutism), and infertility.

Epileptics whose condition is sparked by a focus in the cerebral cortex sometimes feel as if they are having an orgasm just before a seizure, due to electrical stimulation of the brain cells in that area. A woman with a tumor of the cerebral cortex experienced sensations in the genital region as if she were having sexual intercourse. Another woman with an epileptic focus in the cerebral cortex refused both medication and the option of an operation, as she enjoyed the feeling of orgasm that preceded her epileptic attacks.

Patients whose epilepsy is sparked in the prefrontal cortex have been shown to display all kinds of sexually tinged automatic movements, like rhythmic grinding of the pelvis or masturbation. Similarly, temporal lobe epileptic seizures can be accompanied by sexual feelings, sometimes even by the experience of orgasm and spontaneous ejaculation. They can also trigger automatic movements of a sexual nature. But temporal lobe epilepsy is often associated with reduced sexuality between attacks. Surgical removal of the temporal pole can normalize sexual function but can sometimes also result in hypersexuality and Klüver-Bucy syndrome. These opposite effects call for better research into the exact location of the focus and the borders of the operative lesion and better study of any changes to patients' sexual behavior.

Considering the interrelation of brain structures, sexuality, and disease, it's remarkable that patients' medical files rarely contain any information about their sexual behavior. We're programmed to regard sex as an extremely private matter and apparently don't shed these feelings of embarrassment when we put on white coats.

5

Hypothalamus: Survival, Hormones, and Emotions

Here in this well-concealed spot, almost to be covered with a thumbnail, lies the very mainspring of primitive existence—vegetative, emotional, reproductive—on which, with more or less success, man has come to superimpose a cortex of inhibitions.

Harvey Cushing (1869–1939)

HORMONE PRODUCTION BY THE HYPOTHALAMUS AND STREAMS OF URINE

"I suffer from familial neurohypophyseal diabetes insipidus, and in my family's case it started with me. I went through agonies of thirst. In eight hours' time I lost four kilograms and I couldn't stop urinating. . . . My little boy was diagnosed with the same condition last week. He now uses the Minrin nasal spray, and that makes a world of difference."

In the old days, if a patient had to urinate constantly, the doctor would stick his finger in the urine and taste it. If it tasted sweet, the patient was suffering from *diabetes mellitus* (meaning "sweet urine flow"). If it didn't, the diagnosis was *diabetes insipidus* (tasteless urine flow), and that meant something was wrong with the kidneys or the brain.

Every day, large quantities of blood pass through the kidneys in order to be cleaned. During this process, the kidneys recycle around fifteen liters of liquid from the waste fluid. They do this with the help of a brain hormone that inhibits the excretion of water, hence its name, antidiuretic hormone (ADH). It's also known as vasopressin because it raises blood pressure. It's a small protein produced by brain cells in the hypothalamus and transported to the rearmost part of the pituitary gland, where it enters the bloodstream.

The idea that brain cells could produce hormones was first suggested in the 1940s by Ernst and Berta Scharrer. Under the microscope they saw droplet-like structures in large brain cells in the hypothalamus and put forward the theory that they were packaged hormones ready for release into the bloodstream. This controversial notion met with great resistance from their colleagues. "They dismissed the idea emphatically, often viciously," Berta wrote to me. Though by then an old lady, she was still angry at the memory. Opponents claimed that the droplets were just indicative of disease or of changes that occurred after death or had been produced by the staining process. The Scharrers proved these claims wrong by finding similar "nerve gland cells" throughout the animal kingdom, from worm to human. They proved to be a universal cell type that regulates many bodily processes by means of hormones. The Scharrers' observations founded the discipline known as neuroendocrinology.

Their hypothesis that these cells had something to do with the body's water management system was visionary. A hereditary defect in the DNA for ADH makes this hormone's function immediately apparent. People who inherit this condition produce fifteen liters of urine a day. I have followed a family that has had the condition for five generations. It was during my internship in 1968 in Amsterdam's Binnengasthuis Hospital that I first met them. At that time, the family's life was largely dominated by urinating and drinking. One of the family members told me that her mother, who also had the condition, grew so annoyed with her children (with whom she shared a room) getting up constantly during the night to get a drink or go to the toilet that she strictly forbade them to do so, though she herself always kept a kettle under the bed to quench her own thirst. This proved simply unbearable for the children. While their mother slept, they secretly crept under the bed to suck on the kettle's spout. If they woke her up, she would slap them.

When one patient with this syndrome was admitted to a sanatorium as a child, the nursing staff were so irritated by her constant drinking and visits to the toilet that they deprived her of water for long periods. Desperate for fluid, she would drain the flower vases at night. She became so dehydrated that she would have died had it not been for the timely arrival of her parents, who restored her with a large bottle of water. If she went cycling with her sister, they always took plenty of water with them. At gas stations, each one would drain a large bottle and then refill it for the road, leaving the station attendant slack-jawed.

In 1992, working with a research group from Hamburg, we found the tiny defect in the DNA of this Amsterdam family. A single building block of the DNA on chromosome 20 was responsible for the fifteen liters of urine per day. Now people with this condition can administer long-acting ADH in the form of a tablet or nasal spray, reducing the drinks and visits to the bathroom to almost normal proportions. Some refuse this medication, though. They don't see the condition as a disease but rather as a special family trait.

SURVIVING WITHOUT A HYPOTHALAMUS

The hypothalamus is crucial to the survival of the species, because it regulates reproduction, and it's crucial to the survival of the individual, because it directs many bodily processes. Surviving without a hypothalamus can only be done with constant help from others.

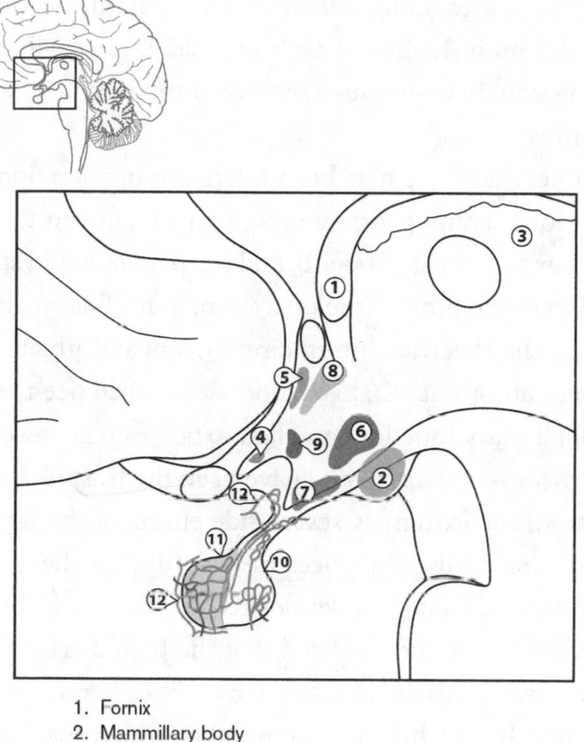

1. Fornix
2. Mammillary body
3. Thalamus
4. Suprachiasmatic nucleus
5. Preoptic region
6. Tuberomammillary nucleus
7. Arcuate (or infundibular) nucleus
8. Paraventricular nucleus
9. Supraoptic nucleus
10. Posterior pituitary
11. Hypophyseal portal system
12. Anterior pituitary

FIGURE 18. The human hypothalamus. Memory information is transmitted from the hippocampus via the fornix (1) to the mammillary bodies (2), and then routed to the thalamus (3). The suprachiasmatic nucleus (4) is the biological clock. Thermoregulation and sexual activity are governed by the preoptic region (5), while the tuberomammillary nucleus (6), the only place where histamine is produced in the brain, is important for focusing our attention. The areas that regulate appetite and metabolism are the infundibular (arcuate) nucleus (7) and the paraventricular nucleus (8). The paraventricular nucleus and the supraoptic nucleus (9) send fibers to the posterior pituitary (10), where oxytocin and vasopressin are released. The infundibular nucleus sends fibers to the capillaries of the hypophyseal portal system (12), where neuropeptides are released that regulate the anterior pituitary (11).

He came to see me together with his extraordinarily devoted mother. Since his operation she had not left his side. It gradually dawned on me that she couldn't—because she was functioning as his external hypothalamus.

Years earlier the young man had undergone an operation to remove a brain tumor, a craniopharyngioma, that threatened to destroy his sight. The surgeon did a good job, and the patient could still see well. However, up to the time of the operation, in his final year of secondary school, he had been an outstanding student and athlete. Now, after not only the tumor but also his hypothalamus had been removed, he had lost all pituitary functions and had to be given hormones to regulate those processes. That would have been the least of his problems, were it not for the extremely severe side effects of the large doses of growth hormone. His joints became painfully swollen, and he suffered from muscle pains and developed breasts. The hormones were reduced when his mother realized that the high dosage was causing the side effects, and his breast tissue was surgically removed.

His mother had to help him remember which medicines he had taken and when it was time for the next dose of hormones. In that way she replaced his memory, which had been very much impaired (see 1–3 in figure 18). She also stood in for his absent biological clock (4), asking that he be given the sleep hormone melatonin to combat his sleep disorders. However, nothing could be done about his complete lack of sexual activity, regulated by the front part of the hypothalamus, which had been removed along with the rest of that structure (5). There was also no solution for his memory and concentration problems, which were caused not just by his impaired memory circuits (1–3) but also by the lack of the histamine system (6), without which he could not focus his attention. The talented boy was unable to continue his studies, as his mother couldn't replace those systems. She also had to regulate the amount and composition of his meals, thus preserving him from the morbid obesity and diabetes that would otherwise inevitably ensue from the removal of the hypothalamus (7, 8). However, by far the greatest danger threatening him was his total lack of thermoregulation (5). If he engaged in physical activity and the temperature then dropped somewhat, in a very short space of time he could develop life-threatening hypothermia, whereas if the sun came out or he warmed up, for instance by walking to the hospital for his appointment with me, his temperature very quickly rose to the level of a fever. Once, prior to an operation, the fifteen-minute wait in a hospital gown was enough to make him cool down so drastically that he nearly went into a coma, and he had to have a hot shower to get his temperature back to normal before the operation could begin. Once again, it was his mother who took over this function of the hypothalamus, following her son around with a thermometer so that whenever the surrounding temperature changed, she could take action.

The boy's situation reminded me of how many vital functions are automatically regulated by that tiny scrap of brain tissue, the hypothalamus. There was still one thing missing, though, from this story of friendly symbiosis between mother and son. "Do you ever get angry?" I asked cautiously. "No," he said, and then suddenly stood up and shouted, "But if I ever get my hands on my brother, he'd better watch out!" Aha, I thought, a classic case of ventromedial hypothalamus syndrome. His mother, now taking on the role of the inhibiting prefrontal cortex, calmed him by putting an arm around his shoulders, saying, "That's what scares me. He flies into such rages I'm scared he might really attack his brother. He's left him in the lurch, you see. It's so tough—it makes it extra hard to cope with all his limitations." His brother, a medical professional on a fat salary, was oblivious to his brother's terrible handicaps and the intolerable burden on his mother. The symptoms of the syndrome in question include attacks of rage, sometimes verging on the homicidal, emotional instability, bulimia, and mental deterioration. Alas, the complete removal of this young man's hypothalamus meant that he suffered from the entire spectrum of symptoms. Without the constant support of his mother and his own extremely disciplined behavior, he would have died much earlier.

The anatomical terms, for enthusiasts (fig. 18): Memory information is transmitted from the hippocampus via the fornix (1) to the mammillary bodies (2) at the rear of the hypothalamus. (These structures had been removed in the operation.) The information is then routed to the thalamus (3). The suprachiasmatic nucleus (4) is the biological clock. Thermoregulation and sexual activity are governed by the preoptic region (5). The memory and concentration disorders were caused not only by the loss of the mammillary bodies but also by damage to the tuberomammillary nucleus (6), the only place in the brain where histamine is produced, which is important for focusing our attention. The areas that regulate appetite and metabolism are the infundibular (or arcuate) nucleus (7) and paraventricular nucleus (8). An MRI of the patient described in this chapter showed that the hypothalamus had in fact been entirely removed. An endoscopic fenestration (keyhole surgery at the base of the brain) confirmed the damage. Given the loss of the mammillary bodies, the input of the fornix, which could still be seen on the MRI scan, wouldn't have been much use to the patient. Moreover, the base of the hypothalamus, the infundibulum, or tuber cinereum, was entirely missing.

DEPRESSION

> The truth lay in this, that life had no meaning for me. Every day of life, every step in it, brought me nearer the edge of a precipice, whence I saw clearly the final ruin before me. To stop, to go back, were alike impossible; nor could I shut my eyes so as not to see the suffering that alone awaited me, the death of all in me, even to annihilation. Thus I, a healthy and a happy man, was brought to feel that I could live no longer, that an irresistible force was dragging me down into the grave.
>
> Leo Tolstoy

> All those who have become eminent in philosophy or politics or poetry or the arts are clearly melancholics.
>
> Aristotle

Many creative minds and famous politicians suffered from depression. Some famous examples are Goethe, Isaac Newton, Ludwig van Beethoven, Robert Schumann, Charles Dickens, Christiaan Huygens, Vincent van Gogh, Charles de Gaulle, Willy Brandt, and Menachem Begin. From his early youth, Abraham Lincoln suffered from severe mood disorders. He's even credited with the authorship of a poem called "The Suicide's Soliloquy," published in 1838.

A doctor must be aware that depression can also be an early symptom of a tumor, an infectious or autoimmune disease, or a hormonal or metabolic disorder. This has led to the theory that in the case of physical illness, depression might confer an evolutionary advantage. The theory goes that when you are depressed, you withdraw, you eat less, you lose interest in everything, and you avoid activity so that all your energy can be devoted to physical recovery. Depression is thought to have another possible evolutionary advantage, being a potentially beneficial response when someone with a dominant status is forced to occupy a lower place in the pecking order. Behavior such as avoiding eye contact or sexual contact would reduce the likelihood of attack by more dominant individuals. Whatever the case, bodily diseases can provoke depression, so you need to start by giving someone with depression a proper physical examination.

Brain disorders, like Alzheimer's, can also start with depression. Prince Claus, the Dutch king's late father, was treated for depression for many years in expensive Swiss clinics before it emerged that his depression was an early symptom of Parkinson's disease. Depression is also often commonly linked to other psychiatric conditions like eating disorders and borderline personality disorder. Patients with schizophrenia often suffer from depression, sometimes committing suicide. But depression does arise as a condition in its own right.

Causes

In the Netherlands, around five hundred thousand people are diagnosed with depression every year. It can be triggered by a very stressful event, like the death of a partner or failing an exam. In many cases, however, it's impossible to find a clear cause. Some individuals develop a strong predisposition for depression very early on. Others are able to rise above the most terrible events with apparent ease. The memoirs of the former Dutch cabinet minister Marcel van Dam clearly show how much adversity some people can take without suffering depression:

> The war broke out and in 1943, when I was five years old, we were warned that the officers of the *Sicherheitsdienst* were on their way to arrest my father. He had been one of the organizers of an uprising by the police in Utrecht against the arrest of Jews. The entire family had to flee instantly, and I ended up living with a farmer and his family who I didn't know at all. My sister wasn't able to escape in time and was put in a concentration camp. I was told, probably for fear of me revealing something, that my father was dead. In 1944, my mother and the younger children were reunited, but my younger brother Leo died shortly afterward, probably of meningitis. My mother's distress was heartrending, and was made worse because my father, who turned out to be alive after all, couldn't be at the funeral. The Germans were waiting to intercept him at the cemetery. When the Netherlands was liberated, it seemed as if we were making a fresh start. Until my brother Wim, who was 12 years old at the time, went with me to the Galgenwaard Stadium in Utrecht to retrieve fire extinguishers from the tanks and army trucks parked there. When we set off for home, he ran across the street without looking and was knocked down and killed in front of my eyes. . . . I've relived all these memories many times. . . . It never disrupted my life. None of these events ever gave me any

> nightmares. I never became depressed, never disproportionately fearful. . . . Why was that? Why do some people become traumatized after what would seem to be much less extreme experiences?

The answer to Van Dam's question lies in a combination of genetic factors and development in the womb and early childhood, which program the activity of the stress system for the rest of our lives. There are different forms and subforms of depression, but they all have a common feature: overreaction of the stress axis. We respond to stress by activating nerve cells in the hypothalamus, which send a substance called corticotropin-releasing hormone (CRH) to the pituitary gland and the brain. The pituitary gland in turn stimulates the adrenal gland to produce the stress hormone cortisol. CRH and cortisol equip our brains and bodies to cope with stressful situations. But if the stress axis becomes overactive, a stressful event can lead to the overproduction of both CRH and cortisol, and these substances can affect the brain so strongly that depression results.

Genetic factors can cause the stress axis to be put into a higher gear during development. This is why certain tight-knit groups, like the Amish in the United States, have a higher than average rate of depression. Depression has been seen to run in certain famous families, like that of Virginia Woolf. Studies of families with this condition helped to locate the first variations in genes (polymorphisms) that increase the risk of depression. Many tiny variations in genes for chemical messengers in the brain also predispose individuals to this condition. People whose mothers were pregnant during the Hunger Winter of 1944–1945 (fig. 9), when famine struck Dutch cities, have a higher incidence of depression. Food shortages are a thing of the past now, but the stress axis can also be permanently activated by a malfunctioning placenta, causing a child to be malnourished and underweight at birth. Exposure to nicotine or certain medicines (like DES) in the womb also increases the risk of depression. After birth, the stress axis of a child who is seriously neglected or abused can be permanently set on red alert. Lasting changes of this type caused by external factors affecting gene expression—in this case of the stress axis genes—are known as epigenetic programming.

We have also found that female hormones (estrogens) stimulate the stress axis, while the male hormone (testosterone) inhibits the stress axis, which would explain why women are twice as likely as men to suffer from depression.

So depression is basically a developmental disorder of the hypothalamus. If a person's genetic background and development cause their stress axis to be switched to a high setting, it overreacts to stressful life events, which can lead to depression. In adulthood, depression can also be caused by other factors, for instance certain types of medication. Prednisone, a much-prescribed synthetic corticosteroid drug, frequently causes mood disorders when taken in high doses. Brain infarcts or damage from MS lesions, particularly on the left half of the brain, also increase the risk of depression, due to the stress axis becoming hyperactive when it's no longer controlled by the cerebral cortex.

Different Types of Depression

We all have a greater sense of well-being in summer than in winter. Some people suffer from extreme seasonal mood swings, though. In summer they can develop hypomania (a state approaching mania) or even become truly manic, while suffering severe depression in winter. This is known as seasonal affective disorder, appropriately shortened to SAD. The former West German chancellor Willy Brandt would tend to become depressed as autumn progressed and the days grew shorter. During these bleak periods, he couldn't bear to see anyone, not even his wife. Seasonal depression is triggered by a lack of sunlight in winter; exposure to sunlight alleviates the symptoms. In the United States there are even insurance companies that send SAD patients from the north of the country to a southern state in order to speed up their recovery. Information on the amount of light in our surroundings is relayed directly to the biological clock, which isn't just a day-night clock but also a seasonal clock, which plays an important role in SAD. Tiny variations in the genes that make the biological clock function pose a risk factor for depressions of this type.

A bipolar disorder with periods of mania and depression can also occur without any clear seasonal link. When Mrs. De Vries came back from walking the dog in the nearby dunes, she found her husband, who had just retired, slumped over the breakfast table, apparently lifeless. She immediately called the emergency services, and the medics tried to revive him, but without success. In the days that followed, her husband's body lay on a bier in the living room. His widow was full of energy, quite tireless in fact, and after the cremation this became even more pronounced. She developed hypomania and then became truly manic a few days later. She went around to all her acquaintances with mementos of her husband, laughing excitedly. In the middle of the night she would call the police, only to menace them with a hockey stick when they arrived at her house. When she threatened her adult daughter with a knife because she didn't want to be incarcerated in a psychiatric institution and be given electroshock treatment, as had happened to her father, the situation became untenable. At length she was persuaded to enter a clinic. There her mania became even worse, despite the medication. She told everyone excitedly that she'd always wanted to stay in this fabulous hotel, and after each hospital meal she would leave a one-guilder tip for the excellent waiters. She would skip through the clinic, singing, arm in arm with a friend who visited her every day, time and again introducing her to a "former classmate," an unfortunate man who had never seen her before and was being driven mad by her constantly harking back to their mythical school days. Her condition improved briefly, and then she fell prey to a terrible depression. The story ended happily, however: She made a full recovery and is enjoying life to the fullest with her eight grandchildren.

The Dutch cabinet minister Ger Klein also suffered from a bipolar disorder (see chapter 15). Hypomanic periods can be very productive. The composer Robert Schumann composed over twenty works in his hypomanic phases, in 1840 and in 1849, while he was unable to compose at all during his periods of depression, in 1844 and 1854. In the winter of 1854 he tried to commit suicide by jumping into the icy Rhine river. He was rescued and spent the last two years of his life in a mental institution. Johannes Brahms, deeply shocked by the illness and death of his friend, started work on *Ein Deutsches Requiem,* which he dedicated to Robert Schumann, his mother, and humanity.

The Russian leader Nikita Khrushchev suffered from alternating depressive and hypomanic phases. After he was deposed in 1964, he fell into a deep depression. In the case of government leaders, especially, there's a strong tendency to deny the existence of a bipolar disorder, because of doubts about the ability of individuals with this condition to make well-informed decisions. Winston Churchill suffered from terrible bouts of depression, which he called his "black dog." According to his private secretary, he could also become madly excited and experienced extreme mood swings with bursts of energy. The terms *hypomania* and *mania* weren't used at the time, but it seems from eyewitness accounts that he must have been bipolar. Lyndon Johnson, who became president after John F. Kennedy was assassinated in 1963, was afflicted by such severe depression after surgery to have his gallbladder and a kidney stone removed that he considered resignation. He, too, had a bipolar disorder, characterized by episodes of profanity and temper. Whether he received medication for it isn't known.

When depression doesn't involve hypomanic or manic interludes, it's classified as unipolar depression or "major depression," of which melancholic depression is a subtype, involving severe disruption of the day-night rhythm and loss of appetite. The type of depression induced by prednisone or similar substances, called atypical depression, is marked by an increased appetite and need for sleep.

The Various Brain Systems and Areas Involved in Depression

Various cell groups become hyperactive in the hypothalamus of depressive patients. In many cases, the stress axis (the hypothalamus-pituitary-adrenal axis) is strongly activated. In postmortem brain material from donors with lifelong periods of serious depression we found a considerable increase in the number of cells in the hypothalamus producing CRH, even if the patient hadn't died during a depressive episode. This ties in with the theory that the stress axis has been set in a higher gear during the development of such individuals. We know that the activation of CRH neurons contributes to the symptoms of depression, because if CRH is injected in the brains of laboratory animals they develop the same symptoms: diminished appetite, changes to the motor system, sleep disorders, fearfulness, and loss of sexual interest. An increase is also found in other stress hormones, like the hormones vasopressin (produced in the hypothalamus) and cortisol (produced in the adrenal gland), which contribute to such symptoms.

Located in the brain stem are three more stress systems that produce the chemical messengers noradrenaline, serotonin, and dopamine and regulate many brain areas, including the hypothalamus. Their possible link with depression was discovered by chance as the result of a side effect of reserpine, a common medication for high blood pressure. The drug was shown to reduce the amount of noradrenaline and serotonin in the brain stem, and depression was a not uncommon side effect. Conversely, the first antidepressants, MAO inhibitors, increased noradrenaline and serotonin levels. The most commonly used antidepressants nowadays are selective serotonin reuptake inhibitors (SSRIs), compounds that increase the level of serotonin by preventing its reabsorption into the body. This gave rise to the theory—now immensely popular—that depression may be caused by abnormally low concentrations of noradrenaline or serotonin. However, patients whose depression can definitely be ascribed to low serotonin levels are in the minority. The mere fact that it takes a few weeks for SSRIs to become effective, even though they raise the serotonin level almost instantly, shows that the link between serotonin and depression isn't that clear-cut. In the case of depressive patients who are very fearful, however, serotonin systems can be disrupted. Low serotonin and noradrenaline levels have also been found in the cerebral fluid of patients who had ended their lives violently, for instance by jumping in front of a train. This group was also found to have high levels of the stress hormone cortisol. Dopamine, the chemical messenger of the reward system, is also involved in the symptoms of depression. When depressive patients are unable to enjoy life anymore, it is probably because of a dip in dopamine.

Our examination of postmortem tissue from the brains of depression sufferers also revealed that the circadian clock, the suprachiasmatic nucleus, is less active in people who suffer from depression. This explains not only their disturbed day-night rhythms but also the effectiveness of light therapy.

Functional scanning studies of depressive patients showed changes in the activity of the temporal and prefrontal cortex as well as the amygdala. The latter may explain the increase in fearfulness. The reduced activity in these brain areas may partly result from raised cortisol levels.

To sum up, an entire network of brain systems and various chemical messengers are involved in the onset of depression. The systems that are the prime cause of depression vary from individual to individual, but in all cases the stress axis is central to the pathological process.

Therapies

Many therapies are used to treat depression that would appear to have nothing in common with one another, but ultimately they all normalize stress axis activity.

SSRIs are very commonly prescribed for depression. In the Netherlands, around nine hundred thousand people take antidepressants.

In around 75 percent of these cases, the patient is indeed very down but not suffering from severe depression, so they won't help very much. Indeed, SSRIs aren't very effective at all. They start to work only after a couple of weeks, during which period there's a real risk of suicide. (And that's not a negligible problem. Around fifteen hundred people kill themselves every year in the Netherlands, and ten times as many attempt to do so). Moreover, these drugs have a placebo effect of 50 percent. Indeed, it's not so strange that the placebo effect is so marked in the case of depression. The expectation that a placebo will alleviate one's pain is linked to increased activity in the prefrontal cortex (see chapter 16). This inhibits the hypothalamus, thus normalizing the activity of the stress axis. Stimulating the inhibitory effect of the cerebral cortex on the stress axis explains why transcranial magnetic stimulation of the cortex is effective. Cognitive therapy and online treatment for depression are successful for the same reason; they produce the same inhibitory effect on the stress axis as transcranial magnetic stimulation, but by different means. We don't know why electroshock therapy is so effective in the case of very severe depression. Perhaps it's a bit like when your computer seizes up: You switch the power off, switch it back on again, and hey presto, it works again. A disadvantage of this therapy is that it can impair memory.

Lithium, the classic medication for bipolar disorder, affects the circadian clock, inhibiting the overactive stress axis and stabilizing mood.

Light improves mood among depressive patients, through its effect on the circadian clock. The latter becomes more active, inhibiting the CRH cells of the stress axis. In the northern United States, seasonal depression is more common than in the sunny southern states. Physical activity can also stimulate the clock, so walking the dog is doubly effective because of the extra light and the extra activity. We found that increasing the amount of light in the living areas of patients with dementia also improved their mood (see chapter 18). Antidepression lamps work just like sunlight, although they're not as efficient. Even on a cloudy day, you're exposed to more light outside than you can obtain from a lamp of this type. (Incidentally, using a light box can get out of hand, and it can very occasionally induce mania or psychosis, so light therapy needs to be carried out under the supervision of a doctor.) In older people, lack of vitamin D can also increase the risk of depression. Vitamin D is made in the skin in response to sunlight. That's why people who live in towns are more likely to have this deficiency than people in rural areas. So sunlight protects you in two different ways. Disrupting the day-night rhythm (for example by going without sleep for a night) has also been found to improve mood, but the effect is unfortunately brief.

When considering all these therapeutic options, we must however bear in mind that depression is essentially an early developmental disorder and that the cause, disrupted brain development, can't be remedied by these therapies. That's why depression frequently recurs.

PRADER-WILLI SYNDROME

> "I'm a social worker at an institution in western Iowa. There I met a man who has been diagnosed with Prader-Willi syndrome. He is forty-two years old, and over the last few years we have seen him deteriorate rapidly, both mentally and physically. My question is, do you know a doctor or psychiatrist in or near Omaha, Nebraska, whom we could contact, so that he could help the man in question? He is an exceptionally pleasant person and it is sad to see him struggling with his mental problems. I appreciate your help."

A wealthy Japanese businessman, the CEO of a car parts factory, married a biologist, and they had two sweet little daughters. But it was unthinkable in Japan for a daughter to inherit his business, so they decided to have a third child. During the pregnancy, the wife felt far fewer signs of life than on the previous occasions. The child was born three weeks prematurely, and the birth was much more prolonged and difficult than before—but it was a boy! However, the baby was so floppy that he couldn't suck, so he was given tube feeding. When he was eighteen months old he started to eat—and seemed to be making up for lost time. However much he ate, he was never satiated; he invariably cried for more and became grossly overweight. When he was four years old he was diagnosed with Prader-Willi syndrome. The parents were also told that the child would always be mentally disabled and that they would face a lifelong struggle to prevent him from becoming obese and getting diabetes, with all its attendant dangers. The mother put electronic locks on the kitchen food cupboards and devoted all of her time to teaching him, stimulating him, and giving him new experiences to take his mind off food. As a result, the child had a strikingly normal build for a young Prader-Willi patient. Yet all of the mother's efforts couldn't prevent the boy from occasionally falling prey to terrible bouts of rage. She joined the Japanese Prader-Willi Association and took him with her to a biennial international conference at which scientists and parents meet to learn from each other. Parents often take their Prader-Willi children to these gatherings, and you can spot them on the flight on the way there: lots of obese children from Europe, Japan, India, and North Africa overflowing their seats, with their disproportionately small hands and feet and their typical almond-shaped eyes. You need only follow them to find the conference.

It was at that conference that the mother of the Japanese boy heard about a new growth hormone therapy that normalized the metabolism of Prader-Willi children, allowing even the fattest children to regain a normal build and ending the incessant battle against hunger. She was fortunate in being able to afford the expensive new therapy, because her Japanese insurance company wasn't yet prepared to foot the bill.

In the United States, Prader-Willi syndrome is known as H3O syndrome (hypomentia, hypotonia, hypogonadism, and obesity). The symptoms are largely due to a disturbance of the hypothalamus. The abnormal birth is actually the first sign of the child's defective hypothalamus, because that brain system plays an active role in timing the start of the birth and in speeding up its various stages (see chapter 1).

Most Prader-Willi patients lack a small piece of chromosome 15; in others, that section of the chromosome doesn't function at all. It's located in the section that they inherited from their father. The opposite strand—the one inherited from their mother—was chemically silenced at the very earliest stage of development and so can't compensate for the absence or malfunctioning of the paternal part. This process whereby the expression of genes is determined by the parent who passed them on is known as imprinting. When we examined the hypothalamus of Prader-Willi patients, we found that the paraventricular nucleus, the center of autonomic and hormonal regulation, was a third smaller than normal and contained only half the normal number of oxytocin neurons. The latter act as your "satiation neurons," signaling to your brain when you have eaten enough. Disabling these neurons in laboratory animals has been shown to cause increased appetite and obesity, and the fact that Prader-Willi patients have fewer oxytocin neurons may account for their inability to feel satiated no matter how much they eat. We're still searching for a link between the Prader-Willi genes at chromosome 15 and the malfunction of the hypothalamus.

Through a network of Prader-Willi parents and scientists, we received a request from a mother in New Zealand who was working as a nurse in a nursing home. She detected in her thirty-nine-year-old son symptoms that she recognized from elderly people with dementia. Was it possible that Prader-Willi patients ran the risk of early aging and Alzheimer's? That question had never come up before, because until recently, Prader-Willi patients died relatively young. In tissue taken from the brains of the few Prader-Willi patients who had lived beyond the age of forty, we indeed found changes that were typical of Alzheimer's (see chapter 18). Since then, reports of early dementia in this patient group have been coming in from all over the world via the Prader-Willi network. Some believe that it sets in very early, before the age of thirty; others speak of a dramatic decline at around the age of forty. Systematic research of this phenomenon is now under way. Does early-onset Alzheimer's form part of Prader-Willi syndrome, or might it be caused by morbid obesity? We know, after all, that certain symptoms of obesity are risk factors for Alzheimer's, such as diabetes mellitus, vascular disorders, high blood pressure, and high cholesterol. If the latter is the case, we can expect to face an explosion of premature brain aging and Alzheimer's as a tidal wave of obesity rolls across the globe.

OBESITY

> What goes into someone's mouth does not defile them, but what comes out of their mouth, that is what defiles them.
>
> Matthew 15:11

The hypothalamus regulates our body weight within very strict limits. Yet on average we all gain about one gram per day. That doesn't sound like much, but obesity has now ballooned into a world health problem: Around 300 million people are obese, and one billion are overweight. Being overweight greatly increases the risk of diabetes, heart and vascular diseases, high blood pressure, certain forms of cancer, and dementia. In the Western world, around 60 percent of adults are overweight, and 30 percent are obese. The recent rapid increase in childhood obesity is particularly alarming. In the United States, 30 percent of children are overweight or obese. For years I was amazed by the increasingly gigantic pairs of jeans I saw in America. But now obesity is everywhere, from China and Japan to Mexico.

We find food tasty, something that used to have a huge evolutionary advantage. Our ancestors spent millions of years in the barren savannas, where every calorie had to be tracked down and consumed. These long periods of scarcity meant that we failed to develop a protective mechanism against eating too much. Food was seldom available in overabundance and then had to be stored as fat, a necessary reserve to get us through the next lean period, which always came. Our autonomic nervous system, guided by the hypothalamus, ensures that in women, fat is stored on hips, breasts, and buttocks, while men develop bellies. Obesity results from a permanent food surplus, less physical labor, and a lack of physical exercise. These days we also eat more carbohydrates and fats and less protein than formerly. But that so many people are fat isn't just due to a lack of self-control. Predisposition is certainly a factor. Obesity has a strong genetic component. Studies of twins, adopted children, and families indicate that around 80 percent of the variation in body weight is determined by genetic factors.

Some people become so fat that their hearts can no longer cope, and they die prematurely. Some are too fat even to leave their homes and have to be hoisted out of the window when hospital admittance becomes necessary. Certain rare genetic factors for extreme obesity that regulate appetite and metabolism in the hypothalamus have now been discovered. Prader-Willi syndrome is one such genetic type of obesity (see earlier in this chapter). Patients with this condition can be so fat that the apron of flesh hanging over their genitals prevents you from knowing whether they are male or female. Normally, the hypothalamus registers how much fat our body has stored by measuring the amount of leptin, a hormone produced by fat tissue. If there are mutations in the leptin gene or the leptin receptor, the hypothalamus will conclude that there's no fat tissue and continually prompt you to eat, resulting in morbid obesity. Mutations have also been identified in which the brain no longer produces α-MSH, a substance responsible for hair pigmentation and appetite inhibition, or no longer receives the chemical message transmitted by α-MSH. Mutations of the α-MSH system often produce extremely fat children with red hair. Such children also don't enter puberty. Reduced sensitivity to α-MSH is found in 4 to 6 percent of obese people. Obesity can also be caused by a mutation in the receptor for corticosteroids as well as by hormonal disorders (like a lack of thyroid hormones, growth hormones, or sex hormones) or an excess of the adrenal hormone cortisol.

Certain drugs, such as antipsychotics, can have the extremely unpleasant side effect of massive weight gain. A boy treated with drugs of this kind gained 150 pounds in a very brief time and subsequently refused any further medication.

Patients suffering from psychiatric disorders like depression also risk becoming obese, as do people with disorders like bulimia and a syndrome that causes nighttime binge eating. Only very rarely are neurological processes in the hypothalamus the cause of obesity. When I was an intern in a pediatric department, I was assigned an eight-year-old patient who was massively overweight but who claimed to eat very little. It turned out, however, that she ate constantly, consuming vast quantities of candy. The cause of her obesity was a tumor in her hypothalamus, which meant that she never felt satiated and so believed that she wasn't eating enough.

Children whose early months in the womb coincided with the famine that struck the Netherlands in the last winter of the Second World War tended to become obese as adults. In a situation like that, a fetus's hypothalamus registers scarcity and calibrates systems so as to retain every calorie that's consumed. If an individual then encounters a surplus of food in later life, he or she runs a great risk of becoming obese. The same problem still affects children who are born underweight because of a malfunctioning placenta or because their mother had high blood pressure or smoked during pregnancy. Maternal obesity during pregnancy, high cholesterol, and the overfeeding of newborn babies also increases the likelihood that a child will later become obese.

Social, cultural, and environmental factors that appear to promote obesity include the mass marketing of candy and chocolate, the universal availability of cheap fast food, and the prevalence of "comfort eating" in response to difficulties. A lower socioeconomic status increases one's risk of obesity. Certain industrial substances that enter the environment have also recently been discovered to cause obesity even in low concentration. Dubbed obesogens, they include estrogens and substances that obstruct the normal functioning of sex hormones during development—endocrine disruptors, for instance (which come from compounds found in certain plastics), and the toxic organotins (which are present in plastic and paint).

Although our gluttony used to have an evolutionary advantage and fat was considered desirable in Rubens's day, nowadays we discriminate against fat people. The common prejudices are that fat people are stupid and lazy, have no self-discipline, and lack initiative. Extremely fat people, moreover, cause feelings of physical repulsion. The dangers of obesity to health are incontrovertible, so there are plenty of reasons to lose weight. But the most effective remedy for obesity—eating less and exercising more—is very difficult to sustain. According to a recent study, retraining eating behavior, using scales that provide feedback by weighing everything you eat, appears to be effective. Rimonabant, a cannabis antagonist, was heralded as a miracle drug, effective against both nicotine addiction and obesity. Unfortunately, however, it not only made people's weight go down but also had the same effect on mood. The drug increased the risk of depression and suicidal thoughts, and the registration application at the FDA has been provisionally withdrawn. The European Medicines Agency made the manufacturer, Sanofi, send all doctors a letter stressing that the drug should no longer be prescribed to patients with depression or at risk of depression. Experiments are being carried out with depth electrodes in the hypothalamus, but so far that hasn't proved effective; one side effect of such stimulation has been to call up memories of the distant past (see chapter 11). And in certain cases, even depth electrodes can cause weight gain. Parkinson's sufferers who have electrodes inserted in their subthalamic nucleus to treat their motor disorders often become overweight.

All in all, in this society of surplus, it's something of a miracle if you manage to keep reasonably trim!

CLUSTER HEADACHE

"as if a red-hot poker were being stuck in my eye."

Functional MRI scans (fMRI) are crucial tools for brain research, but the clinical significance of the technology is limited. There is, however, one shining exception: It has provided fresh clinical insights and a therapeutic strategy in the diagnosis and treatment of cluster headaches. This syndrome (which is luckily rare, affecting fewer than one in a thousand people), causes excruciating headache attacks, mainly for periods of several months; these are the "clusters" after which it is named. The attacks last between fifteen minutes and three hours. Between those periods the patients experience no headaches at all. But around 10 percent have attacks every day or nearly every day, year in, year out, without remission. The unilateral pain around and behind the eye was described by a patient to be "as if a red-hot poker were being stuck in my eye." Cluster headaches are so horrendously painful that the condition is also known as "suicide headache." During a cluster period attacks can be brought on by drinking alcohol or exposure to an environment with low oxygen pressure, for instance in mountainous areas above six thousand feet or in planes with low cabin pressure. More men than women suffer from cluster headache.

Various factors indicate that cluster headaches are a disease of the hypothalamus. The first such indicator is that the patient experiences autonomic symptoms generated by the facial nervous system on the same side of the face that's affected by the pain, such as sweating, tearing up, a runny nose (or, conversely, a blocked nose), and eye redness. The eyelid may droop, and the pupil may constrict. All of these symptoms point to increased activity in the center of the autonomic nervous system, the hypothalamus.

Second, the biological clock in the hypothalamus plays an important role in generating cluster headache attacks. The biological clock (fig. 18) is responsible for all of our day-night and seasonal rhythms, including during periods of illness. The cluster headache attacks often strike at a set time of the day or night, and they show seasonal fluctuation. What's more, the hormonal day and night patterns of cluster headache patients change in a way that suggests alterations in the biological clock, and activity is also noted in that area when attacks are provoked.

Professor Michel Ferrari, headache specialist at Leiden University Medical Center, finds that cluster headache responds better than other headache disorders to treatment; medication often proves extremely effective. Attacks can often be effectively treated with oxygen or sumatriptan and prevented with calcium antagonists or lithium. Only 20 percent of patients don't respond to treatment. The main problem, however, is that the disease often goes unrecognized or is only identified far too late, sometimes after decades of pain, during which desperate patients often undergo extremely drastic therapies, ranging from severing the facial nerve and undergoing major sinus operations to having all of their teeth removed.

Scans carried out during attacks to establish the source of the pain have revealed an increase in grey matter at the rear of the hypothalamus, bordering on the thalamus, indicating that patients have more than the normal number of brain cells on the side of the attacks. And fMRI scans have shown increased activity at this location during attacks (activity ceased during the period of recovery). These patients then had a depth electrode inserted into the rear of their hypothalamus, which was continually stimulated. Eight years of experience have now been amassed using this technique on over forty patients. Electrical stimulation of the area where the headache attacks are generated stops the attacks in 60 percent of patients and makes them sleep better. This treatment doesn't provide instant relief; it usually takes effect after about a month.

Long-term stimulation appears to be safe, although we don't know exactly how it works. PET scanning, a technique that shows changes in brain activity, reveals that the electrode treatment stimulates far more brain areas than just the hypothalamus. In fact, a change of function is induced in a whole network of brain structures involved in pain processing.

The operational mechanism is interesting, but the most important fact is that patients are helped by the therapy. After all, we don't even quite know how aspirin works. But whether using depth electrode stimulation to treat cluster headache is truly effective can only be shown by a controlled clinical trial. A French group of researchers decided to carry out such a trial on a group of eleven patients (randomly selected and split into two subgroups, without the patients knowing which subgroup they belonged to). In the first month after the electrodes had been implanted by a neurosurgeon, one subgroup had their electrodes stimulated, while the other half did not. After a week's rest, the patients' treatment was then reversed (crossover). After this two-month period, however, no difference was seen between the month with and the month without electric stimulation. In other words, the experiment didn't suggest that the treatment was effective. One could of course argue that the group was small and that the method of stimulation wasn't optimal. Whatever the case, the electrodes in the brains of all eleven patients were subsequently activated for the period of a year. Six of the eleven patients benefited from the treatment. This result was in line with expectations, but the effectiveness of the method had not been proven in a controlled trial. So after receiving permission from the ethical committee, the researchers asked the patients if they were prepared to undergo another experimental period in which their electrodes were alternately stimulated or switched off. The patients refused, frightened that the cluster headaches would come back with their old severity. So far, therefore, nothing has been proven. In the meantime it appears that stimulating a subcutaneous nerve at the back of the head—a much less drastic process—can also be effective. So it's extremely debatable whether one should go on implanting electrodes in the brains of cluster headache patients. Clinical research is far from simple.

NARCOLEPSY: WEAK WITH LAUGHTER

> All the narcolepsy patients began to laugh as soon as he came in, until they fell to the ground, completely weak with laughter.

Narcolepsy is a sleep disorder. People who suffer from it are excessively sleepy and less alert during the day, while at night, paradoxically, their sleep is disturbed. Although serious, these symptoms aren't in themselves specific to this brain disorder. Narcoleptic patients also tend to be overweight, but this is of course very common in itself and isn't necessarily connected with narcolepsy.

Narcolepsy does have a number of distinguishing symptoms. In many cases, an emotional experience can make sufferers suddenly lose muscle tone in their arms and legs and fall to the ground (fig. 19). They are so completely weak from laughter or shock that they appear to have lost consciousness, but after the event they can relate everything that happened. This characteristic symptom is called cataplexy. One of our PhD students was so cheerful and full of fun that the narcolepsy patients would begin to laugh as soon as he came into the room, causing the ones with cataplexy to collapse on the floor. This attribute made him the secret weapon of our research. If patients with a history of cataplexy are shown funny cartoons while lying in a functional MRI scanner, their scans show overactivity in the emotional brain circuits and, possibly in response to this, activity in an inhibitory system of the prefrontal cortex. The hypothalamus, where the condition originates, in fact shows reduced activity during the period of cataplexy that follows the laughing fit.

People with narcolepsy can also suffer from sleep disorders that aren't typical of their condition. Some often find themselves unable to move for several minutes after waking up. Such sleep paralysis can be very frightening. Moreover, narcolepsy patients often have extremely vivid dreams—or more frequently nightmares—in the transition between waking and sleeping. These are known as hypnagogic hallucinations and can be so powerful that patients lose touch with reality. A woman who had often hallucinated that all her teeth were being pulled out just as she was waking up didn't realize that a later visit to the dentist was real and not a hallucination. She jumped out of the chair and ran for dear life, just as she did in her hallucination, leaving a very puzzled dentist behind. Other people have hallucinated that dwarfs were stabbing them, that they were being sucked into a corpse, or that they were dying in terrible, violent ways. Hypnagogic hallucinations can be similar to schizophrenic hallucinations and sometimes provoke out-of-body experiences that resemble near-death experiences. Functional scanning does in fact reveal changing activity patterns in the temporal lobe, just as in near-death experiences caused by lack of oxygen (see chapter 16).

The symptoms of narcolepsy are caused by the absence of a chemical messenger in the hypothalamus called hypocretin (or orexin). Degeneration of the cells in the hypothalamus (a phenomenon that has yet to be explained) can induce the symptoms of narcolepsy. This process can be tracked by measuring hypocretin levels in the cerebrospinal fluid. Narcolepsy with cataplexy can also arise when the brain becomes unreceptive to the message of hypocretin. This is caused by a tiny mutation in the DNA of the gene producing the protein that receives this message: the hypocretin receptor. These mutations are rare in humans, but when I was a guest lecturer at Stanford University in the United States, I encountered a dog who had it. It wasn't easy to get to see the dog: My team had to fill out endless forms at checkpoint after checkpoint and comply with all kinds of safety measures and clothing requirements. The animal quarters at Stanford are so costly that, as the head of the research team sighed, "The dog is more expensive to maintain than a post-doc." We had taken along a tin of the dog's favorite food, and we were not disappointed by the effect. The massive Doberman wagged his tail and went into paroxysms of joy. Then his back legs suddenly collapsed beneath him (fig. 19), followed by his front legs, after which he fell asleep on his side. He recovered very rapidly and began to munch contentedly on his favorite meal. We crept away, closing the door of his pen behind us, and returned to the checkpoints. Suddenly I heard the patter of footsteps behind me. I turned around to find myself looking right into the eyes of the Doberman, who was following me, his head on a level with my shoulders. He wanted some more of that delicious dog food and had somehow managed to get the door open. Clearly, the mutation in question doesn't impair the intellect.

FIGURE 19. Narcolepsy is a sleep disorder that causes excessive sleepiness during the day but fragmented sleep at night. When overcome by certain emotions, such as laughter or fright, narcolepsy patients can suddenly lose muscle tension in their arms and legs and collapse on the ground. They aren't unconscious, even though this may appear to be the case (see photos above). This is known as "cataplexy" and results from the lack of a chemical messenger (hypocretin or orexin) from the hypothalamus. Narcolepsy with cataplexy can also arise if the brain ceases to be receptive to hypocretin. The bottom row of photos shows a large Doberman pinscher in ecstasies of excitement at the prospect of being given a tin of his favorite meat. His back legs collapse, followed by his front legs, and he then seems to fall asleep on his side, having cataplexy for a couple of minutes. S. Overeem et al., *Lancet Neurol.* 1 (2002): 437–44.

FITS OF LAUGHTER WITHOUT EMOTION

> A reverend gentleman brought his wife to this city to seek the advice of Messieurs Le Grand, Duret, and myself [physicians], hoping that we could ascertain why she cried and laughed for no cause, but none could cure her. We physicked her in many different ways, but with very little success. In the end he took her away in the same condition in which she had arrived.
>
> Ambroise Paré

In 1996 I was invited to take up the Emil Kräpelin guest lectureship at the Max Planck Institute of Psychiatry in Munich, a renowned institution that combines research and treatment in an excellent way. Trainee psychiatrists spend the mornings in the polyclinic and the afternoons in the laboratory. The institute's research happened to focus on depression, one of the main strands of my own research at the time. My team had carried out a great many studies of postmortem brain tissue of depressive patients and had shown that their stress axes were hyperactive, potentially laying the foundations for symptoms of depression (see earlier in this chapter). Blood tests of depressive patients carried out by the Max Planck Institute also showed stress axis hyperactivity. So the invitation to take up the guest lectureship should have been most welcome, tying in perfectly with my research while at the same time being a great honor. Yet I felt extremely ambivalent about it. This was because the institute had been prominent in the field of eugenics during the Nazi era, euthanizing psychiatric patients and the mentally handicapped. During the Second World War, over 220,000 patients with schizophrenia were sterilized or murdered. It's thought that three out of four psychiatric patients then living in Germany were put to death. Some sources even estimate that they were virtually all killed.

I visited my father and explained my dilemma: on the one hand an extremely attractive invitation, on the other an institute with a deeply tainted Nazi past. He considered the question for about two seconds and then said, "You must go and tell them that we are still here." So that's what I did. Talking about the German occupation of the Netherlands, the Holocaust, and the institute's past proved much less difficult than I'd feared. The director, Florian Holsboer, was Swiss, the heads of the working groups represented a mix of nationalities, and the organization's working language was English. Moreover, the German members of the staff were themselves very much preoccupied with the institute's past. The cellar contained the records of all the murdered patients, neatly archived with German precision, and they were used for academic publications on the institute's terrible history.

After my lecture I met a woman at the polyclinic who had a habit of bursting into stereotypically shrill laughing fits several times a day for no reason. What made it so chilling was that the emotions associated with laughter were completely absent. While knowing that you shouldn't leap to the conclusion that someone is suffering from an extremely rare condition, I couldn't suppress the cautious suggestion that this might be a case of hypothalamic hamartoma. Not long before, I had by chance encountered a growth of this kind in a hypothalamus, and I'd read up on the subject. The patient in Amsterdam had displayed no symptoms, and I'd never yet seen someone in real life who went about bursting into laughter because of a rare little lump in their hypothalamus. But just such a lump turned out to be visible on her scan, right at the back of the hypothalamus, near the mammillary bodies (fig. 18). The lump doesn't grow; it isn't a tumor but a developmental defect. It consists of clumps of nerve cells that failed to find their proper place in the hypothalamus at an early stage of development. In 50 percent of cases, the lump causes epileptic activity, provoking attacks of laughter. These are known as gelastic seizures (from the Greek *gelos,* meaning "laugh"). Some patients alternate between laughing and crying. The local epileptic activity can spark classic epileptic fits with seizures and loss of consciousness. Although the symptoms produced by a hamartoma have led to the hypothesis that there's a "laughter center" at the back of the hypothalamus, it's more likely that the location of the lump activates various brain circuits that cause the abnormal behavior.

The hamartoma can produce all kinds of hormones and can even induce puberty much too early. It can also cause psychiatric problems in children, such as ADHD, antisocial behavior, and intellectual deterioration. The cognitive malfunctions are probably due to damage to the mammillary bodies, which play a crucial role in memory (see chapter 14). Hamartomas can also be responsible for obesity and anger attacks. Abnormal hormone production resulting from these growths can be treated with medication, but the lumps can also be surgically removed or irradiated to cure patients with epileptic activity or abnormal behavior.

In investigating cases of spontaneous laughing fits, however, it's important first to rule out other possible causes, such as enlarged pituitary glands and other tumors, MS, and various developmental malfunctions in the brain.

It doesn't often happen that a researcher like me is able to correctly diagnose a rare condition in practice. After the event I caught myself smiling discreetly—a smile that was accompanied by all the appropriate emotions.

ANOREXIA NERVOSA IS A DISEASE OF THE BRAIN

> The exact nature of this illness hasn't been established, but it must be situated in the hypothalamus.

The French parliament recently considered draft legislation that made the promotion of anorexia a crime punishable by a maximum sentence of three years' imprisonment and a €50,000 fine. The bill in question didn't just target the skeletally thin models in the fashion world but also the "pro-ana" (pro-anorexia) websites that a French minister claimed were disseminating "messages of death." The

French fashion sector signed a charter pledging to promote healthy body images and to stop using ultra-skinny models. The British doctors' association also established a link between abnormally thin models and the onset of eating disorders. And in the Netherlands, there were newspaper reports of a sixteen-year-old girl with anorexia weighing only forty-six pounds who had been expelled from secondary school. People suddenly seemed to buy into the myth that you catch anorexia by seeing it, similar to the way that homosexuality was previously regarded—completely erroneously of course—as a contagious condition (see chapter 3). In neither homosexuality nor anorexia is this view supported by any evidence whatsoever. A British public information campaign costing millions turned out to be a complete waste of money. And that was to be expected, because although an eating disorder can lead to your getting a job as a stick-thin model, no one has ever proved that anorexic models cause a spate of eating disorders.

The absence of a copycat factor in anorexia is demonstrated in the case of a woman who had been blind from the age of nine months who developed a classic form of anorexia nervosa at the age of eighteen. And contrary to popular assumption, there's actually no evidence that anorexia is on the increase. More women, however, are coming clean about their eating disorders following similar confessions from prominent individuals like Princess Diana, the Swedish crown princess Victoria, Jane Fonda, and many other celebrities.

No one denies that anorexia is a dangerous disease. Indeed, in around 5 percent of cases it proves fatal. Some 93 percent of patients are women, suggesting that you have a higher risk of anorexia if your brain has differentiated along female lines (see chapter 3). In Sweden, a cognitive therapy has been developed to teach anorexia patients to regain eating skills; it's known as the Mandometer Method. Of course, the therapy doesn't explain how anorexia is triggered.

All of the symptoms indicate that it is a disease of the hypothalamus. Besides the eating disorder and the loss of weight, symptoms include the cessation of menstruation, lower sex hormone levels, reduced libido, impaired functioning of the thyroid, hyperactive adrenal glands, and malfunctions in the water balance and day-night rhythms. Women who lose a great deal of weight stop menstruating. This is a protective mechanism with a huge evolutionary advantage: Women who don't have sufficient food for themselves certainly shouldn't become pregnant. But in 20 percent of the women who develop an eating disorder, menstruation stops before weight loss occurs. This is an indication of a primary pathological process in the hypothalamus. Some of the symptoms, like the malfunctions in the thyroid and adrenal glands, remain even after the weight has been regained. An extreme preoccupation with calories, precise food selection, and every other aspect of the eating process can persist long after someone has regained their normal weight. Take one patient who, having recovered from acute anorexia, became employed as a recipe writer for a women's magazine. These persistent symptoms also show that a pathological process is at work in the brain and that anorexia symptoms aren't just due to weight loss. Even when anorexia patients start to eat normally, it's debatable, in the case of a great many of them, whether the condition has really disappeared.

The final argument for locating the disease in the hypothalamus is that all of the symptoms of anorexia nervosa can also be caused by a cyst, a small tumor, or some other abnormal process in the hypothalamus. Indeed, autopsies on anorexia patients sometimes reveal lesions in the hypothalamus. There is also the case of a patient who had been undergoing long-term psychiatric treatment for anorexia nervosa and eventually developed other neurological symptoms. Tests revealed a tumor in her hypothalamus. Of course, these comparatively rare findings don't indicate that all anorexia patients have tumors of the hypothalamus, but they do show that a primary pathological process in the hypothalamus can cause all of the symptoms of anorexia and could explain the disease entirely. Indeed, if an MRI scan is carried out in a late stage of anorexia nervosa, the brain is seen to have shrunk, and a wide range of behavioral and cognitive disorders can be expected to result.

We haven't yet established the exact nature of anorexia. However, it's clear that in addition to being more common among women, there are also genetic factors that predispose a person to the disease. A number of the genes in question have already been identified. An extremely stressful life event can appear to be the direct cause of anorexia, but the factors that make someone vulnerable to the disease in the first place come into play at a much earlier stage, probably as far back as during the development of the brain before birth. It seems likely that anorexia sufferers maintain the process of voluntary starvation because they become addicted to the diet-triggered opiates released by their brains, which activate the reward center at the base of the striatum (fig. 16). But the origins of the condition remain a mystery. I favor the theory that anorexia is the result of an autoimmune process. Antibodies are indeed found in the blood of anorexia patients, directed against chemical messengers in the hypothalamus involved in regulating eating and metabolism. The only way to find out more about the condition is to carry out microscope studies on the brains of patients who have died of anorexia. But the prospect of the necessary autopsy meets with a great deal of resistance among those treating the disease as well as among patients and ex-patients. An ex-patient firmly concluded that anorexia couldn't possibly be an illness of the brain, as she was now cured. I must confess that I never quite followed her reasoning. Fortunately, many diseases simply disappear. Another patient also dismissed the notion of a disease of the brain, because it was all about "your attitude to life"—as if that had nothing to do with your brain.

6

Addictive Substances

CANNABIS AND PSYCHOSES

Cannabis has lost its innocence.

Addictive substances have been around throughout humankind's history, but every society and every group has different views on which are permissible and which are beyond the pale. I remember how amazed I was in the 1960s when people would stand with a glass of wine in one hand and a cigarette in the other, complaining loudly about the pot-smoking habits of long-haired layabouts. Addictive substances affect the brain by mimicking its own chemical messengers. Brain cells themselves produce a whole range of opiates and cannabinoids. The nicotine in a cigarette has the same effect as the chemical messenger acetylcholine. Addictive substances also affect the availability or the action of natural chemical messengers. The drug ecstasy, for instance, increases levels of serotonin, oxytocin, and vasopressin. That's why your brain no longer functions optimally when you stop taking a particular addictive substance: You feel terrible and get frequent uncontrollable urges to take more of it. All such substances have a direct or indirect effect on the brain's dopamine reward system (fig. 16), whether or not via an opiate system. Both systems are crucial to the rewarding effect of many normal stimuli, including that of sexual behavior. Not for nothing is the brief euphoria following an injection of opiates often described in sexual terms; after all, it activates the same reward system.

Cannabis has been used for recreational, religious, and medical purposes since time immemorial. The medicinal effects of marijuana were first described in China around five thousand years ago. Nor is its medicinal use in the West new. It is said that Queen Victoria used it as a remedy for menstrual pain. Cannabis has recently been rehabilitated, and medicines with its active ingredient Δ^9-tetrahydrocannabinol (THC) can be obtained in the Netherlands with a prescription. Studies are being carried out on its potential efficacy against pain, fearfulness, sleep disorders, the nausea caused by chemotherapy, and glaucoma (because it reduces eye pressure). Marijuana is also thought to reduce spasticity among patients with MS, although a controlled experiment with THC couldn't confirm such an effect.

Cannabis affects the brain because brain cells themselves produce cannabis-like neurotransmitters. The first such compound to be identified was christened anandamide, *ananda* being Sanskrit for "bliss." The proteins that transmit anandamide's message to the brain, the receptors, are mainly located in the striatum (hence the blissful feeling) and in the cerebellum (hence the unsteady gait after taking marijuana), in the cerebral cortex (hence the problems with association, the fragmented thoughts and confusion), and in the hippocampus (hence the memory impairment). But there are no receptors in the brain stem areas that regulate blood pressure and breathing. That's why it's impossible to take an overdose of cannabis, as opposed to opiates.

The quality of cannabis in the Netherlands has improved so much over the years that it's rapidly changing from a soft drug into a hard drug. Daan Brühl was a nineteen-year-old boy from Amsterdam who was a champion rower. Returning from a disappointing tournament, he was seized by heart palpitations. To calm down he smoked a couple of joints with his girlfriend. Suddenly his expression altered.

He walked to the kitchen, grabbed a knife, and stabbed himself in the heart. He died that evening. Suzanne, a twenty-year-old, became psychotic after smoking large amounts of cannabis. She began to hallucinate and in an extremely fearful state was taken to the hospital, where she was diagnosed with schizophrenia. Schizophrenia is a developmental brain disorder that begins in the womb (see chapter 10). However, the first psychosis doesn't usually occur until around the age of sixteen to twenty because the sex hormones that circulate after the onset of puberty place an enormous burden on the adolescent brain and can bring on the symptoms of schizophrenia. The same applies to cannabis. It's quite possible that teenagers who are admitted to the hospital with symptoms of schizophrenia after smoking cannabis would have suffered a psychosis a few months later anyway. On the other hand, studies show that cannabis users are twice as likely to develop schizophrenia. Patients who already have schizophrenia but are past the worst stage can suffer from a relapse if they use cannabis. The link between cannabis use and psychosis is all the more interesting in the light of the recent discovery that the brain's own cannabis system is activated in schizophrenic patients. Research is now being carried out to find out whether this system could be a new target for schizophrenia medication.

Adult males who smoke marijuana daily for a period of years have been found not only to have a smaller hippocampus (important for memory), a smaller amygdala (affecting fear, aggression, and sexual behavior), and damage to the fiber tracts of the corpus callosum (left-right connections) but also to be more likely to develop a psychosis. Again, though, these characteristics may have been present before using cannabis.

Not every psychosis after cannabis use is the first sign of schizophrenia. Gerard, a Dutch student, was mountain climbing in New Mexico. In the evening, when his group was sitting around the campfire, one of his fellow climbers offered him a joint. After four puffs he experienced hallucinations with flashes of light, became extremely hot, felt dehydrated, developed a burning spot in his heart region along with cardiac arrhythmia, lost all power in his legs, and fainted twice. His friends took him to a hospital in El Paso, where he was told, not very helpfully, that he had had a "bad trip" and could go home again. Afterward he found out that the previous day, one of the other climbers had had a similar problem with that batch of cannabis, so they threw the rest away. The next day Gerard was left with the feeling that there was something odd about his sensory perceptions as well as the world in general (derealization), and he felt very tired. After a few days he had recovered sufficiently to join the next climbing trip. A month after the bad trip, after three days of climbing and considerable physical effort, he suddenly had the same symptoms as before, including the sensation of derealization, a feeling of acting on automatic pilot and observing himself from a distance (depersonalization), a tendency to faint, and serious difficulties in concentrating—something he found particularly worrying. Back in the Netherlands, doctors tried to get to the bottom of his extreme reaction to cannabis. Fortunately, he hadn't had the classic psychosis resembling a schizophrenic episode but had perhaps been intoxicated by a high dose of THC. In the Mexican border region, cannabis is said to be often far stronger than elsewhere. Someone put forward the theory that THC can remain in fat tissue for a very long time, to be released in response to intense physical effort. Moreover, Americans appear to smoke "pure" cannabis on the whole, unmixed with tobacco, causing a greater risk of adverse effects. (When on vacation in Amsterdam, they also put much too much THC in a single joint.) Another possible explanation was that the cannabis Gerard smoked had been sprayed with chemical agents or adulterated, possibly with a toxic weed killer or angel dust. None of these theories could be proven. Luckily, he made an excellent recovery. He's back at university and recently scored high grades on his exams.

Cannabis has lost its innocence. It's much stronger than it used to be and much less harmless than we all thought it was back in the 1960s. For some people, taking cannabis can have disastrous consequences. But compared to the enormous widespread damage that alcohol and cigarettes cause in our society, the problem remains a proportionately small one.

ECSTASY: BRAIN DAMAGE AFTER PLEASURE

> Take me, I am the drug; take me, I am hallucinogenic.
>
> Salvador Dalí

Ecstasy is now known as the "love drug," but it was originally patented as an appetite suppressant in 1914. Its recreational use can be extremely dangerous. A year ago, a trainee nurse decided to write a thesis on ecstasy. She didn't want to do so without any experience of the drug, so she tried it during the festival on Queen's Day, an important Dutch holiday. Just to make sure she wouldn't harm herself, she drank four liters of water. The experiment proved fateful. She lost consciousness, was in a coma for several days, and has been left with lasting brain damage, especially to the cerebral cortex.

When you take ecstasy, you make quite a lot happen in your brain, as a pioneer in this field, the radiologist Liesbeth Reneman, established many years ago, and the risks are considerable. Twenty minutes after swallowing ecstasy, increased amounts of the chemical messengers serotonin, oxytocin, and vasopressin are released. All tiredness vanishes and you feel happy and want to hug everyone. The pleasurable feelings of love and wonderful social interaction last for an hour. If you take a fair amount of ecstasy every weekend, you destroy the brain cells that make serotonin. Less of this messenger is produced, and you function less well. You also need more ecstasy each time in order to get the same pleasant effect. Users are more likely to develop psychiatric and neurological problems like mood disorders, aggression, impulsiveness, and memory impairment. Brain scans of ecstasy users indeed show permanently reduced activity in structures that would be involved in such problems, like the amygdala, hippocampus, thalamus, and cerebral cortex. A recent follow-up study showed that even brief, extremely limited ecstasy use amounting to a couple of pills taken over an eighteen-month period impairs memory and blood flow in the thalamus and cerebral cortex. Scans have also shown that ecstasy can cause blood vessels to constrict or, conversely, to dilate on a long-term basis, depending on the brain region in question. The neurological consequences can include brain infarcts or bleeding on the outside of the brain, resulting in serious and lasting neurological damage.

If you take ecstasy when it's hot and you don't drink enough, you can dehydrate, resulting in organ failure. Some people develop palpitations or have a sudden fatal heart attack. Unity, a voluntary project sponsored by the Jelinek addiction treatment center in the Netherlands, therefore rightly advises ecstasy users to drink a glass of water or a sports drink every hour to prevent dehydration and overheating. But apparently people also need to be explicitly warned that drinking too much water in combination with ecstasy can also be dangerous. Ecstasy causes the pituitary gland to release much more vasopressin (the antidiuretic hormone; see chapter 5), so your kidneys retain all the liquid you drink. As a result, drinking too much can cause water poisoning and serious brain damage.

This is exactly what happened to the trainee nurse researching her thesis: She developed brain and vascular edema—essentially a swelling of the brain—causing serious brain damage. After three days she gradually emerged from the coma, suffering subsequent epileptic attacks. A scan showed severe swelling of the brain, mainly on the left-hand side. In the early weeks she couldn't talk (Broca's area, bottom left of the frontal cortex, fig. 8) or walk (motor cortex, fig. 22) and her sight was impaired on one side (visual cortex, at the rear of the brain, fig. 22). She's still recuperating with therapy. She can't yet read and finds writing difficult. When she speaks, she regularly struggles for words. Recovery is a slow, demanding process, and she's continually tired. It's not yet clear what permanent impairment will result from her brush with ecstasy, but an MRI scan shows damage to the cerebral cortex. Instead of writing a thesis on ecstasy, she now wants to warn schoolchildren about its dangers, particularly in combination with a lot of water. Her own ecstasy party is over for good.

SUBSTANCE ABUSE BY POLITICIANS

> Drunkenness is nothing more than voluntary madness.
>
> Seneca

The terms *addiction* and *substance abuse* mainly conjure up images of neglected, homeless, and often schizophrenic people. But the former British cabinet minister David Owen, a neurologist by training, makes clear in his startling 2008 book *In Sickness and in Power: Illness in Heads of Government During the Last Hundred Years* that the problem of substance abuse can be found even in the highest government circles, with potentially dramatic consequences for the course of history.

Teenagers experiment with pills, magic mushrooms, marijuana, and other potentially dangerous substances. This seems to be a normal aspect of adolescent behavior (see chapter 4). Some world leaders have been forced to confess to similar youthful peccadilloes, which isn't always easy. During his 1992 campaign, Bill Clinton was forced to admit that he had smoked pot as a student, adding rather feebly, "But I didn't inhale." But America has become more tolerant of these things in recent years. When Barack Obama sought the presidential nomination, he was questioned about his book *Dreams from My Father,* in which he wrote about smoking pot and drinking as a troubled teenager, and he owned up to the occasional hit. "I inhaled frequently—that was the point," he said, poking fun at Bill Clinton. This open admission never worked against him.

Before his presidency, George W. Bush was an alcoholic, and certainly not just as an adolescent. At the age of thirty he was arrested for drunk driving and lost his driver's license for two years. During his 2000 campaign he stated that in 1986 he woke up with a hangover after celebrating his fortieth birthday and that he had not drunk a drop since. That claim seems a little questionable. In 2002, Bush fell off a couch while watching a football game and was taken to the Johns Hopkins Hospital in Baltimore. A British doctor there told David Owen that Bush's blood contained significant amounts of alcohol. Bush has refused to answer any questions about his alleged past cocaine use.

There's no doubt at all, however, about Richard Nixon's alcoholism during his presidency. In 1969, when an American spy plane was shot down in North Korea, Nixon stormed about in a drunken rage, telling his national security advisor, Henry Kissinger, "Henry, we'll nuke 'em!" Tapes released after the Arab-Israeli conflict in 1973 also revealed that Nixon was too drunk to discuss the situation with his British counterpart. Russian president Boris Yeltsin suffered severe back pain after being in a Spanish plane that was forced to execute a crash landing. Beginning in 1994 he took more and more painkillers and drank increasingly. At an official ceremony that year in Berlin, in honor of the departure of the last Russian troops, he was clearly drunk. There was another incident that same year when his plane stopped over at Shannon International Airport. The entire Irish government stood ready to welcome him at the foot of the aircraft steps, but he was sleeping off the effects of a drinking bout and failed to appear.

Not that alcoholism is confined to presidents, of course. The American Communist hunter Senator Joseph McCarthy, who destroyed many lives, had a serious alcohol problem and died in 1957 of cirrhosis of the liver. And alcohol certainly isn't the only substance abused by world leaders while carrying out their duties. In 1956, during the Suez crisis, British prime minister Anthony Eden, who was suffering from severe pain, took pethidine, an opium derivative, shortly before chairing a cabinet meeting. He took barbiturates to sleep and amphetamines to stay alert. Indeed, in the final weeks before his resignation he lived on amphetamines, a fact that he didn't conceal from his cabinet.

Some leaders have been successful despite mental health problems and substance abuse. Winston Churchill not only suffered from severe depression but also had hypomanic and manic phases (see chapter 5) in which he drank phenomenal amounts of champagne, brandy, and whiskey. John F. Kennedy had numerous health problems, including Addison's disease, caused by the failure of the adrenal glands, for which he had to take cortisol. At some point during his election campaign he forgot to take his cortisol tablets with him and fell into a coma. In 1938 he was involved in a car crash, after which he suffered from back pain. His treatment involved procaine injections that were administered three times a day or more. (Procaine is a synthetic cocaine derivative that can leak into the brain, causing central effects.) Kennedy was also a recreational user of substances like amphetamines, whose effects he described as delectable and euphoric. He took cocaine both before and during his presidency. He was also being given testosterone, allegedly because of his adrenal gland problem, but you might wonder whether that could have contributed to his irresponsibly macho behavior during the Bay of Pigs invasion of Cuba in 1961. He experimented with marijuana and LSD in the company of his mistresses at the White House. Kennedy also took sleeping pills, painkillers, and phenobarbital, a sedative. On top of that, he was given injections of a mixture of corticosteroids and amphetamines concocted by a doctor. Various doctors prescribed medication for him without knowing what other treatment he was receiving. It has been said that Kennedy played around with doctors even more than with women!

Shouldn't we at least impose the same requirements on those who govern the country as those who drive a car or fly a plane? When are we going to test politicians, on whom we're so dependent, for alcohol, drugs, and medication?

7

The Brain and Consciousness

NEGLECT: HALF A LIFE

"If it isn't there, then I can't be ignoring it."

We are conscious of our surroundings and of ourselves. Some brain structures are crucial to consciousness, like the cerebral cortex, the thalamus (where we receive sensory information), and the white matter, the nerve fibers linking these structures (fig. 20). After a brain infarct on the right side of the brain, both self-consciousness and consciousness of surroundings can be impaired. It's possible for a person not to be aware that they are paralyzed on the left side and to ignore everything on their left, not just in terms of their surroundings but also of their own body. This is known as neglect. If you approach such a patient's bed from the left, even though he can turn his head and see you, he won't notice you. When reading a newspaper, such patients will look only at the right-hand page, and when drawing, will draw only the right side of an object. They eat only what is on the right half of their plates. If you then turn the plate around 180 degrees, they will eat the other half. Neglect can extend to the left half of their bodies. They no longer regard their left arm or their left leg as part of themselves. They will dress or wash only their right side, and comb their hair only on the right side of their heads.

Neglect patients frequently make up extremely imaginative stories to explain the bizarre situation in which they find themselves. Some claim that the hospital is their home and that they have chosen the furniture themselves. One patient remained convinced that her left side was fully functional and that she was physically independent. Yet in her drawings, the left half was completely lacking. "If it isn't there, then I can't be ignoring it," she countered. When she was asked to move her left arm, she answered, "I could do that, but it would be better if I rested it now." When she was asked to walk a few steps after she had claimed that there was nothing wrong with her, her response was, "Of course I could, but the doctors said I should rest."

The mother of a good friend of ours suffered a severe brain infarct on the right side when she was eighty-five, being left paralyzed on the left side. Her mind was still sharp, though, and she hadn't lost her sense of humor. Indeed, her conversations with relatives, friends, and the nursing staff were perfectly normal, with one striking exception. One day she told me that she'd had a really odd dream in which she had a third arm. I carefully took hold of her paralyzed left arm and asked her, "Is this that third arm?" "No," she said, "of course not, that's Kees." Kees was her fifty-five-year-old son. "What is Kees doing here?" I asked. "He's sleeping in my bed, as he always does." This was nonsense; I knew the family very well. "But tonight I needed him," she went on, "and I couldn't wake him up. The same thing happened the night before, when Kitty [her daughter's friend, who visited her nearly every day and was very close to her] was sleeping here, and I couldn't wake her up either," she continued somewhat huffily. She asked if she could have something to drink and went on to speak perfectly normally about all kinds of matters that she wanted to arrange.

The fantasies that arise in the case of neglect in fact spring from a very general principle. If something gets in the way of the brain's information supply, it starts to make up information to fill the gaps. A damaged brain deprived of its customary input invents bizarre stories. It does similar things to compensate for the lack of oral or visual information or information from the memory or limbs (see chapter 7). Unconsciously filling up the little holes in our memory is something the brain does on a daily basis, even when it's intact. We're convinced that events happened exactly as we remember them and will state as much under oath in court. In fact, our brains are just knitting neat stories out of the countless scraps of information they receive, leading to all kinds of consequences.

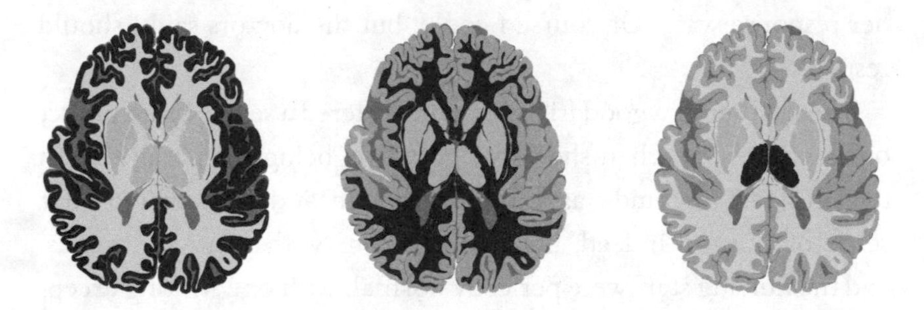

FIGURE 20. Three intact and properly functioning structures, shown here in black, are crucial to consciousness: an intact cerebral cortex (left), the thalamus (right), and the white matter (center), with pathways connecting the cortex and thalamus.

COMA AND RELATED CONDITIONS

> It's as if he has been abandoned twice, first by his brain, then by the people who knew him. Because no one visits him.
>
> Bert Keizer, *Inexplicably Inhabited,* 2010

A coma is a situation in which a patient can't be woken and doesn't respond to external stimuli. It can result from damage to the cerebral cortex, the thalamus, the connections between these two brain structures (fig. 20), or the brain stem (fig. 21), which activates the cerebral cortex and the thalamus. But it can also be caused by a metabolic disorder, drugs, or an excessive amount of alcohol. Some people recover from comas. One boy was driving back from a night out with friends when he crashed into a concrete post at high speed; the accident left him in a coma for six weeks. His family had already been approached about the possibility of his kidneys being donated for transplant. But they thought they detected slight signs of a return to consciousness, so they put off the decision. They were right, because he came out of his coma and completed his education at technical school. He wasn't quite the whiz at math he had been before the accident, but otherwise he was his old self. He got a good job, married, had children, and is now a grandfather. But things don't always go so well. People regularly awake from comas with serious and permanent brain damage or don't awake from them at all.

Vegetative State

Functions crucial to survival, like breathing, heart rate, temperature, and sleep-wake cycles, are regulated in the brain stem (fig. 21), which also contains the centers for coughing, sneezing, and vomiting reflexes. So, as long as your brain stem is still intact, you still go on breathing, even if the rest of your brain no longer functions. This tragic situation arises when people wake up from a deep coma after a serious brain injury, but instead of gradually getting better, they merely exist, like a vegetable. The same actually applies to patients in the final stage of Alzheimer's. They lie in a fetal position, their cerebral cortex no longer functions, and they no longer respond to the world around them (fig. 31).

We need the cerebral cortex to think, speak, hear, feel emotion, and move our limbs. When someone is in a vegetative state, also known as a coma vigil, their brain stem functions are still intact, while the rest of their brain, particularly the cerebral cortex, no longer works. Most patients in this situation gradually regain consciousness after a few weeks, but if the cerebral cortex has been irreparably damaged, they merely progress to a "persistent vegetative state." They can breathe independently and their hearts function normally, so according to the classic definition, they are "alive." Their eyes may be wide open, and they may groan, cry, or laugh, but without the attendant emotions. They appear to be "awake" but don't show physical responses demonstrating any level of consciousness, either of their surroundings or of themselves. Since they seem awake and occasionally grimace or make a sound, it's extremely difficult for their families to accept that they aren't conscious. Parents of newborn children who have had a massive cerebral hemorrhage face the same terrible problem. The child looks normal, but most of its brain has been destroyed.

As long as they are hydrated and fed, patients in a vegetative state can be kept alive for years, as was shown in the case of Terri Schiavo, a brain-damaged American woman who entered a vegetative state in 1990. In 1998 her husband (who was her legal guardian), believing that recovery was no longer possible, petitioned for her feeding tube to be removed. Her parents, however, opposed her euthanasia. For many years the case went from one court to another with much legal saber rattling, while the pro-life movement made the husband out to be a murderer. (And that kind of threat needs to be taken seriously, because most pro-lifers are in favor of the death penalty, and the movement has murders on its conscience.) It took seven years before she was allowed to die, after her feeding tube had been removed by order of the court. The subsequent autopsy confirmed that there was little left of her cerebral cortex and that all that time she had indeed had no prospects at all of a dignified existence.

Eluana Englaro was an Italian woman who entered a vegetative state in 1992 after a car accident in which she suffered irreversible brain damage. Seven years later, her father started a legal battle to have her feeding suspended, because his daughter had expressly said that she would never want to live like a vegetable. On July 8, 2008, nine years later, the Italian Supreme Court awarded her father the right to have her feeding stopped—a remarkable ruling, as euthanasia is illegal in Italy. Eluana was transferred to a clinic that was prepared to receive her and allow her to die, but the Vatican and the government tried to prevent it. "Stop this murderer!" was the predictable reaction of the cardinal who was acting as the Vatican's health minister. The decree issued by Prime Minister Silvio Berlusconi to force the continuation of her treatment wasn't signed by the president, causing Berlusconi to seek an emergency decree. Fortunately for those directly concerned, Berlusconi's decree wasn't approved in time, because several days after the tube was removed, Eluana died.

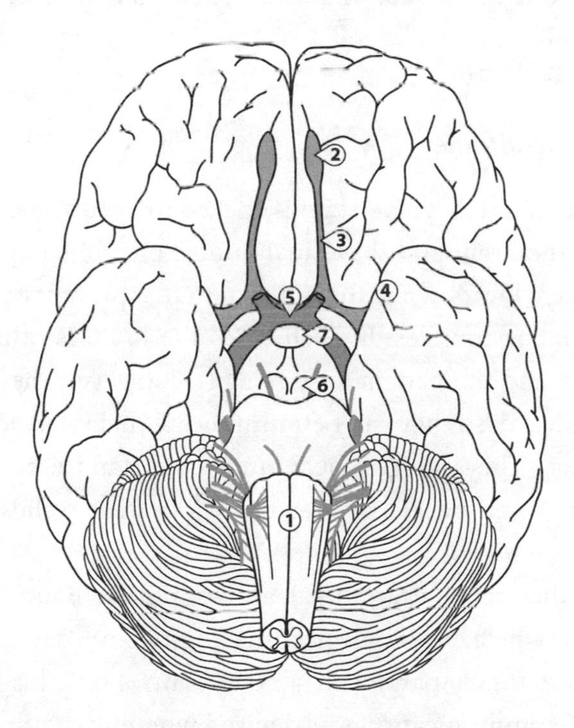

FIGURE 21. The brain seen from below. Breathing, heart rate, temperature, and sleep-wake cycles are regulated in the brain stem (1). The olfactory (smell) system consists of the olfactory bulb (2), the olfactory nerve (3), and the uncus, part of the temporal lobe (4). Also shown are the optic chiasm (5), where the optic nerves partially cross, the mammillary bodies (6), and, in between, the pituitary gland (7).

In the Netherlands, life in a persistent vegetative state isn't regarded as a dignified existence. Keeping patients alive in such situations is deemed medically futile and, in consultation with the family, the decision is usually made to stop treatment. Since the treatment being withheld is medically futile, these aren't, formally speaking, cases of euthanasia. Yet in the Netherlands, too, there are people who are in a coma for long periods of time. It's scandalous that Internet sites exploit the desperation felt by the families of such patients. The organization CWUBS (Coma Wake Up Brain Stimulations), for instance, offers therapy costing over €10,000 to awaken people from a persistent vegetative state. But even giving €100,000 worth of therapy to patients whose cerebral cortex is irreparably damaged won't bring them out of a coma; the sole beneficiary of such therapy is CWUBS itself.

Locked-In Syndrome

The reverse of a vegetative state is locked-in syndrome, which occurs when the brain and the spinal cord are completely separated due to damage low down in the brain stem that prevents nerve fibers from controlling muscles. In such cases the brain is otherwise entirely intact, and the patient is fully alert. However, they are completely paralyzed, so they can't communicate their alertness to their surroundings. They can see, hear, and understand everything, but they can't move or speak. They can only close their eyelids and move their eyes.

In 1995, the French journalist Jean-Dominique Bauby suffered a stroke after which he lay in a coma for twenty days. When he awoke he was totally paralyzed, able to control only his left eyelid. A means of communication was devised whereby the alphabet was read aloud and he blinked when the reader got to the right letter. In that way, letter by letter, he was able to write a memoir, *The Diving Bell and the Butterfly,* describing his full awareness of his surroundings, himself, and the appalling situation in which he was trapped. An impressive film with the same title was made in 2007. In it, extracts are also read from *The Count of Monte Cristo* (1844), by Alexandre Dumas, describing a character with locked-in syndrome, M. Noirtier de Villeforte, who ends up paralyzed and mute after suffering a stroke but is able, by moving his eyes and eyelids, to avert a poisoning and an undesirable marriage. A more recent instance is that of Nick Chisholm, a New Zealander who was knocked out during a game of rugby in 2000. At first he appeared to be simply concussed, but he later had a series of epileptic fits and brain stem infarcts. He was thought to be in a coma until his mother and girlfriend managed to convince doctors that he was aware of what was going on around him. He has since made a partial recovery. In the case of locked-in syndrome, families tend to be aware of consciousness before the doctors, but in the case of a coma families are more likely than doctors to be mistaken in thinking that consciousness exists.

Brain Death

Before the era of transplantation, diagnosing a patient as dead was simple: In the doctor's opinion, a patient's heartbeat and breathing had ceased and couldn't start again. As a doctor, there were always a few minutes of doubt, but then the irreversibility of the process would become clear. Every now and then, a skier is dug out from an avalanche, no longer breathing and with no discernible heartbeat, but subsequently recovers. Cases in which someone is apparently dead but then revives are scarce enough to be well-known. King Louis IX is said to have moved in his coffin during a requiem mass being held for him. The funeral was halted and the king recovered, afterward going on a crusade to Egypt, where he repaid the debt he owed to death. In French, the term for undertaker is *croque-mort* (someone who bites the dead), alluding to the medieval custom of biting a corpse's big toe in order to make sure that the person really was dead. A few years ago, a doctor in the Netherlands mistakenly pronounced an eighty-three-year-old woman dead. When the undertaker's men came to pick up her body from the bathroom floor, it suddenly emitted a quiet "ouch." The old lady suffered no ill effects. (The same can't be said of the doctor.)

From the moment that patients with severe brain damage could be hooked up to breathing apparatuses, the classic diagnosis of death became obsolete, because heart rate and breathing were artificially sustained while a patient was unconscious or "brain-dead." This state can persist interminably. The former Israeli prime minister Ariel Sharon has been on a ventilator since suffering a severe stroke in 2006. His sons want his treatment to continue. In situations like this, the diagnosis is "brain-dead" rather than "dead."

The original definition for the diagnosis "brain-dead" was the "irreversible end of *all* brain activity." But the brains of a quarter of brain-dead patients go on producing the antidiuretic hormone ADH (vasopressin), which ensures that the kidneys reabsorb many liters of water a day from urine (see chapter 5). You can instantly tell when the brain cells that make ADH are dead because the patient produces between ten and fifteen liters of watery urine each day. In brain-dead individuals whose ADH-producing brain cells are still intact, their urine bag fills with a mere one and a half liters of properly concentrated urine.

Various other groups of brain cells may remain active in brain-dead patients, but they won't contribute to the recovery of consciousness. The term *brain-dead* was later redefined by a committee at Harvard Medical School (HMS) as entailing irreversibly fixed pupils, the absence of brain stem reflexes, and the permanent absence of "higher brain functions" like cognition and consciousness. The latter is actually a logical reversal of Descartes's famous "Cogito, ergo sum" (I think, therefore I am). If you can no longer think, because your brain no longer functions, you have also ceased to exist as a person.

Transplantation

Determining brain death is also important for organ transplantation. Besides ascertaining that the HMS criteria apply, the Dutch Health Council also recommends establishing that there's no longer any electrical activity or blood circulation in the brain. Finally, the ventilator is temporarily switched off to check that there's no question of spontaneous breathing. This provides an extra guarantee that the potential organ donor is indeed brain-dead. If these conditions are met and the person in question has previously given permission for their organs to be used, they can be transplanted. Since under the circumstances a person is no longer conscious of their own body, we of course shouldn't interpret the reflex reactions of the spinal cord that occur when a surgeon removes organs from a brain-dead patient as an expression of pain. That's easily said, but it's quite a different thing for the surgeon who sees the body respond when he makes an incision to remove its organs. In the United Kingdom, anesthesia is administered for this procedure. The Dutch association of anesthetists finds this nonsensical, and scientifically speaking they're right. In such cases an anesthetic is given to preserve not brain-dead patients but rather transplant surgeons from discomfort.

BRAIN STRUCTURES CRUCIAL FOR CONSCIOUSNESS

> The cerebral cortex and the thalamus are crucial for consciousness—as is a functional link between these brain areas.

There are two aspects to consciousness. First and foremost, we are *conscious of our surroundings*. A rudimentary form of consciousness is found in every living organism. Even single-celled organisms creep toward food and away from poisonous substances and are thus aware of their surroundings. But it's unlikely that they are conscious in the way that we are. For that, you have to go quite a way up the evolutionary ladder. The second aspect of consciousness is that we are *conscious of ourselves*. Self-awareness is certainly not unique to humans and has been demonstrated in young children and animals using experiments with mirrors. Various species prove to have a highly developed sense of self-consciousness, providing the basis for complex social relationships. Some chimpanzees, orangutans, and possibly also gorillas can recognize themselves in mirrors. A dolphin can see ink marks on its body in a mirror, and an ape can wipe a spot of paint off its own face when it sees it in a mirror, just as a child starts to recognize itself in a mirror between the ages of one and two. An Asian elephant can also recognize itself in an enormous mirror, inspect its own ear, and discover that its face has been marked, as Frans de Waal has shown. And self-consciousness isn't exclusive to mammals. Magpies can also recognize their reflected image. This was shown by placing stickers on their bodies in a place that they could see only in a mirror. The birds removed the stickers from their bodies without touching the mirrors, showing that they recognized their own image.

Some brain structures are crucial for consciousness, such as the cerebral cortex and the thalamus, along with a functional link between the two areas (fig. 20). An individual whose brain stem functions are still intact (meaning that they can breathe and regulate their blood pressure and body temperature independently) but whose cerebral cortex or the connections to it have been destroyed can no longer be said to possess consciousness. This situation applies to patients in a vegetative coma. They don't need to be hooked up to a ventilator, and their heart rate is normal. They can open and close their eyes and groan as well as cry and sometimes laugh uncontrollably. Their sleep-wake rhythm is still maintained by the brain stem (fig. 21), so they sometimes appear to "wake up," but without showing any consciousness of their surroundings or themselves.

While the cerebral cortex is essential to consciousness, its functioning isn't in itself sufficient for consciousness. Under anesthesia, for example, light stimuli still arrive in the visual cortex with a delay of one hundred milliseconds, but we aren't conscious of them. Studies have also shown that even when fully sedated, patients can be influenced by verbal suggestions, music, or the sound of the sea, despite their unconscious state. For consciousness to be intact, the area of the cerebral cortex where the stimulus arrives must also be able to communicate actively with other areas of the brain, which is impossible under anesthesia.

For normal consciousness, you also need an intact thalamus (figs. 2 and 19). The thalamus lies in the center of the brain and plays a crucial role in consciousness, because it's there that all sensory information (except smell; see fig. 21) is received and rerouted to the cerebral cortex. Damage to the thalamus disrupts consciousness. Conversely, a case is known of someone having been restored to consciousness through electric stimulation of the thalamus. Following an accident, a thirty-eight-year-old man spent around six years in a minimally conscious state, between coma and consciousness. He could occasionally communicate through eye or finger movements, but never through speech. Stimulation electrodes were implanted on both sides of his thalamus; within forty-eight hours of stimulation commencing, he woke up. Over the six months of stimulation that ensued, his attention improved, as did his response to assignments, his control of his limbs, and his speech. From a scientific point of view this is an intriguing experiment, but it's debatable whether he will be able to lead a dignified existence in the wake of this heroic remedy. It created an ethical dilemma, in that stimulation of the thalamus made him aware not only of his surroundings but also of himself and his terrible predicament.

THE IMPORTANCE OF FUNCTIONAL LINKS BETWEEN BRAIN STRUCTURES FOR CONSCIOUSNESS

> The reactions of a vegetative patient showed that he must at least still possess a residual degree of higher brain functions, like cognition.

Some areas of the brain, like the cerebral cortex and thalamus, are crucial for full consciousness (figs. 19 and 20). For consciousness to arise, not only must these areas and their links be intact, but they must also be able to communicate well with each other. Functional scans (fMRI) of patients in a vegetative coma show that parts of the cerebral cortex still function. A strong pain stimulus can even activate the brain stem, thalamus, and primary sensory cortex (fig. 22) of such patients. But these areas have been shown to be functionally disconnected from higher-order areas of the cortex that are necessary to consciously register pain. In the same way, a sound stimulus can activate the primary auditory cortex, but the functional disconnection means that such signals don't reach the higher-order areas necessary to register sound consciously. In other words, neuronal activity in the primary sensory or auditory cortex is necessary but not sufficient for consciousness. For consciousness, a functional connection is needed with the network of the prefrontal and lateral cortex (frontoparietal network). So recovery from a vegetative state goes hand in hand with recovery of the functional links between this network's components.

A series of spectacular observations was made by researchers from Cambridge and Liège using fMRI, starting with a twenty-three-year-old woman who had been in a vegetative state for five months following a traffic accident. Under the circumstances her brain was remarkably undamaged. When someone spoke to her, her middle and superior temporal gyrus (part of the temporal lobe, fig. 22) would show normal activation. Ambiguous sentences caused Broca's area (language processing, fig. 8) to light up. When asked to "visit" all the rooms in her house in her mind, activity was seen in the parts of the brain that control spatial orientation and locomotion: the parahippocampal gyrus (fig. 26), the parietal cortex (fig. 1), and the lateral premotor cortex (fig. 22). When instructed to play tennis, the area for motor coordination (supplementary motor area) lit up. A study was then carried out of fifty-four comatose individuals who had sustained severe brain damage (thirty-one of whom were in a minimally conscious state). In five cases, appropriate responses were shown to commands. Four of these individuals were in a vegetative state. Although changing activity patterns appeared to show that they were aware of themselves and their surroundings, one wonders to what extent one can really speak of "consciousness" in such a situation. A subsequent experiment involved asking one vegetative patient, a twenty-nine-year-old man, simple questions like "Is your father named Thomas?" or "Do you have brothers?" He was asked to think of one type of activity if the answer was yes, and another type of activity (giving a different brain image) if it was no. The different patterns of brain activity showed his answers to be correct in five out of six questions. His responses revealed that he must have at least retained a residual degree of higher brain function, like cognition. But whether these experiments show a form of consciousness in which you're aware of your own situation, for which communication between intact brain areas is essential, remains unclear. And says nothing of whether the patient would wish to continue living in such a situation.

"Absence seizures" are a form of epileptic seizure in which there's a break in consciousness for around five to ten seconds. During that time, patients look blank and don't respond. They often blink their eyes and smack their lips. Their awareness is impaired, and their frontoparietal networks (fig. 1), crucial to consciousness, are much less active. In complex partial seizures, sometimes lasting several minutes, consciousness is also impaired. Patients are awake but can no longer respond. They make automatic movements with their hands and mouths, and frontoparietal activity is again much reduced. These changes in activity in the cerebral cortex aren't found in patients with temporal lobe epilepsy whose consciousness isn't impaired (see chapter 15).

Indeed, the distinction between a vegetative state and minimal consciousness revolves around the functioning of the frontoparietal network. In the former case, the network is disconnected. In the latter, speech and complex auditory stimuli do spark general activity of the network that is crucial to consciousness, as shown in fMRI and PET studies. This also means that in principle, an entire network can be recruited in these patients, as was shown in the case of the minimally conscious man who was aroused when his thalamus was electrically stimulated using brain electrodes. It's claimed that music or electrical stimulation of a nerve in the arm (the median nerve) can speed up the process of reviving someone from a state of minimal consciousness, but very few controlled studies have been done, and so far there have been no spectacular results.

ILLUSIONS AND LOSS OF SELF-CONSCIOUSNESS

> The "I" is the body's rather untrustworthy partner, which cheats on it whenever it gets a chance.
>
> Victor Lamme, *There's No Such Thing as Free Will,* 2010

For self-consciousness you need a combination of sensory input and an intact cerebral cortex. The premotor cortex is important for the feeling that a certain body part belongs to us. It is there that various types of sensory information, like input from eyes, ears, organs of balance, muscles, tendons, and joints (proprioception) and the sense of touch are put together. You can fool your premotor cortex with the following little trick. Put a rubber hand on the table where your own hand would be, meanwhile putting your hand under the table where you can no longer see it. If someone then repeatedly strokes the rubber hand and your own hand at the same time, your brain combines the sight of the rubber hand being stroked with the sensation of your real hand being stroked. After about ten seconds you begin to regard the fake hand as your real one. If someone suddenly hits the rubber hand with a hammer you jump out of your skin. It seems that the combination of touch (coming from your own, hidden hand) and visual information (coming from the fake hand) is needed for the illusion that this is your real hand. Scans of people experiencing this illusion show activity in the premotor cortex and cerebellum. The feeling that a body part belongs to you appears to be based on nothing more than the activity of the few groups of neurons in a few very specific brain areas.

Self-consciousness can be impaired or lost for various reasons. In the first stage of Alzheimer's, around 10 percent of patients aren't aware of their degeneration. The percentage increases as the disease progresses. This unawareness that something is wrong with you is called anosognosia (from the Greek *nosos*, "disease," and *gnosein*, "to know"). It's usually the person's partner who notices that something is wrong and makes them see a doctor. Anosognosia is linked to reduced activity in the angular gyrus, near the upper edge of the temporal lobe (fig. 28). It's here that sensory information from the body and the surroundings is combined, making this area essential for self-consciousness. This part of the cerebral cortex is increasingly damaged as Alzheimer's progresses.

Out-of-body experiences or near-death experiences (see chapter 16) are also caused by a malfunction in the angular gyrus. A lack of oxygen prevents the angular gyrus from integrating the sensory information coming from your body, including the organs of balance, disrupting consciousness of your entire body.

Building on the rubber hand experiment, the Swedish scientist Henrik Ehrsson induced out-of-body experiences in experiments using cameras linked to a head-mounted video display. The participants were given goggles with a video screen for each eye. The screens showed images from two cameras filming them from behind, so that participants saw a 3-D image of their own back. Ehrsson then used two plastic rods to simultaneously touch a participant's actual chest and the chest of the virtual body, moving the second rod to where the virtual chest would be according to the camera pictures. This gave participants the illusion that they were in the virtual body, and made their own body appear to be someone else's. When the virtual body was threatened with a hammer, the participants reacted as if the threat were real. Their fearful attempts to ward off the blow were accompanied by physiological responses (notably the level of perspiration on the skin), showing that their emotions were aroused. In Switzerland, Olaf Blankes carried out similar experiments in which participants watched holographic projections of their own bodies. Afterward he blindfolded them and asked them to walk back to the spot where they had been standing. Participants who had had an out-of-body experience during the experiment walked back to the spot where their virtual bodies had stood. So self-consciousness isn't a metaphysical construct. Your brain constantly manufactures the sense that your body belongs to you, using sensory information from muscles, joints, vision, and sensation.

Methods of fooling consciousness can also be used to treat patients with chronic phantom pain, for instance after having had an arm or leg amputated. The neuroscientist V. S. Ramachandran discovered that phantom pain is caused by a conflict in the brain. Each

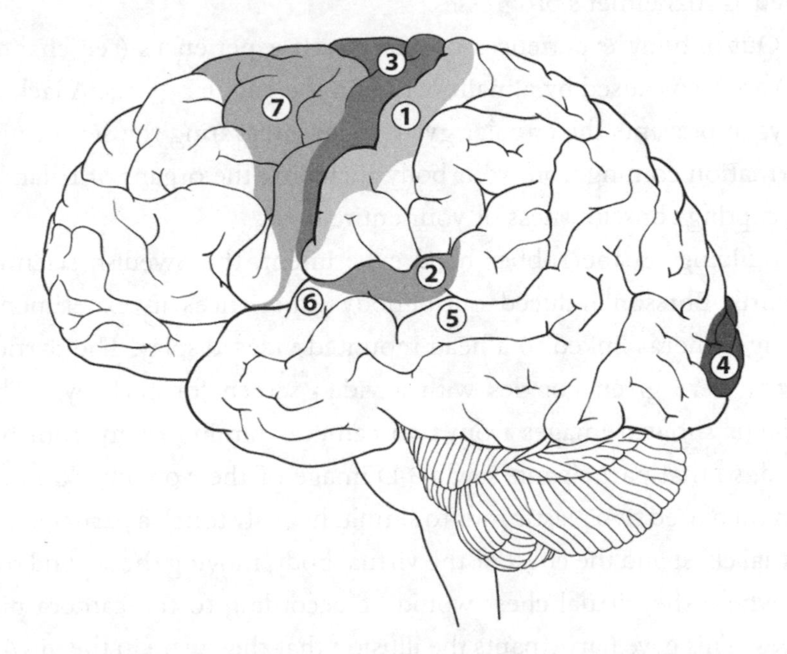

FIGURE 22. Some specialized cortical areas. (1) Primary sensory cortex, (2) auditory cortex, (3) motor cortex, (4) visual cortex. Also: (5) middle temporal gyrus, (6) superior temporal gyrus, and (7) premotor cortex.

time a patient wants, say, to move their (amputated) hand, they receive a signal back that it is impossible. As a result, the brain ultimately forces the phantom hand into an extremely painful, cramped position. Ramachandran's solution was as brilliant as it was simple. He placed a mirror perpendicular to his patients' chests, between their two hands, so that when they stretched out their normal hand and viewed the mirror from that side, it appeared as though both hands were working. The patients were given exercises in which they calmly opened and closed their normal hand while looking at their "phantom hand" in the mirror. Although they knew that it was an illusion, the visual input of a relaxed, calmly moving hand helped their phantom hand to relax and the phantom pain to disappear. One man whose leg had been amputated had been unable to wear his leg prosthesis for eight years due to phantom pain in his stump. After a mere three to four hours of mirror therapy, his pain had gone, and he was able to practice walking on his prosthetic leg for the first time, even though he knew that the leg he had seen move in the mirror no longer existed.

"FILLING IN" MISSING INFORMATION

> If information enters the cerebral cortex via an abnormal route, the patient isn't conscious of this fact.

Alien hand syndrome (see chapter 17) shows that self-consciousness also requires effective communication between the left and right hemispheres of the brain. This syndrome can occur if the bundle of fibers connecting the two (corpus callosum, fig. 2) is damaged. Sometimes surgeons have even deliberately severed this link in a last-ditch attempt to make life bearable for patients with disabling epileptic seizures. After the operation, these patients turned out to have split consciousness. The neurobiologist and Nobel laureate Roger Sperry discovered that one side of the brain wasn't aware of what the other side was seeing. In an experiment, patients could describe only images that reached the left side of the brain, because the ability to speak is located on that side. However, they seemed unaware of images that reached only the right side of the brain. Yet if they were asked to use their left hand (controlled by the right side of the brain) to indicate the image that had just been shown to the right side of the brain, they could do so. So on an unconscious level they had access to information that reached the right side. The left side of the brain then made up a story combining the information from both sides of the brain. The story was logical to the patients but completely incomprehensible to their surroundings. When the right side of the brain registered a written instruction to stand up and walk away, a patient obeyed. When asked why he did so, he didn't say, "You just asked me to," because he hadn't consciously registered the instruction. So he made up a reason to explain his behavior: "I'm just going to get some hot chocolate."

The peculiar situation in which a neglect patient finds himself is also often made "plausible" with a great deal of inventiveness and imagination. A paralyzed patient accounted for her situation as follows: "I'd like to get up, but my doctor won't allow me" (see chapter 7). In fact these kinds of fantasies are based on a very general principle. If the brain doesn't receive the right information in the expected place, the cerebral cortex at that location will work harder to fill in the gap. Such inventions are perceived as real information (see chapter 10). This phenomenon can also arise when there's a lack of auditory information, causing people to hear songs nonstop, or of visual information, causing them to see nonexistent objects in dim light. Lack of memory information resulting from alcohol abuse can lead people to make up events constantly without being aware of it, and lack of information from limbs as a result of amputation can cause phantom pain (see chapter 10). Each brain function has its own local system that enables consciousness (fig. 22). It's due to the differences in the location of increased cortical activity that we "see" things when we lack visual input and "hear" music when we lack auditory input.

The phenomenon of "blindsight" demonstrates the importance of information following the right route to the right part of the cortex. It was always thought that damage to the left primary visual cortex (fig. 22) would result in total blindness of the right field of vision and the other way around. But when individuals who were perceptually blind in a certain area of their visual field had to guess where a light stimulus was located in that area, they were able to do so correctly to an extent that couldn't be due to chance. Seeing something without being aware of it is referred to as type I blindsight, or attention blindsight. It was assumed that this unconscious mode of seeing was due to the receipt of visual information in subcortical areas. A new scanning method showing nerve pathways (diffusion tensor imaging) has revealed that individuals with this form of blindsight do receive information in the part of the cerebral cortex where visual information is processed but that it arrives via an abnormal route through the brain. So even though information arrives in the part of the cerebral cortex where it's normally received, a patient with blindsight isn't conscious of it, apparently because it travels by an unusual route. This would also explain why neglect patients can see something but not be conscious of it, because the information arrives in the cortex by another route, as a result of the damage caused by their stroke.

NOTIONS ABOUT THE MECHANISMS OF CONSCIOUSNESS

> Consciousness can be seen as an emergent characteristic generated by the joint functioning of the enormous network of nerve cells.

Throughout history, many metaphors have been used to describe consciousness of surroundings, like the "Cartesian theater," "the film in your head," and "a TV screen." But they are all based on the dualist notion that there's a little man in your head watching the images. It is a curious idea, not least because it raises the question of what is in the head of the little man. Another little man? No, we just have an amazing network of neurons.

John Eccles, who in 1963 won the Nobel Prize for his research into synaptic transmission, simply could not accept the idea that the neural network was responsible for consciousness. Instead he devised a theory (philosophical rather than neurobiological) that the neural units of the cortex were linked to mental units called "psychons." He believed that these psychons acted on the cortex in "willed" actions and thought and that their common activity gave rise to consciousness as an integrated mental process. No one actually knows what a psychon is. That makes it untestable as a theory and therefore, from a scientific point of view, an unacceptable hypothesis. It is, moreover, entirely redundant. All recent research suggests that the joint activity of enormous numbers of neurons in communication with a number of brain areas provides the foundation for consciousness.

Consciousness can be seen as an emergent characteristic generated by the joint functioning of specific areas of the huge network of neurons in our heads. Brain cells and areas have their own separate functions, but their functional links with one another jointly endow them with a new, "emergent" function. There are many examples of emergent characteristics. For instance, we know hydrogen and oxygen as gases. But when these molecules bind, a substance with entirely different characteristics emerges, namely water. The question of what exactly is needed from a neurobiological point of view to enable this new characteristic, consciousness, to emerge from neural activity is something that preoccupies many brain researchers. The Amsterdam neuroscientist Victor Lamme is looking for an explanation in the functioning of neurons. His theory is that for consciousness to exist, neurons in the prefrontal and parietal cortices have to relay information back to the cerebral cortex. One of the routes involved is via the thalamus. This recurrent processing extends from the purely sensory to the motor areas. Lamme believes that the selective attention crucial to our consciousness emerges because only a few of the objects that we perceive undergo recurrent processing. So we report on the stimuli on which our attention is focused while being unaware of the rest. There's no reason to assume that basic mechanisms like recurrent processing and attention aren't common to all animals, albeit to varying degrees. The philosopher Daniel Dennett seeks to explain consciousness as a purely bodily, chemical phenomenon, a view I share. However, he also believes that humans have a different *kind* of consciousness than animals because of the far-reaching impact of our linguistic development. I think it's more logical to assume that animals have a different *degree* of consciousness. And although there are differences between species in this regard—a magpie's ability to recognize itself in a mirror is far removed from a dog's ability to distinguish the smell of its own urine from that of another dog—animals can be said to possess rudimentary self-consciousness. In humans, consciousness doesn't depend on language, by the way. People whose language areas have been disabled after a stroke are still fully conscious of their surroundings and of themselves. By nodding or shaking their heads they can make considered decisions, even if they can no longer verbalize them.

The importance of being conscious of your surroundings and of yourself is primarily expressed in social interaction, which involves observing and constantly interpreting your situation compared to others and learning from the mistakes that you make in this process. And that brings us back to Charles Darwin and Frans de Waal, who pointed out the enormous evolutionary importance of individuals being able to function well in the complex social interaction of the group (see chapter 20).

8

Aggression

BORN AGGRESSIVE

> I have heard of cases in which a desire to steal and a tendency to lie appeared to run in families of the upper ranks.
>
> Charles Darwin, *The Descent of Man,* 1871

Humankind is an aggressive species, just like chimpanzees. It's not for nothing that we share ancestors. In the 1960s and 1970s there was a universal belief in social engineering. Give everyone a good environment in which to live, and aggression and crime would disappear overnight. Anyone who thought differently was publicly reviled. Now that it's once again permissible to consider the biological background to our behavior, we can also look at the question of why one person is more aggressive than another and why some are more likely than others to commit crimes.

Boys are more aggressive than girls. That's something that's determined before birth. The peak in testosterone that male fetuses produce halfway through pregnancy makes them more aggressive for the rest of their lives. Girls with adrenal gland abnormalities that cause them to produce too much testosterone before birth are also much more aggressive later. And hormone-like medicines taken during pregnancy can raise the aggression levels of both boys and girls.

Some children are, however, markedly more aggressive than others, and they are more likely to commit crimes: 72 percent of young offenders in Dutch prisons have been sentenced for crimes of aggression. A strikingly high incidence of psychiatric disorders was found among this group—as high as 90 percent in the case of adolescent males. Besides antisocial behavior, there is a strong link between delinquency and addictive substance abuse, psychoses, and ADHD. Genetic factors are also influential, as has been shown from studies of twins. Tiny variations in the DNA (polymorphisms) of the gene for proteins that break down chemical messengers in the brain can lead to more aggression, alcoholism, or violent suicides. A reduction in the activity of the chemical messenger serotonin is linked to greater aggression, impulsiveness, and antisocial behavior. Some Chinese men have been found to carry a tiny variation in a gene involved in processing serotonin that is linked to extremely violent crime, antisocial personality disorders, and addiction to alcohol and other substances. Another variation of the same protein increases the likelihood of borderline personality disorder, which can also be marked by impulsiveness and aggression. So our genetic background can contribute significantly to our aggressive and criminal behavior later.

A fetus's surroundings also affect its later propensity for aggression. Tests establishing fitness for medical service showed that men who had been severely malnourished in the womb during the Dutch famine in the winter of 1944–1945 were two and a half times more likely to have an antisocial personality disorder (see chapter 2). Malnourishment in the womb still occurs, even in our affluent society, when a placenta malfunctions. A combination of genetic factors and a mother's smoking during pregnancy can increase a child's risk of ADHD by a factor of nine, and ADHD is associated with greater aggression and a greater likelihood of delinquency (see chapter 2).

It's not just our level of aggression that's largely determined before birth. This isn't a new idea, simply one that was regarded as taboo when faith in social engineering was at its highest. Charles

Darwin (1809–1882) came to the same conclusion in his autobiography, writing that he was "inclined to agree with Francis Dalton [his cousin] that education and environment produce only a small effect on the mind of anyone, and that most of our qualities are innate." That puts the potential influence of parents and a host of well-meaning social organizations in the right perspective (see later in this chapter).

YOUNG AND AGGRESSIVE

> The Ministry of Justice is now looking beyond social factors as determinants of aggression and crime.

We're born with different propensities for aggressive behavior depending on our gender, our genetic background, the amount of nourishment we received from the placenta, and our mother's consumption of nicotine, alcohol, and medication during pregnancy. The likelihood of our displaying uninhibited, antisocial, aggressive, or delinquent behavior increases in puberty as testosterone levels rise. And there are considerable gender-based differences in such behavior. Men are five times more likely than women to commit a murder. Moreover, men murder a relative or acquaintance only in 20 percent of cases, as opposed to 60 percent of cases for women. The age at which men commit murders follows a stereotypical curve. As testosterone levels rise during puberty, so too does the incidence of murder. This peaks around twenty to twenty-four years, then declines to low values around fifty to fifty-four years. An identical age pattern for murders has been found in very different parts of the globe, from the United States to England, Wales, and Canada. The decline in criminal behavior among people in their late twenties doesn't mirror declining testosterone levels but is attributed to the late development of the prefrontal cortex (fig. 15), which restrains impulsiveness and promotes moral behavior. A logical consequence would be to apply adult criminal law only when this brain structure is mature, around the age of twenty-three to twenty-five. However, politicians take no account of this development pattern, preferring to drum up votes from a fearful electorate by urging just the opposite—that is, lowering the age of criminal responsibility. The functioning of the prefrontal cortex is inhibited by alcohol, which can lead to sudden, mindless violence after a night out. Damage to the prefrontal cortex in the first years of life can also disrupt social and moral behavior later in life.

Testosterone stimulates aggression. Some men have higher testosterone levels than others and are therefore more likely to become aggressive. Men imprisoned for rape and other violent offenses have been found to have higher levels of testosterone than other types of offenders, and these levels are higher in prisoners in general and in military recruits with antisocial behavioral tendencies than in the rest of the population. The same link between higher testosterone levels and greater aggression applies to female prisoners too, by the way. The aggression shown by hockey players during games can easily be measured by the incidence of blows with sticks. Here, too, a link has been found between aggression and blood testosterone levels. So it's worrying that such huge amounts of anabolic steroids are currently taken in the world of sports to increase muscle mass, because this hormone also increases aggression.

Environmental factors play a role, too. Violent films and computer games have recently been shown to heighten aggression. Interestingly, the same effect is produced by reading biblical passages in which God sanctions killing (but only in people who are religious). What's more, physical factors like temperature and light greatly affect our actions. Everyone knows that long, hot summers can spark violent incidents. The influential factor in decisions to go to war turns out not to be military strategy but the amount of daylight and the temperature. This emerged from Gabriel Schreiber's study of 2,131 conflicts in the last 3,500 years, which found a pattern of annual rhythms. For centuries, the decision to declare war has largely been made in the summer in both the northern and southern hemispheres, while season didn't play a role in equatorial regions.

Factors such as a deprived background and lack of education have of course long been known to contribute to aggression and delinquent behavior. Indeed, these are the only factors to have been researched in previous generations. When the Italian criminologist Cesare Lombroso (1835–1909) was accused of paying too little attention to the social causes of crime, he answered that this had already been done by countless academics, adding that it was "pointless to prove that the sun shines." Until recently, the Dutch Ministry of Justice also focused only on the social causes of crime, but it's now showing interest in other factors that increase levels of aggression and the likelihood of criminality.

AGGRESSION, BRAIN DISORDERS, AND PRISON

How often does our criminal justice system violate the principle that penal law shouldn't apply to people with a brain disease?

Rules on criminal liability have been in place ever since Daniel M'Naghten killed the British prime minister's secretary in 1843 and—to the general shock of Victorian England—wasn't jailed but put in a lunatic asylum. Under the "M'Naghten rules," as they came to be known, criminals with mental disorders could be judged "guilty but insane" and placed in a secure hospital facility rather than sent to prison. Yet although we agree that such individuals can't be held criminally liable, prisons today are full of people with psychiatric or neurological diseases. According to the Dutch forensic psychiatrist Theo Doreleijers, 90 percent of young people in prison have a psychiatric disorder, and 30 percent of individuals detained under hospital orders have ADHD.

In the case of brain disorders associated with aggressive behavior, two areas of the brain working together are of special significance: the prefrontal cortex (fig. 15) and the amygdala (fig. 26). The front of the brain, the prefrontal cortex (PFC), inhibits aggressive behavior and is crucial for moral judgments. Children whose PFC has been damaged often have difficulty learning moral and social rules. Vietnam veterans with damage to the prefrontal cortex became more aggressive and violent, and impulsive murderers also show reduced activity in their PFC. Brain disorders that affect the PFC tend to be associated with aggressive behavior. A surgeon who carved his name in a patient's abdomen at the end of an operation turned out to be suffering from Pick's disease, a form of dementia that starts in the PFC. Schizophrenia, also marked by reduced activity in the PFC, can lead to aggressive behavior. John Hinckley Jr. gained notoriety after his attempt to assassinate President Reagan. (The bullet from his pistol hit Reagan under the left armpit and bored through his left lung but stopped an inch from his heart.) Hinckley's brain scan, which went around the whole world, clearly showed the shrinking of the brain that's typical of schizophrenia. He's still in prison. In 2003, another schizophrenic patient, Mijailo Mijailović, murdered the Swedish foreign minister, Anna Lindh, after he stopped taking his medication. He believed that Jesus had chosen him for this purpose and heard voices telling him to commit the murder. Conversely, aggressive behavior can sometimes be the first symptom of schizophrenia.

The amygdala (fig. 26), an almond-sized structure, is located deep within the temporal lobe. When you hold the gelatinous mass of a brain in your hands during an autopsy, you can feel, within the pole of the temporal lobe, the solid little button of the amygdala. Stimulation of the amygdala inhibits or induces aggressive behavior, depending where and how it's done. Its inhibitory effect was convincingly demonstrated by the Spanish physiologist José Manuel Rodriguez Delgado, who was able to stop a bull in mid-charge through the remote electrical stimulation of its amygdala. If you disable this structure on both sides, even sewer rats will become tame. Some psychopaths have a malfunction of the amygdala. This prevents them from seeing from their victims' facial expressions that they are suffering and thus from feeling empathy toward them. In 1966, Charles Whitman killed his wife and mother and then went on to shoot fourteen people dead and wound thirty-one others at the University of Texas in Austin. He was found to have a tumor of the temporal lobe, which was pressing on the amygdala. It makes you wonder how many other people who go on shooting sprees in schools or elsewhere have a brain disorder. Ulrike Meinhof started her career as a critical journalist, later becoming one of the founders of the Rote Armee Fraktion in Germany, a terrorist group that killed thirty-four people. Meinhof committed suicide in her cell in 1976. Doctors had previously discovered that she had an aneurysm, a bulge in the wall of a blood vessel at the base of the brain that was pressing right on the amygdala. This caused lasting damage. When she was operated on for the aneurysm, the neurosurgeon also damaged the prefrontal cortex, so there were two possible causes for her aggressive and lawless behavior.

Other brain disorders that are sometimes linked with aggression are mood disorders, borderline personality disorder, learning disabilities, brain infarcts, MS, Parkinson's disease, and Huntington's disease. Even patients with dementia can be aggressive. In 2003, an eighty-one-year-old Dutch woman who'd been placed in a nursing home because of her dementia murdered her eighty-year-old roommate. She was found on the toilet in a confused state, and it was only when a nurse took her back to bed that the victim was spotted. Fortunately, the Public Prosecution Service decided not to prosecute. In "civilized" countries like the United States and Japan, schizophrenic patients who have committed a murder can still be given a death sentence. I hope that will never happen again in the Netherlands. But how often does our criminal justice system violate the M'Naghten rules?

GUILT AND PUNISHMENT

> The criminal justice authorities should learn from the medical world how to adopt an evidence-based approach founded on properly controlled studies.

Criminal law can only be applied to people with a healthy brain. This principle also has a biological foundation. Rhesus monkeys normally punish any animal that doesn't stick to the group rules. However, the primatologist Frans de Waal observed that a mentally retarded rhesus monkey with Down syndrome was allowed to get away with breaking all of the group's standard rules. Humans should behave similarly, but apparently we find it hard to do so.

Forensic psychiatrist Theo Doreleijers discovered thirteen years ago that 65 percent of underage delinquents brought before a public prosecutor had psychiatric disorders, but medical reports had been requested in less than half of the cases. Can we hold such children liable for their deeds? Child abusers have often themselves been abused as children, so to what extent are they culpable? How accountable is an adolescent for his actions when his brain is suddenly deluged with sex hormones that are modifying the function of almost all of its parts? A child has to learn to deal with a whole new brain during puberty, at a time when the prefrontal cortex, which inhibits impulsiveness and controls moral behavior, is extremely immature. And how accountable are addicts for their condition, which was caused by tiny variations in their DNA or malnourishment in the womb?

In other words, moral condemnation and punishment based on personal accountability rest on very shaky ground. However, our sense of morality is strongly anchored in our evolutionary development, because it affects the survival of the group. It also accounts for the idea that each individual is responsible for his or her own deeds, illusory though this is.

However, contrary to what is sometimes thought, that we're programmed in certain ways doesn't mean that we should do away with punishment entirely. After all, the next time we decide whether or not to do something, our brains can factor an effective punishment into our unconscious deliberations. And punishment also has aspects that have nothing to do with personal accountability. Society requires criminals to atone for their deeds; it also wants them to be locked up for its protection and as a warning to others—though the effectiveness of this latter aspect is debatable.

The knowledge that we possess about the neurobiological risk factors for aggressive or criminal behavior always relates to a *group* of individuals with a certain characteristic. As a result, we can't assure a court that a particular factor has contributed to a particular *individual* committing a crime. Some therefore claim that the practical contribution of neurobiological knowledge to sentencing or detention on remand is of marginal importance. Ybo Buruma, a member of the Dutch Supreme Court, rightly said in an interview in the newspaper *NRC Handelsblad* (November 7, 2000), "Courts, like doctors, deal with individuals." But he went on to draw exactly the wrong conclusion: "I think all this knowledge is terrific, but as long as we can't apply it on an individual level in court cases, it's of no use to us." Thus he reduced law as a science to the level of medicine a hundred years ago, when doctors also treated their patients on an individual basis to the best of their ability but had no idea what the effect would be. Medicine has learned its lesson; evidence-based medicine is always founded on the effects on a well-defined group of patients. You never know whether the one patient you prescribe medicine for will belong to the 95 percent who are cured as a result or the 5 percent who will experience serious side effects and, very occasionally, die. Yet you make the decision to treat that one patient on the basis of good data. And this is how we should look at the factors that determine aggressive and criminal behavior in a particular group and the way in which this group responds to preventive measures and different types and degrees of punishment. Only on the basis of such data can we make pronouncements about an individual that are based on probability, in the knowledge that our judgment regarding that person can't be entirely certain but will at least be correct with respect to the group to which he belongs. Alas, the criminal justice authorities have a very long way to go in this regard. They keep trying out new forms of punishment, from community sentencing to boot camps for young criminals, without a proper control group, which means that the effectiveness of a given punishment will always be controversial.

VIOLENT WHILE ASLEEP

> There is, in all of us, even in good people, a lawless wild beast that emerges when we sleep.
>
> Plato, *The Republic*, 380 B.C.

Dream sleep coincides with darting movements of the eyes, which is why it's also known as REM (rapid eye movement) sleep. It's also referred to as "paradoxical sleep" because EEG scans reveal that the brain is extremely active at this time. This combination of brain activity and rapid eye movement was discovered by Eugene Aserinsky in 1952, when he monitored his small son during REM sleep.

During dream sleep we exhibit many of the characteristics of psychiatric and neurological disorders. Our higher visual centers are activated, and we hallucinate like patients with schizophrenia. We experience incredibly bizarre events in a world in which the laws of physics and of everyday society no longer apply. Dreams often carry an emotional or aggressive charge; not surprisingly, the amygdala (fig. 26), the center of aggressive behavior, is activated at these times. When we dream we make up stories, just as people with alcohol dementia fill up the holes in their memories with stories about events that never took place (see chapter 10). A few minutes later we forget everything that we experienced in our dreams, as if we were suffering from a severe form of dementia. During dreams we lose muscle tension, just as narcolepsy sufferers with cataplexy do while awake.

It's not for nothing that we lose muscle tension while we sleep. Retaining it can lead to activity during sleep; sleepwalkers, for instance, are deeply asleep, but they have normal muscle tension. They can perform automatic, semi-purposeful actions, of which walking is an example. They are unaware of what they do and are afterward unable to remember any of their actions. (A scan of a sleepwalking patient indeed showed that large parts of the cerebral cortex aren't activated during sleep.) The French scientist Michel Jouvet carried out experiments on animals in which he created slight lesions in the brain stem, destroying the nerve cells that make the muscles relax during sleep. The animals in the study were shown to carry out the actions that they were dreaming about. He saw a cat in dream sleep leap on its imaginary prey with open eyes, without having the least awareness of her surroundings. She wasn't at all interested in a bowl of tempting cat food, nor did she purposefully remove the mess that was placed on her coat, though she did automatically clean her coat while asleep. Rats with lesions of this kind played with invisible rats during their dreams, and squirrels dug up nuts.

Humans, too, sometimes perform complicated actions during dream sleep of the type witnessed in the above animal study. They also occasionally become aggressive. One woman told me:

> Three years ago my husband was suffering from nervous tension. One night he made such strange noises in his sleep that he woke me up. I tried to calm him by stroking his head. That turned out to be a bad idea, because he grabbed me by the throat and tried to throttle me. Since I was by now wide awake I was able to free myself and to wake my husband. When I told him what had happened he was dreadfully shocked, so much so that he hardly dared go back to sleep. He told me that he'd dreamed he was being attacked and that he'd tried to defend himself. This dream recurred a few times. Each time I was woken up by the sounds he was mak-

ing. I made sure to put some distance between us before stroking him softly so that he calmed down again. We discussed these events with our children and with friends, and wondered what would have happened if I hadn't been able to free myself. Would he have gone to jail?

That is indeed the question. People who are tried for crimes are sometimes acquitted on the grounds that they had clearly been asleep when they broke the law. Some people can indeed perform very complex actions while asleep without being the least aware of it. In 90 percent of cases these are men, and these events occur in a transitional stage between REM sleep and other forms of sleep. Such actions, like sleepwalking, are completely automatic. People have been accused of robbery, rape, and attempted murder while asleep, and some have even been thought to commit suicide, though such deaths could also simply be attributed to accidents while sleepwalking. Those affected sometimes have brain disorders like narcolepsy or Parkinson's disease, but in many cases there's absolutely no neurological or psychiatric abnormality. Events of this kind can be induced by fever, alcohol, lack of sleep, stress, or medication. Some sleepwalkers who are extremely mild and amiable when awake are shockingly violent when asleep.

In 1987 Kenneth Parks drove fourteen miles while asleep and battered his mother-in-law to death. He woke up just as he was about to kill his father-in-law and gave himself up to the police. He was subsequently acquitted. Julius Lowe was a frequent sleepwalker, and during one of these episodes he killed his eighty-two-year-old father, to whom he was extremely attached. A man by the name of Butler shot his wife dead while in a sleep-confused state; he was later found guilty, however. While on vacation in 2008, a fifty-nine-year-old Briton called Brian Thomas strangled his wife, to whom he had been married for forty years. He told the court that he had been dreaming that he was fighting with a robber who had broken into their caravan. Thomas had suffered from sleep disorders, including sleepwalking and insomnia, from an early age. He was taking medication for this condition, but a side effect was that he became impotent. Since he and his wife were going on vacation and wanted to be "intimate," he had briefly stopped taking his pills. The case against him was dropped when the judge ruled that he couldn't be held responsible for this tragedy on account of his sleep disorders.

A total of sixty-eight such murders by sleepwalkers have now been recorded. To prove that you carried out a crime while asleep, you need to undergo a battery of sleep studies by highly qualified experts and to have an excellent lawyer. But given the obvious impossibility of proving beyond a doubt that a crime was committed while the perpetrator was actually fast asleep, courts are mostly reluctant to acquit such individuals. And you can hardly blame them.

9

Autism

DANIEL TAMMET, AN AUTISTIC SAVANT

"So what are you painting?" I inquired with interest.
"The number pi," he replied.

Daniel Tammet has Asperger's syndrome, a form of high-functioning autism, and is also a savant: He possesses undreamed-of numerical and linguistic abilities. In 2004, he set a European record by reciting pi from memory to 22,514 decimal places in five hours and nine minutes without making a single mistake. It had taken him three months to learn. People with autism often have synesthesia, a condition in which sensory and cognitive pathways are interlinked, causing letters or numbers to be perceived as colored. Tammet sees Wednesday, the day on which he was born (January 31, 1979), as blue, hence the title of his book: *Born on a Blue Day*. He sees numbers not only in color but also in different shapes and sizes, and he can identify every prime number under 9,973 by its crystalline sparkle.

I spent a few days with Tammet when the Dutch translation of his book came out. He told me proudly that he had started to paint. "So what are you painting?" I inquired with interest. "The number pi," he replied. He sees number series like the decimals of the number pi as mountain landscapes consisting of series of differently colored numbers. Synesthetes have unusually strong connections between the various areas of the cerebral cortex. As a result, the visual cortex, which normally occupies itself only with the sense of sight, receives extra information about calculation going on in other areas of the brain. Complex calculations suddenly become easy when they are translated into images. But Tammet also has extraordinary linguistic abilities, being able to learn a new language in a single week, even one as difficult as Icelandic. That's an unusual combination, but what makes Tammet unique as a savant are his well-developed social skills, which tend to be lacking in autistic savants. These skills enable him to convey very movingly in his book how lonely he was as a child, how desperately he wanted to have friends and yet was isolated by his differentness. He also speaks of the many phobias that he combated as a child by thinking of numbers, which he regarded as his only true friends, and of his obsessive need for order and regularity, which has stayed with him ever since. Every day he eats exactly forty-five grams of porridge for breakfast and drinks a cup of tea at exactly the same time; strict regimes of this kind counteract his anxiety. These characteristics, so typical of Asperger's, have never before been communicated so eloquently. What makes his book fascinating is the personal, poignant account of what a child with these gifts lacks, how problematic Tammet's development was, and how he succeeded, step by step, in overcoming his social deficiencies to become a completely independent adult. Tammet earns his living by giving online language courses—online communication being much easier for individuals with autism to deal with than face-to-face conversations.

Over twenty years ago, Dustin Hoffman movingly portrayed the challenges of living with autism in the film *Rain Man,* inspired by the savant Kim Peek. Daniel Tammet regarded his meeting with fellow savant Peek as a highlight of his life. On the way to his meeting with him, which was featured on a BBC documentary, Daniel tried, like Rain Man, to earn money by counting cards in Las Vegas. The experiment failed; he lost heavily. Then he decided to use his intuition instead. That worked perfectly, and he won over and over again. Since then, Daniel Tammet has been known as "Brain Man," which certainly does justice to his phenomenal cognitive capacities but ignores his most special achievement, which was the insight and courage he displayed in overcoming his many handicaps and becoming a socially competent, extraordinarily sympathetic savant.

When you read Tammet's book you're constantly confronted with the blurry boundaries between what is considered normal and what is classified as a psychiatric problem, and you find yourself wondering how different savants are from people regarded as geniuses in the days before the labels *savant* or *Asperger's* had been invented. As a boy, Picasso struggled with reading, writing, and arithmetic. Einstein was slow to talk and would apply picture thinking to complex problems in the field of physics. The dividing line between psychiatric disorders and great gifts is often a very narrow one and strongly depends on how someone is viewed by their surroundings.

AUTISM, A DEVELOPMENTAL DISORDER

> Autism has only fairly recently been classified as a developmental brain disorder originating in the womb.

Autism is marked by severely disrupted social skills and a very confined repertoire of activities and interests. It was first described in 1943 by Leo Kanner in Baltimore and in 1944 by Hans Asperger in Vienna, who both used the same term independently of one another. However, there were great differences between the two descriptions. The children whom Kanner described scarcely spoke, were mentally subnormal, and displayed symptoms that were mostly neurological. The children Asperger described as *Intelligenzautomaten* (intelligence machines) had a precocious grasp of language, could talk about their experiences and feelings, and were normally abled. Asperger's publications made little impact until 1981, when it was suggested that people of normal intelligence with autism be designated as suffering from Asperger's syndrome.

Brain development in autism is atypical. Between the ages of two and four there is too much brain volume, delaying growth in some areas and prematurely terminating it in others. The main cause of autism is genetic. Daniel Tammet has a younger brother, Steven, who also has Asperger's. (He has an encyclopedic knowledge of the band the Red Hot Chili Peppers.) Their father was frequently admitted to psychiatric clinics. Tammet's grandfather had epilepsy and was so ill that his wife was advised by a psychiatrist to divorce him. Paternal age plays a role in autism, too: The syndrome occurs ten times more frequently in people born to fathers in their fifties than in those whose fathers were in their twenties when they were conceived. In addition, the likelihood of autism increases if a child acquires metabolic disorders or infections in the womb, has an older father or mother, or is deprived of oxygen at birth.

The symptoms of autism appear early, around the age of three. Children with autism don't make contact with others and have motor problems due to a developmental disorder of the cerebellum. They are clumsy and display stereotypical autistic behavior, like flapping their hands or walking on tiptoe. Daniel Tammet writes how very much he wanted to have friends but how impossible this was because he was "different." Both he and Temple Grandin, a professor of animal science with autism, invented friends to compensate for their lack of company. Team sports are hugely problematic for people with autism. Daniel hated soccer and rugby, as he was always being picked last for the team. But he was good at trampolining and chess. He was taught chess by his father at the age of thirteen, beating him in his very first game.

People with autism have trouble interpreting emotion and empathy. They don't understand why another child is crying. According to Temple Grandin, her emotional circuitry had simply been disconnected. Indeed, disorders are now being found in the social brains of people with autism, in which the chemical messengers vasopressin and oxytocin play a crucial role. Individuals with autism also tend to shy away from bodily contact, even though they may feel a need for it. Temple Grandin, who invented veterinary machinery, found a professional solution for this problem. She constructed a "hug machine" in which she could lie, with the sides (controlled by air pressure) squeezing against her. People with autism can also be oversensitive to certain sounds. Tammet describes how the sound of cleaning his teeth drove him mad as a child, making him stuff cotton in his ears. (He told me that this troubles him much less now that he has an electric toothbrush.) Sometimes people with autism will concentrate so hard on an idea or task that they can't hear what is being said to them. Tammet tells the story of how he didn't hear the voice of the mayor calling him forward to receive an award because he was concentrating so hard on counting the number of links in the latter's chain of office.

Autism has only fairly recently been classified as a developmental brain disorder. Thirty years ago I remember the parents of a child who had been "different" right from the start being told, after lengthy tests by psychiatrists and psychologists, not just that the diagnosis was autism but also that their method of upbringing was to blame. That was the fault of Kanner, who devised the theory of "refrigerator mothers," maintaining that autism was a response to a lack of close maternal contact. In 1960 he went so far as to claim that the mothers of autistic children "had just happened to defrost enough to produce a child." How many parents' lives were made undeservedly wretched by this ludicrous notion?

SAVANTS

> It isn't uncommon for individuals with autism to have a unique talent, but only in exceptional cases are they as multitalented as Daniel Tammet.

One in ten children with an autistic spectrum disorder has savant qualities, usually a talent that contrasts sharply with the mental disability and handicaps they may also have. However, few of these gifted children become truly creative as adults, either because of the kind of talent that they possess or because of their personality. Half of savants have an autistic spectrum disorder, and the other half have brain damage or a brain disease.

A savant's talent can be very limited. Twins George and Charles, though unable to count, were calendar calculators: They automatically "knew" on what day a date fell in a particular year. Savants are able to make unconscious use of algorithms. But not all known stories about the gifts of savants are reliable. In his book *The Man Who Mistook His Wife for a Hat,* Oliver Sacks describes autistic twins who, when they saw the contents of a box of matches fall on the floor, immediately both cried out "111!" They also saw that the figure 111 was composed of three times the prime number 37. By the time *Rain Man* was made, this number (using toothpicks instead of matches) had grown to 246. Four remained behind in the box. Daniel Tammet doesn't believe this story. According to him, no one, not even Kim Peek, could identify the exact number, if only because the matches at the bottom of the pile would be hidden from sight. The abovementioned twins had an IQ of 60 and were unable to do even simple sums. Sacks describes how the two would exchange prime numbers. When he took along a book of prime numbers and joined in the game, the twins were delighted. But after a while, they continued with twelve-digit prime numbers, whereas Sacks's book stopped after ten digits. Again, Tammett voiced his doubts. He didn't think that anyone knew of such a book, and when Sacks was recently asked its title, he answered that it had disappeared!

The term *idiot savant* is used to describe the combination of an exceptional gift and low IQ (30 to 70). It was coined in 1887 by John Langdon Down (the British doctor who gave his name to Down syndrome), who went on to say that he had never met a female idiot savant. They do exist (see for instance Nadia in later in this chapter), but they are greatly outnumbered by boys.

Leslie Lemke, who was born prematurely, spastic, blind, and with an abnormal left prefrontal cortex, is an idiot savant with unusual musical capabilities. At the age of seven, his mother let him feel the keys of the piano. A year later he was able to play six instruments. When he was fourteen he played Tchaikovsky's first piano concerto flawlessly after hearing it just once on television. He is famous for his ability to improvise. After hearing a piece of music by a particular composer only once, he can effortlessly improvise in the same style. He gives classical concerts, but is mentally retarded, with an IQ of 58.

It's not uncommon for autistic individuals to have a talent, but only in exceptional cases are they as multitalented as Daniel Tammet. These talents are largely possessed by boys and lie in the field of art, music, calendar calculation, and almost instantaneous mental calculation. They almost always go hand in hand with a remarkable memory. A Japanese savant who went on a journey of several months was afterward able to make very detailed drawings of the things he'd seen on his trip. Savants appear to store all the information that enters their short-term memory in their long-term memory, too. They can remember vast quantities of trivial facts, like license plates and railway timetables; it's as if they are unable to forget information. But Tammet said that he would now no longer be able to recite that long series of pi decimals; he would need to practice them again.

But remarkable memory isn't in itself enough to account for the savant syndrome. These individuals possess genuine talent, too. Stephen Wiltshire was an autistic boy with a verbal IQ of 52. He was known for his "London alphabet," a series of pictures showing landmark structures in London, drawn at the age of ten. He went on to draw in New York, Venice, Amsterdam, Moscow, and Leningrad. After a forty-five-minute helicopter flight above Rome, he produced a six-foot drawing showing every house, window, and pillar in the city with photographic precision. He has sometimes been compared to a printer because of the automatic way he draws. Artistic savants always have a strong preference for a particular subject and a particular technique. It's striking that they almost never draw people; the social brain is their Achilles' heel.

BRAINS OF SAVANTS

> Brain damage at an early age appears to foster the development of savant qualities, because at that stage the brain is still fully able to make new connections with other structures.

There are various theories about the neurobiological background of the savant syndrome. The exceptional gifts associated with it almost never develop unless there's brain damage, especially on the left side of the brain. It's thought that the brain damage allows links with other brain structures to be reinforced, enhancing the functioning of the visual cortex. Indeed, there are many examples supporting this theory. The left side of Kim Peek's brain was damaged, and there was no connection between the right and left hemispheres. Peek was able to read two pages simultaneously, using both eyes separately, at lightning speed. He read nine thousand books about the history of the United States and knew them all from memory. But he couldn't look after himself, relying on constant help from his father.

Epilepsy is often associated with autism. Daniel Tammet had his first serious epileptic seizure at the age of four and was treated effectively for three years with Valium. He suffered from temporal lobe epilepsy on the left side, which could explain his compulsive writing around the age of seven and his subsequent religious feelings (see chapter 15). Damage to the left side of the brain might cause right-side compensation and thus promote numerical skills, but the left side of Tammet's brain shows no trace of damage. Indeed, he's also a linguistic genius.

One theory is that everyone possesses potential savant talents localized in the "lower" regions below the cerebral cortex that are suppressed by "higher" processes. The psychiatrist Darold Treffert has dubbed this the "little Rainman" that each of us possesses. The idea is that these hidden talents can only be expressed by switching off the part of the brain that controls higher functions. Cases have indeed been known of people developing savant-like qualities due to a form of dementia that starts in the left frontal region. Some start to paint compulsively, for instance. These outbursts of creativity go hand in hand with a loss of language and social skills. In such cases, brain activity is concentrated in the rearmost region on the right, the visual cortex. When magnetic stimulation is used to disable the left frontotemporal region temporarily in healthy subjects, some improve at tasks like drawing, math, and calendar calculation. However, these improvements are modest, and no exceptional artistic feats have been seen. So the notion of a universal "little Rainman" doesn't satisfactorily explain the syndrome and moreover doesn't take into account its genetic component.

Brain damage at an early age appears to foster the development of savant qualities, because at that stage the brain is still fully able to make new connections with other structures. A Japanese savant caught whooping cough and measles at the age of four. It impaired his speech development, but when he was only eleven he made extraordinarily beautiful drawings of insects.

Some claim that savant talents are entirely due to training. Tammet jokes that he learned to count so well because he was one of nine children. It's true that savants are able to focus strongly and train obsessively, thus honing their skills. But they have to have a talent to start off with. And these talents are expressed at a very early age, both in savants and in child prodigies like Mozart, which conflicts with the training theory. When the young Mozart heard Gregorio Allegri's *Miserere* in St. Peter's in Rome, he made a few notes and then wrote out the music from memory back at his lodgings, in contravention of a papal ban. Stephen Wiltshire was producing remarkable drawings by the age of seven, an ability that didn't markedly improve afterward. Some children are able to perform calendar calculations from the age of six.

Talents sometimes disappear with age. The autistic girl Nadia showed a remarkable talent for drawing between the ages of three and seven. First she drew horses and other animals, later people. By the time she was nine, however, she had lost her unusual ability. It would seem that the improved functioning of the left side of the brain, which is responsible for speech, inhibited her drawing skills. In that respect, too, Daniel Tammet is an exception. Developing his social skills didn't cause him to lose his numerical and linguistic talents. He has a remarkable brain in every respect.

10

Schizophrenia and Other Reasons for Hallucinations

There is only one difference between a madman and me. The madman thinks he is sane. I know that I am mad.

Salvador Dalí (1904–1989)

SCHIZOPHRENIA, A DISEASE OF ALL AGES AND CULTURES

When he arrived at the other side in the region of the Gadarenes, two demon possessed men coming from the tombs met him. They were so violent that no one could pass that way. "What do you want with us, Son of God?" they shouted. "Have you come here to torture us before the appointed time?" Some distance from them a large herd of pigs was feeding. The demons begged Jesus, "If you drive us out, send us into the herd of pigs." He said to them, "Go!" So they came out and went into the pigs, and the whole herd rushed down the steep bank into the lake and died in the water.

Matthew 8:28–32

Schizophrenia has been "treated" in various ways over the centuries. In China, four-thousand-year-old skulls were found with holes trepanned into them, to let out the evil spirits believed to possess schizophrenia sufferers. In some cases bone had started to grow over the hole, showing that the patients had survived the operation for some considerable time. Jesus set a long religious tradition with his banishing of evil spirits (see the epigraph to this section). The Catholic Church ordained exorcists up to around 1970. That profession has since disappeared, though Catholic bishops still appoint priests to carry out exorcisms. In the Protestant church, exorcisms are performed by ministers. Demon banishing goes on in the Islamic tradition, too. When she lived in the Netherlands, the sister of the Somali feminist and activist Ayaan Hirsi Ali received medication for her schizophrenia. But back in Somalia she fell into the hands of Islamic clerics, who put her in an empty room containing only a mattress. They took her medication away and beat her to ritually banish her demons. This treatment proved fatal.

A painting by the medieval artist Hieronymus Bosch in the Prado, in Madrid, represents "the stone operation." A doctor is pretending to remove a stone from the head of a schizophrenia sufferer in a placebo operation. Bosch portrayed the doctor with an upturned funnel on his head to show that he was an impostor, and next to him a nun with the Bible on her head to show that the church was equally culpable.

A picture on the gable stone of the old asylum in the Dutch town of Den Bosch shows how in 1442, schizophrenic patients were locked up in prison. For a few cents, families could go and gawk at the "lunatics" on Sundays. In the 1920s and 1930s, when my mother was a seventeen-year-old trainee psychiatric nurse, the straitjacketed patients were put in baths of alternately cold and hot water. She said she would never forget the sound of them endlessly banging their heads on the edge of the bath, the only movement they could make. Up until the 1950s, schizophrenia was "treated" by means of a lobotomy, an outpatient procedure that involved severing the connections to and from the prefrontal cortex. Opponents called it "partial euthanasia," since it turned patients into robots. This made their care extremely easy, though, and the terrible procedure became popular (see chapter 13). Fortunately modern antipsychotics have made it obsolete.

In China, a relative sits next to every hospital bed in order to help the nursing staff and make sure that their loved one has everything they need. If no relative is available, a co-worker will take their place. It makes Chinese hospitals rather jolly, chaotic places. But in the closed psychiatric wards, the picture is totally different. On my visit to one, I felt as if I were walking onto the set of *One Flew over the Cuckoo's Nest*. The enormous ward contained two seemingly endless rows of beds with identical bedding. Next to every bed hung an identical towel, while each bedside table had on it only a glass, no personal belongings. The patients on this male ward all wore identical striped pajamas. They received no visits at all; their families had disowned them. I was the first visitor for many years, and a foreigner at that. One of them was a sailor who spoke good English. He had seen much of the world, including Rotterdam, and acted as an interpreter for the excited group of patients who crowded around me, tugging at my arms in an effort to draw attention to their personal stories. I was moved and saddened by these accounts. It was very hard to abandon them once more to their isolation.

On that occasion I traveled on from China to Jakarta to give a series of lectures. I was picked up by a young chauffeur whose car radio was blasting out house music at full volume. I asked him cautiously whether he could perhaps turn it down a bit, at which he smiled understandingly and asked what kind of music I liked. "Mozart's *Requiem*," I replied, thinking that it would prove a conversation stopper. The next morning he came to pick me up for the first lectures. To my utter amazement, amid the din of chaotic traffic—Jakarta is one gigantic slow-moving parking lot—he suddenly put on Mozart's *Requiem*. I must confess that I was extremely moved. The next day, when we were stuck in traffic, he asked me what I knew about the treatment and care of schizophrenia sufferers. It turned out that his brother had schizophrenia and lived at home. When he was in a very bad way, they would give him a few drops of medicine. The little bottle of haloperidol had been very expensive, and they'd been making it last for years. I asked him where they kept the bottle. Just in the living room, he told me. Thinking of the average room temperature in Jakarta, I told him that that wasn't such a good idea, because medicine kept that long could lose its efficacy or even become toxic. He fell silent for a while, then said, "Oh, that explains it." Apparently the medicine hadn't worked so well the last time they'd given it to his brother, so they tried a drop on the parrot, which fell down dead on the spot.

Treatment for schizophrenics can be even worse elsewhere in the world, as I saw from an award-winning World Press Photo exhibited in Amsterdam in 2004. Taken in Bangladesh, it showed an eighteen-year-old boy in the completely empty cell of a psychiatric clinic. He was lying on a stone floor, wearing only a pair of shorts. His legs were pinioned in a medieval contraption of wooden blocks. His arms were raised in despair, his fists clenched and his face twisted in a grimace. Apparently there were twenty-four such rooms in the "clinic" in question, and according to its director, thousands of patients had been "cured" in this way since its founding in 1880.

The afflictions of psychiatric patients in the Netherlands may be heartrending, but the sufferings of their counterparts in many other regions of the world are of a different order entirely. Our wealthy nation should never use this as an excuse to sanction cuts in treatment, though, because that would mean putting more patients in isolation cells, and isolation only makes their symptoms worse.

SCHIZOPHRENIA SYMPTOMS

> Our hope for the future lies . . . in organic chemistry or in an approach to [psychosis] through endocrinology. Today this future is still far off, but we should study analytically every case of psycho-

sis because the knowledge thus gained will one day direct the chemical therapy.

Sigmund Freud in a letter to Marie Bonaparte, 1930

Schizophrenia affects 1 percent of the population, but because sufferers have it for such a long time it fills almost half of the beds in psychiatric hospitals. Schizophrenia patients are often depressed and feel that their lives are pointless. Around 10 percent attempt to kill themselves. Suicide makes the burden imposed on their families even heavier.

Schizophrenia is characterized by "positive" and "negative" symptoms. Positive symptoms are abnormal experiences, like delusions and hallucinations. During a psychosis, people see things and hear voices that they experience as completely real. (*"Later, after I'd lost my job, I started to hear voices when I was at home . . . and to be troubled by different voices in my head. Sometimes they're very aggressive, and cut right through me."*). Scans show that during these hallucinations the areas of the brain that normally process auditory or visual input are extremely active. They can't be distinguished from real experiences because they take place in the same areas of the brain where external stimuli are normally processed. Other patients suffer from delusions. They believe that they are being watched or controlled by mysterious powers. (*"In the last week of my work and the two weeks that followed they treated me without my consent, using an extremely advanced system. . . . What's more, they modified my brain with this apparatus, so I can communicate with people in the street by transmitting thought waves."*) During the hallucinations, patients may hear voices giving unwanted instructions. Some are even told to kill someone (see chapters 8 and 17). While in the grip of a psychosis, one woman believed that she could fly. She threw herself out of the window and was killed.

Negative symptoms entail the loss of normal abilities, like taking initiative, organizing one's life, tidying up one's room, and looking after oneself. They also include muted emotions and cognitive deterioration. Many patients end up as vagrants, sleeping on city streets. They often take addictive substances, which might work early on as a form of self-medication against negative symptoms. In the long run, such substances can exacerbate the positive symptoms and cause damage. Negative symptoms are caused by reduced activity in the prefrontal cortex. A current therapy is to apply transcranial magnetic stimulation to that area. Stimulating the areas of the cerebral cortex that are extra active can also reduce hallucinations.

Schizophrenia is more common in men, who are also more affected by it than women. The initial symptoms of the disease can be difficult to identify. A year or two before their first psychosis, young people often show signs of paranoia, start taking drugs, abandon their studies, and become withdrawn. Isolation can exacerbate the condition. Schizophrenia is largely genetic, so having a relative with schizophrenia heightens your risk of this disorder. The first experience of psychosis peaks around the age of twenty. In women there's a second peak that coincides with menopause. Changing hormone levels during puberty and menopause bring on the disease, though a predisposition for it arises in the womb. Female hormones reduce the negative symptoms of schizophrenia if taken together with standard medication.

As the disease progresses, the brain shrinks and its ventricles (cavities) become larger, creating too much space between the convolutions of the brain, just as in the case of many elderly people. This shrinking certainly isn't caused by treatment, because it was shown to exist back in 1920, long before medication for schizophrenia had been developed. Nor is it specific to schizophrenia. It's also seen in the aging process and in various forms of dementia. In fact, there are no brain changes that are specific to schizophrenia, so diagnosis is entirely dependent on psychiatric investigation. It is, however, important to rule out rare brain diseases whose symptoms can mimic those of schizophrenia. But once the diagnosis has been made, early treatment is very important to prevent further brain damage from psychosis.

SCHIZOPHRENIA, A DEVELOPMENTAL BRAIN DISORDER

Schizophrenia is a developmental brain disorder that is caused by a combination of factors and is present at a very early stage—indeed, the main foundation is laid at conception. Studies of families and twins show that the genetic component of schizophrenia is around 80 percent. The genetic factors are many and varied, differing in each family, but all involve tiny variations in the genes that affect brain development or in the production and breakdown of chemical messengers in the brain. The normal development of the fetal brain can subsequently be further disrupted by a host of nonhereditary factors. Maternal malnourishment during the first three months of pregnancy doubles the risk of schizophrenia. This first emerged in studies of children born in Amsterdam after the famine of 1944–1945 (see chapter 2). It was recently confirmed by studies of children born in China during and after the mass starvation in Anhui province in 1959–1961, in the wake of Mao's "Great Leap Forward." The same risk arises if the fetus is malnourished due to a malfunctioning placenta. Toxic substances in the environment, like lead, can also impair brain development in the womb and increase the likelihood of schizophrenia. In addition, you're more at risk of developing schizophrenia if you were born in the winter or if your mother was exposed to flu during the sixth month of pregnancy. How those two factors interact isn't clear. Toxoplasmosis and the Borna disease virus can also be passed on to the fetus, increasing the risk of schizophrenia. Psychological factors, such as stress during pregnancy, play a role too. Moreover, life events, like the death of a relative and pregnancy during wartime, increase the likelihood that a child will develop schizophrenia in later life.

A strong correlation has been found between problems at birth (e.g., forceps delivery, low birth weight, a period in an incubator, and premature birth) and subsequent schizophrenia. Traditionally, it was assumed that these problems at birth affected a child's brain, increasing the likelihood of their developing schizophrenia. For childbirth to proceed normally, however, subtle interaction is needed between the brains of both mother and child. So you could see birth as the first functional test of a child's brain. Disruptions to the birth process can therefore be regarded as the first symptoms of malfunctioning brain development, which are later manifested as schizophrenia (see chapter 1).

After birth, an environment full of stimuli increases the risk of schizophrenia. You're more likely to develop the disease if you live in a city than in the country. Migrants are also at increased risk, probably because of the difficult social circumstances in which they often live. Quite a few adolescents go to their doctor with the first symptoms of schizophrenia after smoking joints. Whether cannabis induces the disease or simply brings forward the moment at which symptoms occur is still a subject of fierce debate.

It is clear from the brains of schizophrenia patients that the disorder arises very early in life. In schizophrenics, a high percentage of cells in the hippocampus are in disarray—something that can only have happened during the first half of pregnancy. Abnormal patterns of brain convolution are also found, as well as groups of cells that have failed to migrate to the right place in the cerebral cortex. This too can only happen during early development.

So although most people with schizophrenia are admitted to clinics as young adults, the foundation for the disease is laid in the womb. It's terrible to think that as late as the 1970s, psychotherapists were spreading the pernicious message that schizophrenia was caused by a mother's coldness and mixed messages (the double bind theory). Family therapists were given the task of reeducating mothers or even of "rescuing" children from the clutches of their pathological environment, a situation that caused extra pain for parents struggling to do their best for their children. The Dutch psychiatrist Carla Rus was so alienated by this approach that she stopped training as a family therapist. My mother, on the other hand, had her own views on what caused schizophrenia. She had a button printed with the words, "Madness is inherited, you get it from your children."

HALLUCINATIONS DUE TO A LACK OF STIMULI

> I doubt if a single individual could be found among the whole of mankind free from some form of insanity. The only difference is one of degree. A man who sees a gourd and takes it for his wife is called insane because this happens to very few people.
>
> Desiderius Erasmus (1469–1536)

If brain structures stop receiving information in a normal way, they start making up information. This applies both to sensory information—from ears, eyes, and limbs—and memory information. A fifty-seven-year-old man who had been suffering from a disease of the inner ear for twenty years found that in the space of twelve months his hearing had greatly deteriorated despite his two hearing aids. During that year he was plagued by nonstop music in his head. Day and night he heard the national anthem, Christmas carols, psalms, and sometimes children's songs. Though distorted, he could always recognize the tune and sometimes sang along. These musical hallucinations are a form of tinnitus better known to patient associations than to the average doctor.

When your brain manufactures information on the spot where it's normally processed, it's interpreted as if it had entered from outside, via the normal route. If the auditory cortex (fig. 22), for example, stops receiving the information it normally gets from the ears, it starts to work overtime, producing something that that part of the brain normally processes: music. You would therefore expect the maddening tunes to disappear if you stimulated the auditory cortex. However, it wasn't easy for the man I met to find a doctor prepared to try that. In the end, he was treated by Dirk de Ridder of Antwerp. A short test involving electromagnetic stimulation of the auditory cortex caused his tinnitus to vanish, only to return gradually after a few days. He then splurged on a pair of €4,000 Varibel "hearing glasses" developed by Delft University of Technology, which greatly improved his hearing and reduced his tinnitus. This shows that the brain stops producing old information once it receives fresh input again, and it makes no difference whether the input is meaningful (the hearing glasses) or has no information content (the electromagnetic stimulation).

Charles Bonnet syndrome is another phenomenon wherein the brain manufactures information to compensate for a lack of input. The condition provokes colorful visual hallucinations in individuals with impaired sight, typically older people with cataracts, glaucoma, or retinal bleeding. The hallucinations—often complex, vivid images of people—tend to occur in dim light and quiet surroundings. Sufferers of Charles Bonnet syndrome are aware that their hallucinations aren't real and find that they usually disappear if they shut their eyes. An eighty-three-year-old woman who had played an active role in the Dutch resistance during the Second World War and who had become practically blind due to glaucoma confided anxiously to her daughter that whenever she blinked her eyes she saw swastikas.

In the case of Charles Bonnet syndrome, the visual cortex (fig. 22) receives insufficient information from the eyes and starts to produce its own pictures. A similar phenomenon occurs in the case of memory loss. People who suffer from Korsakoff's syndrome, a dementia that results from alcohol abuse, produce fake memories of events that never took place, known as confabulations. Phantom sensations following amputation appear to be based on the same principle. Lacking customary input from a limb, the brain "makes up" the presence of a missing arm or leg. Hallucinations can also be a sign of neurodegenerative diseases like Lewy body dementia, which often involves impaired visual perception, and Alzheimer's disease and Parkinson's disease.

In schizophrenia, input to areas of the cerebral cortex is also reduced, so the hallucinations it provokes could be caused by the same mechanism. Depending on the area of the cortex that is overactive, schizophrenia patients see or hear things that aren't there. A group headed by René Kahn in Utrecht has indeed shown, in a series of pioneering experiments, that electromagnetic stimulation of the brain reduces hallucinations in schizophrenia patients. Conversely, isolation cells, in which these patients tend to be confined during acute stages of the disease, diminish brain input even further and can thus make their symptoms much worse.

Mountaineers, especially when alone, sometimes have very vivid hallucinations (hearing voices, seeing people, or having out-of-body experiences) or are overcome by fear. So it's interesting that the revelations received by the leaders of the world's three main religions were preceded by a period of isolation in the mountains. On two occasions, Moses received the Ten Commandments from the Lord on Mount Sinai. On the second occasion he spent "forty days and forty nights" there alone "without eating bread or drinking water." When Jesus took the disciples Peter, John, and James up a mountain to pray, they had a vision of Moses and Elijah. The Prophet Muhammad saw the Archangel Gabriel during his lonely vigil on Mount Hira. These experiences involved seeing bright lights, hearing voices, and experiencing fear, just as in the case of mountaineers. When the brain is very isolated it starts to use stored experiences and thoughts to manufacture things—sometimes even new religions.

OTHER HALLUCINATIONS

> When we remember we are all mad, the mysteries disappear and life stands explained.
>
> Mark Twain (1835–1910)

Delirium

Hallucinations are by no means confined to schizophrenia. They are most common in cases of delirium. In the Netherlands, around one hundred thousand patients a year experience delirium. Most of them are elderly people who have been operated on under general anesthesia (because of a broken hip, for instance). For an old brain, an anesthetic is like a dose of near-lethal poison. In intensive care, up to 80 percent of patients experience delirium. Delirium can also result if the brain's functioning is impaired by pneumonia, dehydration, medication, drugs, or malnourishment. In older people it can even be caused by a simple urinary infection. Then there's the famous delirium tremens, which isn't solely caused by alcohol poisoning—it can result from alcohol deprivation. Brain damage due to lack of oxygen, low blood sugar, or an infarct can also induce delirium.

Delirious patients are extremely confused. They are often restless, have memory problems, and are aggressive, noisy, and sometimes hyperactive to the point of falling out of bed and breaking something, ending up in even worse shape than they were before. But there's also a peaceful type or phase of delirium in which patients simply lie in bed apathetically, staring blankly. Their consciousness is impaired. They don't know where they are or sometimes even who they are. They can't think straight or concentrate. The condition sometimes resembles dementia, but delirium strikes all of a sudden, while the onset of dementia is gradual. Delirious patients will hallucinate, often seeing creepy crawlies everywhere. Some have been known to refuse food or drink because it appeared to be covered with ants. One patient saw beetles coming out of the ceiling. Feverish children can have visions of cartoon characters (one little girl reported seeing Donald Duck riding her father's bicycle along her bedroom wall). The hallucinations and delusions are often frightening. One patient, a Holocaust survivor, thought that he was being sent back to a concentration camp and was terrified of his doctors and nurses, believing that they had come to get him. In his desperate attempts to escape, he tore his drip out of his arm and then the tube through which he was being fed—a dangerous thing to do, because if the food had gotten into his lungs he could have developed pneumonia. A female patient thought that she had been tied up and raped in the hospital. An old friend of mine, refusing to believe that his operation was over, asked the doctor why on earth he had visited him in the middle of the night. He went on to reprimand the doctor for not having the decency, during this nighttime visit, to answer his queries about the results of his blood test. My friend believed that he had then gone to the laboratory himself to get the test results. Both the doctor's visit and his nocturnal trip to the laboratory had in fact been a hallucination. An old lady who fought tooth and nail with the nursing staff explained later that she thought her bed was her grave. She made desperate attempts to climb out, but the nurses kept pushing her back in.

More is gradually emerging about why some people are more prone to delirium than others. Delirium is basically caused by an overdose of the chemical messenger dopamine. There are a great many tiny variations, polymorphisms, in the DNA of the gene that produces the protein that receives the dopamine message in the brain cells. Those tiny variations are what make you more or less susceptible to delirium. Delirium causes brain damage and increases the risk of dementia. And its effects can last a long time. Many people experience long-term problems with reading, writing, walking, and memory and never recover entirely. Around a third of people over sixty-five who experience delirium die within a few months. So it's a serious disorder and one whose risk is determined from the moment of conception.

Hearing Voices

> It often takes quite a while for children to discover that other people don't hear voices, after which they tend to be afraid to talk about such experiences.

There are people who aren't psychotic and yet do hear voices. In fact, between 7 and 15 percent of the population hear voices, yet only a fraction have mental health problems. This phenomenon forms part of a spectrum, with healthy people who hear voices at one end and schizophrenia patients at the other. People who hear voices at the onset or in the wake of a psychosis are somewhere in the middle. In healthy individuals, hearing voices frequently starts at a young age and tends to run in families. It often takes quite a while for children to discover that other people don't hear voices, after which they tend to be afraid to talk about such experiences. Some are very attached to their friendly voices, like the lady whose voices had told her since the age of eleven that there was no reason to be afraid. Patients with mental health problems, however, often hear voices with threatening, negative messages ("Why don't you jump in front of a train today?" or "You must die, Evelyn, you're evil, and evil people must die"). No wonder that voices like that produce paranoia and psychosis. By contrast, the voices that healthy people hear typically provide friendly help and advice, though there are cases of them saying nasty things like, "You're ugly, you're worthless, you're fat."

Unlike psychotic patients, healthy people can control their voices. They can call them up as well as order them to go away at inconvenient times. Functional brain scans show that in the healthy group, brain activity isn't very different from that of patients with psychosis. In both cases, activity is seen in Broca's area (language production) and Wernicke's area (hearing, processing, and understanding language) (fig. 8) and the primary auditory cortex is activated (fig. 22). It looks as if the links between those areas are disrupted, which would tie in with the theory that when input to a particular brain area is reduced, that area starts to produce its own information (see earlier in this chapter). In people who hear nasty voices, the right side of the brain is much more active. Attempts have been made to silence the voices by means of transcranial magnetic stimulation of the overactive area, but up to now that method has proved no more effective than placebos. I wonder how many people who work as TV psychics or claim to have paranormal gifts aren't receiving messages from the other side but are just hearing their own brains.

Olfactory Hallucinations

> Gershwin died at the age of thirty-eight, shortly after partial removal of a tumor of the uncus.

The uncus (fig. 21) is a structure located in the front of the temporal lobe above the amygdala that is involved in smell. While conducting an orchestra, the famous composer George Gershwin suddenly started to smell burnt rubber and blacked out for ten to twenty seconds. He'd had an uncinate fit, which is an olfactory hallucination sometimes associated with epilepsy. Despite consulting a great many doctors, it took six months before Gershwin's fits—which grew increasingly frequent—were shown to be caused by a tumor of the uncus. Gershwin died in 1937 at the age of thirty eight, shortly after the tumor was partially removed.

11

Repair and Electric Stimulation

> I believe that the great diseases of the brain . . . will be shown to be connected with specific chemical changes in neuroplasm. . . . It is probable that by the aid of chemistry, many derangements of the brain and mind, which are at present obscure, will become accurately definable and amenable to precise treatment, and what is now an object of anxious empiricism will become one for the proud exercise of exact science.
>
> J.L.W. Thudichum, 1884

AGE-RELATED BLINDNESS: MACULAR DEGENERATION

> Oh, the endless labor of the intellectual—pouring all this knowledge into the brain through a three-millimeter aperture in the iris.
>
> Irvin D. Yalom, *When Nietzsche Wept,* 1992

My father went blind in the last year of his life. He lived to be eighty-nine. Fifteen years ago we would go to Leiden together every week, where he was undergoing laser treatment for degeneration of his retina. The retina, which evolves during fetal development from a bulge in the brain, converts light into electrical signals that the optic nerve transports to the back of the brain, where we see. When I arrived at the teaching hospital in Leiden for the first time with my practically blind father on my arm, I said, "We have to make a right here." "How do you know that, have you been here before?" he asked. "No, but there's a great big sign with an eye on it and an arrow pointing to the right," I explained. Without missing a beat, he said, "I'd love to know what their signs for gynecology and obstetrics look like." He had the most common form of age-related blindness: macular degeneration. It's caused by new blood vessels growing right under the yellow spot (the macula), the part of the retina where you see the best. This destroys the retina, and you begin to lose your sight, starting in the middle of your field of vision. At first the objects you see are distorted, then a black spot forms in the middle of your central vision, which gradually becomes larger. In the wet form of macular degeneration, the new blood vessels leak blood and fluid. Reading and writing soon become impossible, and in the long run you can't even make out large objects. On the journey back to Amsterdam after my father's first laser treatment in Leiden, he asked me what month it was. "January," I answered. "That's odd," said my father, "surely it's much too early for the bulb season?" It turned out that the injection of fluorescent substance he had been given in preparation for the laser treatment made everything look yellow, creating the illusion that the tulip fields were in bloom. The laser therapy, alas, didn't help preserve what little remaining sight he had.

After my father's death, laser treatment improved, and an effective therapy has recently been devised for wet macular degeneration. Antibodies like Avastin have now been developed that inhibit the vascular endothelial growth factor (VEGF), the molecule that prompts the growth of new blood vessels that destroy the retina. In order to halt the degenerative process, this substance has to be injected into the eye every month using an extremely thin needle. A similar substance called Lucentis has since been devised specifically for optical use. It is said to stabilize the condition in over 90 percent of patients and even improve vision in a third of this group. Other forms of therapy for macular degeneration are under development as well. In a short space of time, wet macular degeneration has gone from being an incurable to a treatable eye disease. The breakthrough in treating it came from Avastin, which was originally developed as a medication for intestinal cancer. That's often the way in medicine: However carefully you target your approach to a particular disease, the breakthrough often comes from a completely unexpected sector.

SERENDIPITY: A LUCKY ACCIDENT

> It's not uncommon for medical findings to be made by accident, but you must keep an open mind and be able to make important deductions from apparently insignificant facts.

If medication for Parkinson's disease stops being effective, electrodes are sometimes implanted in a patient's brain. They are then stimulated electrically by a pacemaker, which temporarily switches off activity in a small part of the brain. It's very impressive to see how violent tremors suddenly stop when the patient switches on the stimulator. Yet this treatment is the result of serendipity, a lucky chance. In fact, it's not uncommon for advances in medicine to be made by accident, when a doctor or researcher is exploring an entirely different avenue. In 1952, a Parkinson's patient was set to undergo a drastic brain operation in a desperate attempt to stop his exceptionally severe tremors. The plan was to sever motoric pathways so as to paralyze the patient. During the operation, the surgeon, Irving Cooper, accidentally tore open a blood vessel. To stop the bleeding, the blood vessel was tied up, and the operation was abandoned. To everyone's surprise, the patient's tremors vanished. Cooper later went on to cauterize that blood vessel deliberately in other Parkinson's patients, thus disconnecting a small region of the brain. He succeeded in reducing tremor in 65 percent of Parkinson's patients and muscle rigidity in 75 percent. Subsequent experimental disconnection of different brain regions proved most successful in the area below the thalamus, the subthalamic nucleus (fig. 23). This is where most electrodes are still inserted in the brains of Parkinson's patients. The nice thing about electrodes is that disconnection is reversible; as a result, you can see where they are most effective and can keep adjusting the way in which they are stimulated. This treatment alleviates symptoms like slowness of movement, muscle stiffness, and tremors, though it can't slow the progress of the disease.

Around thirty-five thousand people worldwide are going through life with depth electrodes in their brains. As in the case of any effective therapy, side effects have emerged. Some Parkinson's patients with depth electrodes have problems interacting with their partners and co-workers. Although most patients were personally very satisfied with the quality of their lives, their relatives sometimes reported that they were more irritable and prone to mood swings. Around 9 percent experienced psychiatric side effects like increased impulsiveness or crying fits. Electrode stimulation can exacerbate depression. We have even seen cases where the patient committed suicide, despite the electrodes having been implanted at the right location in the subthalamic nucleus. (Ten years ago, neurologists wouldn't have been interested in this correlation, but the boundary between neurology and psychiatry is fast becoming blurred.) Sometimes an implant can cause dementia symptoms because of bleeding or brain damage. But these symptoms occasionally disappear if the stimulator is configured differently. Cases have also been reported of psychosis, sexual disinhibition, and gambling addiction. One patient who had been a typical tightfisted Dutchman prior to the operation couldn't resist the lure of slot machines afterward. It was only years later, when his mounting debts forced him to sell his house, his wife demanded a divorce, and he tried to commit suicide, that the problem was brought to the attention of his doctors. (Gambling addiction can also be a side effect of the classic medication for Parkinson's, L-dopa, the dopamine system [fig. 16] being central to addiction.) Another patient treated with electrodes became manic and started buying houses in Spain and Turkey, something he really couldn't afford to do. But he absolutely refused to have the stimulator switched off.

Electrode stimulation can also impair cognition, memory, and speech. However, most of the psychiatric side effects are temporary and easy to treat, and can even be prevented. They also tell us something about the function of brain structures and circuits in psychiatric symptoms, like the role of the dopamine system in addiction. The success of depth electrodes in treating Parkinson's has led to their application in many neurological and psychiatric disorders, like unbearable pain, cluster headache, depression, phobia, muscle spasm, self-mutilation, and obsessive-compulsive disorders. Studies are being carried out on the treatment of obesity and addiction using depth electrodes. At present, the possibilities seem endless. Irving Cooper could never have imagined all this when he accidentally caused a brain hemorrhage in 1952.

DEEP BRAIN STIMULATION

> Electrical stimulation deep in the brain appears not only to have useful clinical applications but also to provide basic information about how our brains work.

Depth electrodes were used to spectacular effect on a thirty-eight-year-old man who had been in a state of minimal consciousness for six years following an accident. He could occasionally communicate with eye or finger movements but never with speech. Under normal circumstances, recovery after twelve months in a state like that is virtually impossible. A breakthrough was made, however, when stimulation electrodes were planted on either side of the thalamus in the center of his brain, where sensory information is received. Within forty-eight hours of stimulation he woke up and would turn his head when someone spoke his name. Over a four-month period of electrical stimulation he began to recover his speech, eat, drink, and comb his hair.

Depth electrodes have also recently been used to treat obsessive-compulsive disorder, a condition that causes people to perform certain actions compulsively, like washing their hands hundreds of times a day or pulling out their hair. They become fearful if they don't give in to their obsessions, rendering a normal social life impossible. Since performing compulsive actions makes people with OCD feel good, it's thought that the brain's reward system is involved. The good feeling is caused by the release of the chemical messenger dopamine in the nucleus accumbens (fig. 16). Research by Damiaan Denys shows that when traditional treatments fail, patients can benefit from electrodes on both sides of the nucleus accumbens. The theory is that stimulating the region causes dopamine to be released, creating the same reward as the compulsive behavior. As a result, compulsive hand washing can be reduced from ten hours a day to fifteen minutes a day, allowing patients to lead a normal life again. One patient who was successfully cured of compulsive behavior ended up thinking obsessively of sex instead. The electrode turned out to be close to the bed nucleus of the stria terminalis (figs. 10 and 11). Whether this was really the cause of the obsessive thoughts has yet to be established.

Along with new breakthroughs, new side effects of depth electrode treatment are also being reported, as in the treatment for tinnitus, which sometimes affects people who are hard of hearing. Because their brain isn't receiving normal auditory information, it starts to produce its own sound sensations, causing some people to hear constant music (see chapter 10). Stimulating the part of the brain that stopped receiving auditory information would seem a logical way of curing this condition. However, a patient who was treated for tinnitus with depth electrodes in the temporal cortex (fig. 28) not only continued to hear the annoying tunes but also had an out-of-body experience as a side effect. He felt as if he was standing a couple of feet to the left of and behind his body (see chapter 7).

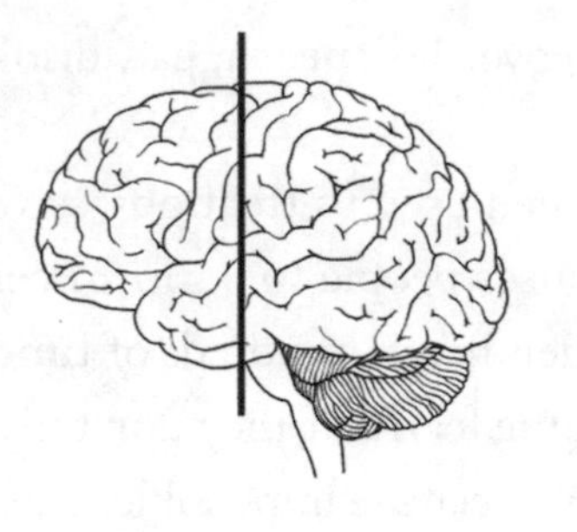

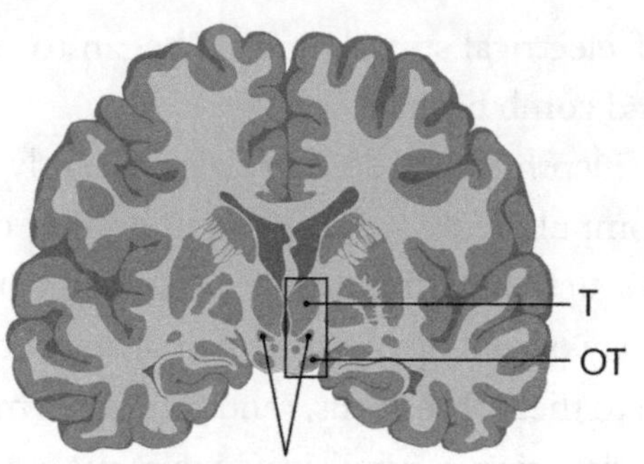

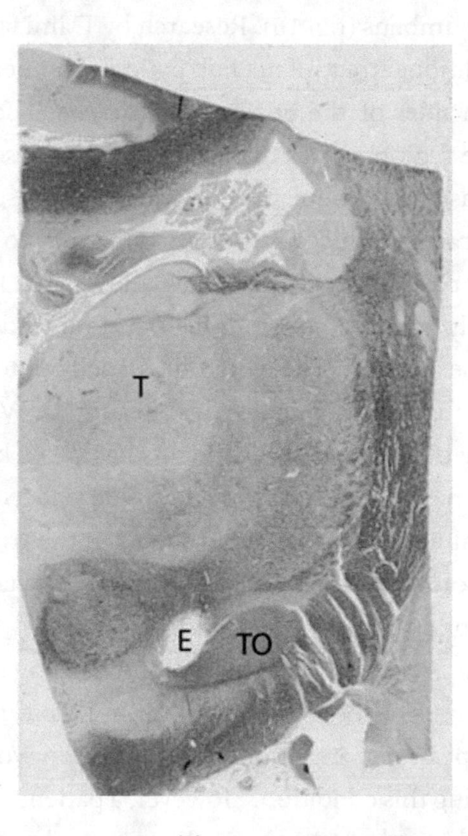

FIGURE 23. The depth electrode (E) was inserted at the correct location in the subthalamic nucleus of the brain of a Parkinson's patient. T = thalamus, OT = optical tract.

Another unexpected side effect occurred in the case of a man who'd had depth electrodes implanted in his hypothalamus to inhibit his binge eating. He was so obese that he didn't even fit into the scanner. Stimulation from the electrode didn't help him to lose weight, because he had a tendency to switch it off at night so that he could eat. But when it was on, he suddenly experienced events that had taken place thirty years earlier, like walking through the woods with friends. Each time he could remember more and more details. This side effect seemed to be due to activation of the temporal lobe. Episodic memories of this type also occur when people see their lives flash past in near-death experiences (see chapter 16). The temporal lobe is important for memory, and since stimulation has been shown to improve memory, studies are now under way to see if it can be used to help patients with memory problems. So this technique not only appears to have great clinical potential, but it can also teach us a great deal about how our brains work by showing us what happens when depth electrodes aren't quite in the right place.

BRAIN STIMULATION AND HAPPINESS

> Happiness is nothing more than good health and a poor memory.
>
> Albert Schweitzer

I was once told by Ruut Veenhoven, a professor of "social conditions for human happiness," that happiness doesn't depend on having a goal in life. That didn't surprise me at all, given that life came into being and evolved by accident and has no purpose. But enjoying life is useful because it's closely connected to eating and reproducing and therefore promotes survival. Indeed, the pursuit of pleasure is such a strong impulse that it causes overpopulation and obesity. Positive emotions like falling in love, maternal love, and pleasure in social contacts also benefit the survival of the species.

Human cognitive development has enabled feelings of pleasure to be elevated to the "higher" orders of art and science, altruism, finance, and transcendent activities, culminating in happiness. Happiness is contagious. When someone is happy, their friends, partners, and family are more likely to be happy.

Positive feelings can also be disturbed, as in the case of some psychiatric patients. Mania can go hand in hand with intense happiness. Conversely, anhedonia, the absence of any feeling of pleasure, is a characteristic of depression and is associated with schizophrenia, autism, and addiction. The ventral striatum plays a crucial role here. Parkinson's patients with lesions in that area sometimes suffer from flat affect (lack of emotional expression) or even anhedonia. The increased levels of corticosteroids associated with depression inhibit dopamine release in the ventral striatum, thus apparently blocking all pleasurable feelings. Conversely, stimulation of the area can alleviate depression.

Feelings of pleasure and happiness are linked to activity changes in a great many areas of the brain. Activity in the prefrontal cortex increases in response to both tasty food and financial reward. This area also controls whether or not you give in to temptation. But the prefrontal cortex is not where pleasure is *generated*. Patients who have undergone leukotomy, an operation in which the prefrontal cortex is disabled, can still derive pleasure from food and sex. The pleasurable feeling is generated in reward systems located lower down in the brain.

Addictive substances make use of existing brain systems to induce pleasurable feelings. Sigmund Freud took cocaine for a while, writing in 1895 that the feelings the drug generated couldn't be distinguished from normal feelings of well-being. Tests in which laboratory animals were given small amounts of opiates in the areas of the brain that are considered "hedonistic hotspots" show that those brain structures are *sufficient* to induce pleasurable feelings. However, it's only possible to say whether an area is *necessary* for pleasurable feelings if the feelings disappear when the area is disabled. Similarly, stimulating a brain area (the ventral striatum or nucleus accumbens) is *sufficient* to create a rewarding effect, but disabling the area does very little to impair the rewarding effect of food, showing that it isn't *necessary* for the creation of such an effect. Our taste for sweet food depends on a single hedonistic hotspot at the base of the brain. Disabling that area makes sweet food taste repellent. Likewise, the hypothalamus is necessary for infatuation, maternal love, and pair forming. The other brain areas that show changes of activity in response to pleasure or happiness aren't necessary for the pleasurable feeling, but they are necessary for processes linked to it, like learning, memory, decision-making, or behavioral effects. The dopamine reward system is involved in anticipatory pleasure, motivation, and attention relating to enjoyment. When someone is clinically depressed, the stress hormone cortisol inhibits the dopamine reward system, preventing them from feeling pleasure. Cocaine, on the other hand, extends dopamine's availability to receptor cells in the brain. The brain's own opiates are also involved in the sensation of happiness. Oxytocin and vasopressin play a role in infatuation, orgasm, pair forming, and maternal love. Deficiencies in those two chemical messengers are associated with autism.

Some people can generate feelings of happiness themselves. Brain scans of nuns instructed to reexperience their ecstatic love of God showed changes of activity in reward structures. A brain tumor in the temporal lobe can also generate ecstatic experiences, for instance the feeling of direct contact with Jesus. The ecstatic experiences cease when such tumors are removed.

Alas, it is impossible to induce extreme happiness by means of a stimulation electrode in a single brain location, but there are "self-stimulation hotspots." Rats with stimulation electrodes placed in certain areas of their brains can be stimulated to eat, drink, and have sex many times in the space of a minute. But whether they actually enjoy it is debatable, given the findings emerging from stimulation studies of humans. One young man who'd had an electrode implanted in the accumbens/septum region constantly engaged in self-stimulation. He protested vehemently when the electrode was removed. It had induced feelings of alertness, warmth, arousal, and an urge to masturbate, but never orgasm or clear evidence of pleasure. A young woman who stimulated herself constantly while experiencing erotic sensations never brought herself to orgasm. Moreover, the constant stimulation led her to neglect herself. It seemed as if she, too, wasn't experiencing any kind of real pleasure. So for the time being we'll have to rely on the old-fashioned method of obtaining pleasure and happiness. And there's nothing wrong with that.

PROSTHESES IN THE BRAIN

> The garage called. Your brain's ready.
>
> W. W. Tourtelotte

The brain gets information from the outside world through the senses and then takes action by means of motor control. Until recently, if you lost one of your senses you were doomed to be blind or deaf or, if your spinal cord was damaged, to be paralyzed for life. At the Netherlands Institute for Neuroscience's International Summer School of Brain Research in 2008, however, new developments in brain computer interfaces or neuroprostheses appeared to offer future hope of returning sight to blind people and enabling paraplegics to walk again. By far the greatest advances are being made in the field of hearing. Since 1960, bionic ears in the form of cochlear implants have been implanted in patients whose deafness is caused by an inner ear disorder. The implants stimulate the nerve cells connected to the nonfunctioning hair cells in the inner ear. Since 1980 it has been possible to augment this technique by implanting twenty-two electrodes, and over a hundred thousand people with cochlear implants are now able to hear surprisingly well or even normally. However, implants do not work when deafness is due to failure of the auditory nerve. Deafness of that sort has been successfully treated by implanting twelve electrodes in the brain stem (rather than in the inner ear), enabling auditory information to reach the brain, thus improving communication.

Millions of people around the world are blind because the light-sensitive cells in their retinas, the photoreceptors, have been destroyed. Gerald Chader, an ophthalmologist from Los Angeles, described experiments with three such patients who were totally blind. Information was sent from a tiny camera on a pair of spectacles to a miniature receiver that had been attached to the patients' retinas during an operation. A microprocessor translated the visual signals into electronic signals. Sixteen electrodes made contact with the layer of nerve cells in the retina that was still intact, which passed on the information to the brain via the optic nerve. After considerable training, the patients could distinguish between objects such as a cup and a plate. The number of electrodes is being gradually increased to one thousand; with any luck facial recognition will become possible within five to ten years. In a study by another research group, the miniature camera transmitted visual information to a device in the patient's pocket, where it was electronically processed and then sent to a receiver connected to a large number of microelectrodes implanted in the visual cortex (fig. 22).

It is increasingly possible to determine from the electrical activity of large numbers of cells in the motor cortex what motion they intend on carrying out, making it feasible to control a prosthetic arm, for instance. Such advances also open up possibilities for the future treatment of paraplegia. Paraplegic animal studies show that electrical stimulation of the spinal cord, three months of training, and supporting medication can generate a walking pattern that isn't directed by the brain. Gregoire Courtine of Zurich estimates that within five years' time he'll be able to use this approach to treat patients. A spectacular experimental result was obtained in the case of twenty-five-year-old Matthew Nagle, who was left fully paralyzed after being stabbed in the neck with a knife. A 4-by-4-millimeter plate with ninety-six electrodes was implanted in his motor cortex (fig. 22).

When connected to a computer, the plate created an interface powered by signals from the cells controlling his motor system. He was able to use the computer after just a couple of minutes, moving a cursor on the screen by simply imagining the action. He could draw a circle on the screen, read his email, play computer games, and even open and close a prosthetic hand. But this experiment, besides revealing the potential of a neuroprosthesis, also showed its limitations. Before the operation, Nagle had been able to operate a computer using speech recognition. After the operation he was hooked up to a large computer, with an assistant always standing behind him. As a result, the added value of the implanted electrodes wasn't great. So when the electric signal from his brain decreased after about nine months, he decided to have the electrodes removed. A great deal still needs to be done, but there are certainly many hopeful advances being made in this field of research.

TRANSPLANTATION OF FETAL BRAIN TISSUE

> If the transplantation of fetal brain tissue becomes effective, what characteristics might you acquire from the donor?

Parkinson's disease is characterized by the death of dopamine cells located in a part of the midbrain called substantia nigra, from the Latin for "black substance" (fig. 24). In autopsies the substantia nigra normally shows up as a black band shimmering through the brain tissue, caused by the dark pigmentation of the dopamine-producing neurons. When the dopamine cells die off, they can no longer innervate and control the motor area in the middle of the brain, the striatum. That is what causes the movement impairments typical of Parkinson's.

It would be logical to assume that one could cure Parkinson's by replacing the dead cells. In 1987, a Mexican doctor by the name of Ignacio Madrazo published an article in the renowned *New England*

Journal of Medicine in which he reported incredible improvements in patients with Parkinson's disease after tissue from their own adrenal gland (which contains dopamine cells) was transplanted into the caudate nucleus (fig. 24). His article sparked a wave of similar transplants, some two hundred in the space of two years. The operation, however, proved ineffective, and 20 percent of those who underwent it during the two-year period died. Autopsies showed that the transplanted tissue from the adrenal gland had not survived. Only scar tissue was found in the striatum of the patients. Dr. Madrazo's promising results were probably due to a combination of poor research and placebo effects (see chapter 16).

Since 1988, in an alternative to the patients' own adrenal tissue, dopamine cells from fetal brain tissue have been transplanted into the striatum of Parkinson's patients. In order to be effective, the tissue must come from six- to eight-week-old fetuses. In around 85 percent of patients operated upon, the transplanted material can be viewed using a PET scan. Indeed, dopamine cells that had been communicating with the host brain cells have still been found in the striatum of deceased patients sixteen years after the operation. However, the new dopamine cells sometimes showed signs of Parkinson's disease. That the condition can infect transplanted tissue might explain why patients who initially appeared to benefit from the procedure subsequently deteriorated. Moreover, tissue from four embryos is needed for a single transplant. That amount is hard to obtain, because the tissue comes from aborted fetuses whose use in a transplant requires maternal consent. As a result, a great deal of hope has been invested in an alternative source of transplant tissue—specifically, embryonic stem cells—from which dopamine neurons can be cultured. But these therapies are still very risky and have many drawbacks. One case has already emerged of a patient who developed a brain tumor four years after having stem cells injected into his cerebellum. Stem cells have the potential to grow into anything, including tumors.

The transplant of fetal dopamine cells into the brains of Parkinson's patients did produce some positive results. The patients were able to cut down their L-dopa medication, and their movement disorders were slightly reduced. But it is still very far from a true cure, and the results vary. Moreover, both the effects and the side effects resemble those of L-dopa. In around 15 percent of cases, abnormal movements (dyskinesia) arise as a complication of the transplant, but the same applies to patients taking L-dopa. Placebo-controlled studies were also carried out in which, in a blind trial, 50 percent of patients underwent the operation but didn't receive a transplant. Two years later there was no longer any appreciable difference in terms of movement disorders between the transplanted patients and those who underwent a fake operation. So all in all, the results are not convincing (see chapter 16).

A second brain disorder that has led to the experimental transplant of fetal brain tissue is Huntington's disease, an inherited condition that causes movement problems and in which brain cells in the striatum waste away. At a later stage, dementia ensues. The mutation that causes the disease is so rare that in all South African patients the disease can be traced back to a single sailor who arrived at the Cape of Good Hope on Jan van Riebeeck's ship in 1652. The first transplants of fetal striatum tissue have been given to Huntington's patients, and monitoring in a multicenter study is showing clinical improvements. Studies of patients who have since died show that the transplants contain living cells that integrate in the network of brain cells. One transplant, however, grew too fast, causing neurological problems. So, here, too, optimism needs to be tempered.

Fetal retinal tissue transplants are being used to treat blindness caused by nerve cell degeneration, like retinitis pigmentosa and macular degeneration. The results are encouraging.

If transplants of fetal brain tissue become truly successful in the future, ultimately enabling brain defects to be effectively repaired using this technique, we will face an important question. After all, many of our characteristics, including our personality, are determined by the development of brain structures in the womb. If fetal

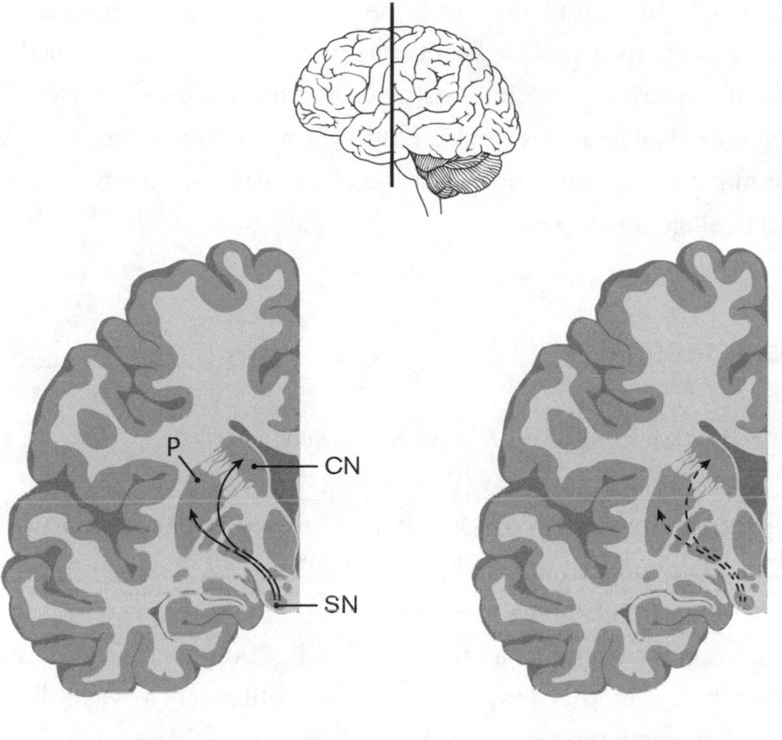

FIGURE 24. In Parkinson's disease, the dopamine-producing black pigmented cells in the substantia nigra (SN) die, and can therefore no longer control the motor area, the striatum (P = putamen, CN = caudate nucleus).

donor tissue is implanted in your brain, what characteristics might you acquire from the donor? These will depend on the area of the fetal brain used and the place where it's implanted. Yet even taking this into account, it's difficult to predict what these transplanted characteristics might be. When this technique becomes effective and is applied to higher brain structures, like the cortex, you might wonder to what extent you're in fact compiling a new person. How much transplanted tissue would it take before a recipient should add the donor's surname to his or her own? It will get even more interesting if we manage to transplant brain tissue from other species. Because of the scarcity of human fetal brain tissue, fetal brain tissue from pigs has been transplanted into the brains of Parkinson's patients, who were then given medication to prevent rejection. So far these operations haven't been successful: Only a few pig cells survived. But if xenotransplantations of this kind ever work in the future, might not human recipients find themselves endowed with the friendliness and intelligence of pigs?

GENE THERAPY

Medication in the form of a piece of DNA . . .

In gene therapy, pieces of DNA containing the code for a particular protein (a gene) are inserted into a cell. The cell then starts to produce medicine in the form of the gene product, that is, a new protein. Brain researchers had thought that it would take an extremely long time before this new therapy, which until recently was still only being used experimentally in cultured cells and laboratory animals, could be applied in a clinical setting to treat disorders of the nervous system. But gene therapy is already being tested on patients with eye disorders and Alzheimer's disease.

In recent years, the research group led by Mark Tuszynski in San Diego has been the first to apply gene therapy to the treatment of Alzheimer's. They are getting cells to produce nerve growth factor (NGF) as a possible medicine, targeting an area of the brain that's important for memory, the nucleus basalis of Meynert (NBM, fig. 25). The cells of the NBM, which is located at the base of the brain, make sure that the chemical messenger acetylcholine—important for memory—is produced throughout the cerebral cortex. NBM cells become somewhat less active with aging, much less active in the case of Alzheimer's. Tuszynski first showed that he could restore NBM neuron activity in aged rhesus monkeys using NGF gene therapy. He did this by removing some skin cells (fibroblasts) and culturing them outside the body. He then inserted the NGF gene in these cells and transplanted them into the brains of old monkeys, close to the NBM. The skin cells were shown to produce NGF in the monkeys for at least a year and to restore the activity of NBM cells.

The same procedure was adopted with Alzheimer's patients. For the first stage of the new therapy, eight Alzheimer's patients were selected who were at such an early stage of the disease that they could still understand the experiment and give formal consent. In this phase 1 study, intended to show how a new therapy is tolerated, patients' fibroblasts were removed, cultured, and genetically engineered to produce NGF. This was done using a virus as a vehicle. The virus had been disabled in such a way that it could still penetrate the cell, along with the NGF gene, but no longer multiply and thus cause disease. The NGF-producing skin cells were then injected into the region of the NBM in an operation involving stereotactic surgery. (This technology—dubbed "cerebral GPS" by the Dutch doctor Bert Keizer—shows very precisely where the tip of the needle is located in the brain.)

In the case of the first two patients, the operations were far from successful. As is customary in stereotactic brain surgery, the patients weren't anesthetized. Although tranquilized, they moved when the cells were injected. Subsequent bleeding in the brain caused paralysis on one side. One patient went on to recover from the paralysis, but the other died five months later of lung embolisms and heart failure, a complication that had nothing to do with the operation or the gene therapy. In subsequent operations, the cells were injected under general anesthesia, preventing any movement. PET scans showed that the cerebral cortex became more active after the procedure. It has been claimed that the memories of Alzheimer's patients who received gene therapy deteriorated only half as rapidly as those who weren't given this treatment. But this was a phase 1 study, so it lacked good controls. The brain of the patient who died after five months showed a robustly stimulating effect on the NBM neurons, giving hope that gene therapy can work.

It will take a while, though, before the effects and side effects of this therapy are known. Previously, three Alzheimer's patients in Sweden had also been given NGF—in their case it was infused into their brain cavities with a miniature pump. But the experiment was stopped because the treatment had little effect on memory function while causing serious side effects in the form of chronic pain and weight loss. We can only hope that the NGF now being produced by the cells that Tuszynski injected into the brain tissue will stay in place better, eliminating the side effects. (We found that sensitivity to NGF was greatly reduced in the NBM of Alzheimer's patients. Whether this will prove problematic isn't yet clear.) The next step that Tuszynski will take is to inject NGF directly into the brain with the aid of another virus, which may prove a more effective technique.

In late 2009 it was reported that in France gene therapy had been used to cure two boys of the fatal brain disease adrenoleukodystrophy (ALD). People with this rare hereditary condition lack the ALD protein that breaks down fatty acids. The latter build up in the myelin sheath, the protective layer that coats nerve fibers in the brain. As a result the nerves lose function, causing progressive physical and mental disability. The disease was brought to international attention by the movie *Lorenzo's Oil,* in which the father of a boy with ALD tries to cure him with a mixture of oils (a method that ultimately proves unsuccessful). In the French study, an intact ALD gene was inserted into stem cells taken from the boys' bone marrow using a lentivirus (a stable virus form) as a carrier molecule, after which the modified cells were replaced in the bone marrow. Exactly how the engineered cells prevent the defects in the brain is unclear, but the two seven-year-old boys in question have been doing well for two years now.

Many laboratories are now working on gene therapy for a wide range of diseases. In our laboratory, Joost Verhaagen is using it to repair damage to adult spinal cords. The day when patients can be cured of spinal cord injuries and brain infarcts is still far away, but the first favorable results with laboratory animals already show the potential effectiveness of such therapy. Experiments are being carried out to repair damaged nerve fibers by implanting cells engineered to produce growth factor at the site of spinal cord injuries. At the same time, proteins that block the regrowth of nerve fibers in the damaged spinal cord are inhibited. New advances have been made in the latter area: Promising animal experiments have prompted Martin Schwab of Zurich to set up a clinical study using antibodies to neutralize a protein that inhibits such regrowth in recent spinal cord injuries.

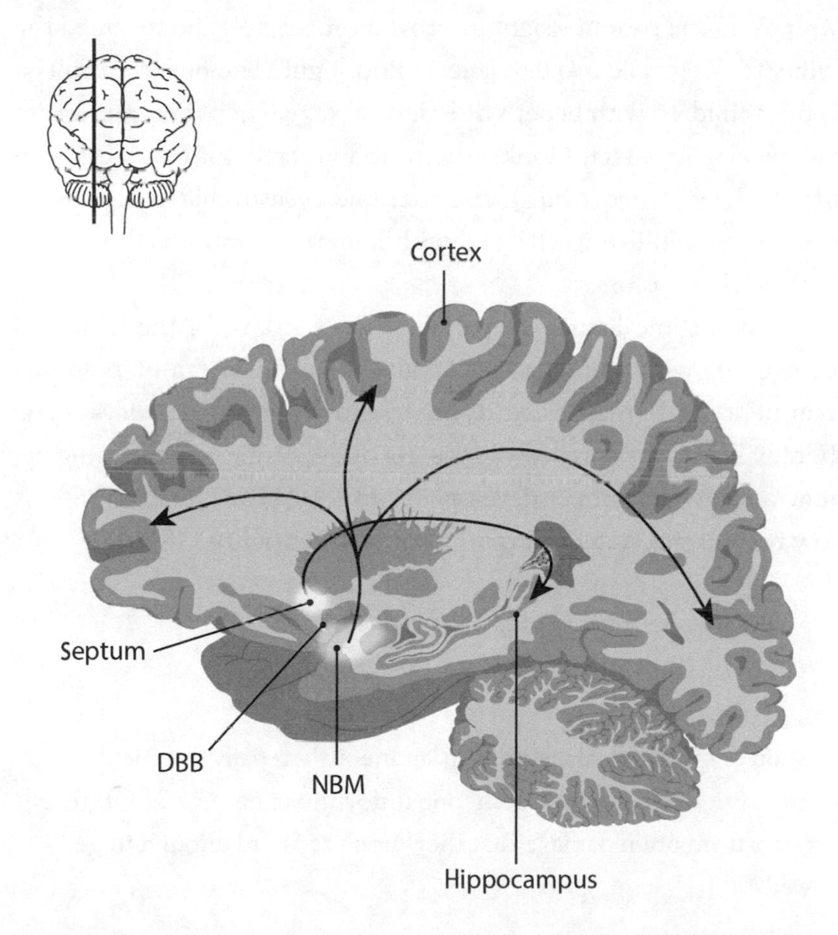

FIGURE 25. The basal nuclei–the nucleus basalis of Meynert (NBM), the diagonal band of Broca (DBB), and the septum–are the source of the chemical messenger acetylcholine in the cortex and hippocampus. This chemical messenger is important for memory (see also fig. 33).

Gene therapy for disorders of the nervous system is most advanced in the field of ophthalmology. Children with Leber's disease—a hereditary condition caused by a genetic mutation—are born with poor sight and go completely blind in adulthood. Experimental gene therapy on dogs proved effective against this disease. A phase 1 study was then carried out on three young adult patients whose retinas were seriously affected to determine whether treatment involving insertion of a piece of DNA coding for the missing protein was safe. The new therapy caused no serious side effects. Moreover, one patient's sight improved remarkably; he regained the ability to detect and avoid objects in poor light. The next step will be to treat children with Leber's disease at a stage when their retinas are still reasonably intact. Monkeys with red-green color blindness have already been cured using gene therapy. Measurable results were achieved within five weeks, and within eighteen months they could distinguish all colors.

The first clinical studies involving gene therapy for the blind and for patients with dementia herald an entirely new era of potential treatment for human brain disorders. In its very early days, gene therapy sometimes proved shockingly disappointing. One young patient died, while others developed leukemia. But gene therapy is now reemerging as a promising form of treatment.

SPONTANEOUS REPAIR OF BRAIN DAMAGE

> Brain damage can sometimes spontaneously repair to some degree. But don't reproach someone unfortunate enough not to recover from brain damage that they didn't try hard enough to get well!

It was previously thought that brain tissue, once lost, could never be regenerated and that the functional improvements that can occur after a stroke are simply due to reduced swelling and, to a limited extent, to functions being taken over by other regions of the brain. People who suffer trauma and go into a coma can come out of it within a matter of days or weeks or progress to a vegetative state also known as coma vigil, in which they are awake without being conscious. Coma vigil can herald improvement, but some patients remain in this state permanently without making any further progress (see chapter 7). After three months in a vegetative state a patient is thought to have no chance of recovery. Yet people occasionally do come out of a vegetative state after a very long period. An exceptional case is that of Terry Wallis, who went into a coma after a car accident and then progressed to a minimally conscious state from which he awoke nineteen years later. He had occasionally responded to external stimuli by nodding and grunting, but he couldn't communicate his thoughts and feelings. Yet nine years after the accident he started to speak the occasional word, and after nineteen years he fully regained the power of speech. He was also able to count and to move his limbs. However, he remained severely handicapped, unable to walk or to feed himself. He was also unaware of the passage of time; he didn't know that his daughter—a baby at the time of the accident—had become a stripper or that his wife had had three children by another man. You might wonder whether this constitutes a worthwhile existence and whether Terry, described by the media as "a modern Lazarus," is himself happy about this "miracle," but his case is truly special from a medical point of view. His recovery is attributed to the fact that new axons—nerve fibers that create connections between different brain regions—formed in his brain. Over a period of eighteen months, MRI scans showed an increase in the volume of nerve fibers at the rear of the cortex as well as in those connecting the various cortical regions. Increased activity was also detected in a part of the parietal lobe called the precuneus, which is important to our consciousness of our surroundings and of ourselves. This area, which doesn't function in patients in a vegetative state or coma or suffering from dementia—or during sleep, as a matter of fact—remains active when the brain is in a minimally conscious state. It was during the period when these activity changes were registered that Wallis regained consciousness. An increase was subsequently measured in the fibers of his cerebellum, at a time when his motor functions had improved strikingly. What distinguishes the few patients who are able to emerge from a vegetative state or a minimally conscious state after such a long period is still a mystery. But their existence does overturn the long-held view that recovery is impossible in such cases.

The story of Jill Bolte Taylor caused a media sensation. She was a brain researcher at Harvard who, at the age of thirty-seven, suffered a massive stroke in her sleep. She woke up with a pounding pain behind her left eye, and when her left arm became paralyzed she realized what was happening. With huge difficulty she telephoned a colleague to try to ask for help. She thought she was speaking clearly but in fact could only utter unintelligible sounds. Luckily the colleague, realizing that something was terribly wrong, called 911. Trying to communicate at a moment like that must be nightmarish. A doctor friend of mine realized that he was having a stroke and called his GP. The GP listened briefly to the incomprehensible sounds at the other end of the line, decided that it must be a prank caller, and promptly hung up. At that moment my friend's wife came home with the groceries, and he called out to her. Annoyed, she yelled back, "How often do I have to tell you to wait until I'm in the same room before you talk to me! I can't hear a word you're saying out here!" and went on unpacking the groceries. Luckily he recovered spontaneously from the damage caused by his stroke and has fully regained the power of speech. The situation in which Jill Bolte Taylor found herself was very different. Two and a half weeks after the cerebral hemorrhage, surgeons removed a golf ball–sized blood clot from her brain. She was left unable to walk, talk, read, or write and could remember nothing of her former life. With her mother's help, she gradually learned how to function again. It took her eight years to recover fully. She has written a bestseller about that period in which she describes how she used her willpower and her knowledge of the anatomy of the brain to consciously stimulate the damaged brain circuits and get them working again. This is pseudoscientific mumbo jumbo, but it proved immensely popular with the general public. "I truly believe that as a patient you're responsible for your own recovery," she said with great conviction. Of course it's important to do your utmost to regain health after a cerebral hemorrhage or stroke. But the danger of Taylor's enthusiastic but unscientific pronouncements is that they can be used to reproach those unfortunate patients who don't recover from such injuries for failing to work sufficiently hard on their recovery. When I first began studying to be a doctor, my father placed the power of medicine in perspective. "There are two kinds of maladies," he said. "One kind goes away by itself, and the other kind you can't do anything about anyway."

12

The Brain and Sports

NEUROPORNOGRAPHY: BOXING

> In various civilized countries, this form of inflicting deliberate neurological damage on one another has been banned for decades.

Witnessing aggression sparks aggression. Measures have rightly been taken to restrict excessively violent computer games. So it doesn't make sense that certain forms of primitive aggression, like boxing, are still permitted. You can watch boxers inflicting permanent brain damage on one another on prime-time television, yet no one seems to get very upset about it. As the audience howls encouragement in the background, you see (repeatedly and in close-up) detailed footage of the onset of neurological damage: an unsteady gait, impaired speech, eyes flicking left and right, an occasional classic epileptic fit, reduced consciousness after being knocked down, unconsciousness after being knocked out, and occasionally coma and death. It's more or less a complete course in neurology, in fact. Since the Second World War, around four hundred boxers have died from injuries incurred under the supervision of the various boxing unions. It's incredible that the most revolting examples of this neuropornography are shown on television, even at times when young children might still be watching.

In boxing, long-term brain damage from repeated blows to the head is much more common than acute damage. In 1928, the term *punch-drunk* was coined for boxers who stood unsteadily, moved slowly, and developed behavioral disorders, varying degrees of dementia, or Parkinson's disease. This was replaced by *dementia pugilistica,* and now the neutral term *chronic traumatic brain damage* is used to describe the condition suffered by 40 to 80 percent of professional boxers. Around 17 percent of professional boxers have Parkinson's. Muhammad Ali, formerly the world's greatest boxer and fastest talker, has become a shuffling Parkinson's patient with a mask-like face who struggles to form a sentence.

If boxing is character-forming, character can't be located in the brain, because there the only observable effect is deterioration. Many brain areas shrink as cells are lost, fibers are torn and lose their protecting myelin sheath, and the typical changes of Alzheimer's and Parkinson's can be seen under the microscope. If boxers die suddenly, it's usually because of cerebral hemorrhage. When someone is knocked out, their brain is slammed into the hole under the skull. This compresses the medulla, the lower half of the brain stem, which regulates vital functions like breathing, temperature regulation, and heart rhythm, with potentially devastating effect. Blows also destroy the hypothalamus and the pituitary gland, causing hormonal deficiencies in 50 percent of boxers. Their sense of smell is also impaired. Despite the use of protective helmets, one in eight amateur boxing matches leads to concussion. The current debate as to whether the brain damage incurred by boxers who are more genetically susceptible to brain trauma should be monitored more closely using psychometric tests is incomprehensible. By the time changes show up, it's too late. Professional boxing has already been banned for decades in Sweden, Norway, Iceland, North Korea, and Cuba. (Since 2001 Norway has in fact banned all martial sports that can lead to concussion, including the popular and spectacular K-1 fights, a form of kick-boxing.) In other countries, doctors have urged a ban on boxing. When you try this in my country, people counter by saying that boxers engage in this activity of their own free will, forgetting that the Netherlands outlawed dueling (in which people also partook of their own free will) centuries ago. You might wonder whether boxers are already showing the first signs of dementia when they elect to take up this barbarian form of "sport." But that's yet another argument in favor of protecting them from themselves and finally banning this embarrassing remnant of our primitive evolutionary past.

SEX AND THE OLYMPIC GAMES

> The collective sex tests for the Olympic Games have caused a great deal of unnecessary misery.

In 1912, Baron Pierre de Coubertin, the founder of the International Olympic Committee (IOC), opposed the inclusion of women in the Games on the grounds that their participation would be "impractical, uninteresting, unaesthetic, and incorrect." When women were allowed to take part, a distinction had to be made between the two sexes because of the biological advantage that testosterone gave men in terms of height and muscle power. At the time of the ancient Greeks, the distinction between men and women was easily made. Sports were performed naked, and if you didn't have a penis they didn't let you join in. But chromosomal gender, internal and external gender, and gender identity (feeling male or female) aren't always identical, and these discrepancies are sometimes linked to differences in testosterone levels. Women with too much testosterone can pose a threat to other sportswomen. Dora (Heinrich) Ratjen, a man who had been persuaded by the Nazis to pose as a woman high jumper, competed in the 1936 Olympics, though he came in only fourth. (The deception wasn't revealed until the 1950s.) It was at those

Olympics that the first serious gender controversy arose, when the American gold medal winner of the hundred-meter sprint, Helen Stephens, was accused of being a man, but turned out on inspection to be a woman. Ironically enough, Stella Walsh, the gold medal winner of 1930 whom Stephens beat in 1936, turned out, when she was murdered during a robbery, to be of indeterminate gender. In 1967, a number of Soviet athletes who had been asked to strip in front of a panel of gynecologists failed to turn up for the inspection. It was assumed that they had too much testosterone, either because of a disorder or injections.

The aim of testing for chromosomal gender is to prevent unfair competition in sports. Yet it hasn't led to any deserved bans, only to personal misery. The test involves screening cells scraped from the inside of a cheek. If someone has two X chromosomes (making them genetically female) a dark spot called a Barr body shows up inside the cell's nucleus. The first athlete to fail this test was the Polish sprinter Ewa Kłobukowska, who was banned from further participation and ordered to return her Olympic medals (Tokyo 1964). She turned out to have a deviant chromosome pattern of which she'd been unaware, and the affair caused her to become clinically depressed. This test has also been used—quite wrongly—to ban athletes with androgen insensitivity syndrome, like Maria Patino. People with this syndrome have a variation in the receptor for testosterone, which prevents testosterone from affecting their body or brain. Genetically male (XY) individuals with this syndrome therefore develop into heterosexual women. Despite having internal testicles (in the abdominal cavity), they don't possess an unfair advantage in sports. Quite the reverse, in fact, because they lack the effect produced on normal women by testosterone from the ovaries and adrenal glands. Paradoxically, girls with a mild form of congenital adrenal hyperplasia weren't banned as a result of chromosomal testing, even though their higher testosterone levels can generate more muscle tissue. SRY, a new genetic test introduced in the 1990s, didn't improve the situation. Patino was initially felled by the verdict from that test but subsequently fought back and in 1988 was the first sportswoman to be reinstated. It's not known what test was used in 1950 to impose a lifetime ban on the runner Foekje Dillema, who, according to recent research by Anton Grootegoed, may well have been a woman but also possessed some testicular tissue as a result of a rare chromosomal abnormality. Whatever the case, it meant that top runner Fanny Blankers-Koen lost her main rival, and there are rumors that she or her husband had a hand in initiating a sex test by the Dutch National Athletics Federation. Foekje, too, was ultimately reinstated, albeit posthumously.

There was an enormous fuss when some male-to-female transsexuals took part in competitions as women—as if one would undergo the protracted misery of a sex change just to win a medal. However, Renée Richards, an MtF transsexual, won a court case in the United States that allowed her to compete in women's tennis. Since 2004, following research by Louis Gooren, a Dutch professor of transsexology, MtF transsexuals have been allowed to take part in sports two years after gender-reassignment surgery if their hormone levels have "normal" values and their sex change has been formalized. The Canadian transsexual cyclist Kristen Worle sought to participate officially in the Beijing Olympics but, alas, failed to qualify.

In 1999 it was decided to abolish the collective sex tests for the Olympic Games. A team of specialists would instead be on constant standby to investigate any problems professionally. This was a distinct improvement on applying a simple but fallible test to a very complex problem. As of May 1, 2011, the International Association of Athletics Federations (IAAF) has opted for the most logical and simple approach. Investigations are now confined to the level of testosterone in the blood. If women's testosterone levels are lower than would be normal in men, they may participate in women's events. An exception is made for individuals with androgen insensitivity syndrome like Maria Patino. At long last a logical solution has been found that doesn't entail someone suddenly being told that they are not a woman but a man. And it's a good solution, because it's all about the effect of testosterone on our muscles. But it will be interesting to see how the IAAF deals with the boundaries of normal values.

DEATH OF THE FITTEST

> How did people ever get the idea that exercise—apart from the mental kind—is healthy?

Over the last hundred years, the average life expectancy has risen from forty-five to nearly eighty years, even though we have simultaneously become much less physically active. You might happily conclude from this that it pays to be lazy. But strangely enough, no one does. There are few things people agree on in this world, but it is a truth universally acknowledged that we don't get enough exercise these days. As a result, you can no longer enjoy a pleasant stroll through the woods without being overtaken by panting and sweating joggers, most of whom look as though they're in agony. Any self-respecting company sponsors athletes; marathons are organized for children with cancer. Amsterdam's Academic Medical Center, which should know better, organizes an annual charity run. At 6:45 A.M. an exercise initiative aimed at Dutch senior citizens bombards the nation with images of elderly people leaping about in leotards.

Where on earth did people get the idea that sports are good for you? Certainly not by working in an emergency room on Sunday mornings, as I used to do. Early January 2009 was a case in point. The Dutch canals froze, whereupon half the people of Holland dug out their skates and hit the ice—in many cases literally. Hospitals were struggling to cope with the ten thousand extra patients being admitted with broken bones, hypothermia, and the like. Hardly a contribution to public health, any more than the special flights made every year to transport Dutch skiers with broken legs back from winter sports regions. There are 1.5 million sports injuries a year in the Netherlands, half of which require medical treatment. If sports were banned, all our waiting lists would disappear overnight. We already know that boxers can end up with lasting brain damage (see earlier in this chapter). The risk run by kickboxers turns out to be ten times as great. Soccer players lose brain cells, too, what with all that heading and the odd elbow in the face. Long-distance runners have been dropping dead since the very first marathon in Greece. Some 15 percent of paraplegics incurred their injuries from sports. The American actor Christopher Reeve, famous for playing the role of Superman, was paralyzed for life after breaking his neck in a fall from his horse.

A compulsion to exercise can also be a sign of disease; it is, for instance, a typical symptom of anorexia nervosa (see chapter 5). People with anorexia often work out obsessively. Decades ago, long before jogging was fashionable, the neurologist Frans Stam was looking out of his window at Valeriusplein, a square in Amsterdam. To his amazement he saw someone emerge from one of the houses opposite his, run around the square several times at high speed, and then go back in again. This process was sometimes repeated several times a day. A few months later, the man in question was hospitalized; he proved to have Pick's disease, a form of dementia in which the prefrontal cortex atrophies, which often first manifests itself in behavioral disorders. Since Stam told me that story, I've always been a bit wary of joggers. No one seems to worry very much about the increased risk that athletes have of contracting ALS, a form of motor neuron disease, or about the fact that up to 150 people a year in the Netherlands suddenly drop dead on playing fields or in gyms. Bodybuilders happily inject themselves with anabolic steroids and in the past took growth hormone preparations made from human pituitaries that sometimes turned out to be infected with Creutzfeldt-Jakob disease, a type of dementia that progresses very rapidly. It does indeed seem, as the Dutch magazine *Vrij Nederland* once jokingly claimed, that half the population takes part in sports, while the other half drives them to the hospital.

You might argue that these are just some of the minor risks of a lifestyle that on the whole makes people long-lived and healthy. But you'd be on shaky ground. The studies and statistics that allegedly support the benefits of exercise aren't based on properly controlled trials but on comparisons of groups of people who themselves opted to take up a sport (or not). This self-selection makes it impossible to draw any valid conclusions. Conversely, as far back as 1924, Raymond Pearl found that extreme physical exertion actually shortened lifespan. This appears to apply to the whole animal kingdom. Comparative research by Michel Hofman at the Brain Institute shows that two factors determine our lifespan: metabolism and brain size. The higher the metabolism, the shorter the lifespan. This ties in with the finding that top athletes at Harvard have shorter lifespans than their non-athletic classmates. So the enormous physical effort involved in exercise might even have a life-shortening effect. The American researcher Rajinder Sohal found that the more flight movements a fly makes, the sooner it dies. If you prevent flies from wasting energy by confining them within two plastic plates, so that they can't fly, they live up to three times longer. A single organ, the brain, also affects length of life: The larger and more active it is, the longer your lifespan. Conversely, brains that are too small, for instance because of microcephaly or Down syndrome, shorten lifespan. Stimulating the brain also appears to delay the onset of Alzheimer's and can mitigate the symptoms of this disease (see chapter 18). Eminent scientists are said to have larger brains and to live longer. You can increase brain size by providing stimuli in the form of new and constantly varying information, for instance by giving children an enriched environment (see chapter 1). So it would appear to be healthier to watch sports than to take part in them. And if you're really bent on taking up a sport, how about chess?

13

Moral Behavior

PREFRONTAL CORTEX: INITIATIVE, PLANNING, SPEECH, PERSONALITY, AND MORAL BEHAVIOR

The functions of the prefrontal cortex came to light as a result of brain damage and disorders.

As with many human brain structures, the functions of the prefrontal cortex (fig. 15) came to light only as a result of accidents, operations, and neural disorders. In 1848, a railroad construction foreman named Phineas Gage was stuffing blasting powder, fuse, and sand into holes with an iron rod when an explosion blew the thirteen-pound rod right through his head, causing half a teacupful of his brain to be spilt on the ground. Amazingly, Gage not only survived the accident, he remained fully conscious in the aftermath of his terrible injury. But he underwent a marked personality change. Previously hardworking and responsible, he became fitful, capricious, aggressive, and foulmouthed, and he ended up losing his job. As his friends put it, "Gage was no longer Gage." Indeed, one of the functions of the prefrontal cortex, which had been damaged in Gage's accident, is to ensure that we conform to social norms.

Another one of its functions was discovered shortly afterward, when Paul Broca, a physician working at a hospital in Paris, carried out an autopsy of a patient who had been nicknamed "Tan," because "tan" was the only distinguishable sound he could make. His inability to speak proved to be caused by an infarction in the frontal lobe of his left cerebral hemisphere (fig. 8). This region of the brain, which is responsible, among other things, for the production of grammatically correct sentences, is now known as Broca's area. Damage to this region after a stroke leaves patients with aphasia, that is, speech disorders.

A hundred years after Gage's accident, during the "heyday" of psychosurgery, lobotomies were performed to destroy the prefrontal cortex deliberately. Misinterpreted findings in animal studies suggested that lobotomies could help schizophrenia sufferers or violently aggressive individuals. The film *One Flew over the Cuckoo's Nest* memorably shows how the operation transforms a rebellious and difficult patient into an apathetic zombie who just sits in a chair, gazing vacantly into space. Of course, he is now much more manageable from a clinician's point of view—and that was deemed important when considering a lobotomy. Eventually, doubts were increasingly voiced about whether lobotomies were an effective treatment for aggressive behavior. It may be that the medical world was alarmed by the fate of the Nobel Prize–winning surgeon who devised the procedure, António Egas Moniz, who was shot—by a disaffected patient, it was claimed—and spent his remaining years in a wheelchair.

In 1951, 18,608 lobotomies were performed in the United States, mainly on schizophrenia patients. The vogue was started by Walter Jackson Freeman, nicknamed "Jack the Brainslasher," who enthusiastically performed the operation all over the United States, touring the country in a van that he called his "lobotomobile." After stunning a patient with an electric shock, he would hammer an ice pick through the back of the eye socket into the frontal lobe, thus destroying the connection with the other parts of the brain. The effect of the procedure wasn't at all understood at the time. Pope Pius XII announced that the Church did not oppose lobotomy "as long as free will is retained, even if there be some loss of personality"; another high-ranking Catholic added, "If the soul survives death, it can presumably survive a lobotomy." The operation was later rightly characterized as "partial euthanasia," because the patients' personalities were blunted, and they became completely apathetic. Doctors eventually stopped resorting to the operation—not from ethical considerations, but because it was rendered unnecessary by the emergence, in the mid-1950s, of psychoactive drugs. The damage caused to so many individuals by lobotomies was never properly documented, though it became abundantly clear from the operations that the prefrontal cortex played a crucial role in expressing personality and taking initiative.

In his book *The River That Flows Uphill,* William H. Calvin tells the following anecdote:

> The famous Montreal neurosurgeon Wilder Penfield had a sister . . . who was one of those cooks who could spend four hours preparing a five-course meal and have everything turn out just right. Nothing got cold or overcooked, because it was always ready to come off the burner or out of the oven just when it was needed. Now that's truly a precision-timing scenario. But Penfield's sister began to lose this ability. Over the course of several years, the holiday family dinners began to distress her because she couldn't get properly organized as she had used to. For ordinary dinners she was still a good cook. Most physicians wouldn't have picked up on such subtle clues. But Penfield's clinical instincts told him that she might have a frontal lobe tumour. She did. He operated. She recovered.

The removal of much of the right prefrontal cortex along with the tumor, however, did nothing to restore her ability to organize. Fifteen months after the operation she prepared a dinner for five, which failed disastrously due to her inability to take initiative and make choices, typical functions of the prefrontal cortex.

The prefrontal cortex also ensures that we conform to social norms. In Pick's disease, a rare neurodegenerative disease, and other forms of frontotemporal dementia, the prefrontal cortex becomes severely damaged. These conditions cause the prefrontal cortex tissues to shrink until they resemble a walnut in its shell (fig. 30). The early stages of Pick's disease aren't so much marked by memory failure as by behavioral disorders, general dementia only setting in many years later. A professor who developed Pick's disease suffered from a loss of decorum in its early stages, famously urinating against a salon piano. Interestingly, microscopic postmortem analysis at the Netherlands Brain Bank of the brains of patients who had been diagnosed with Pick's disease revealed an absence of "Pick bodies" in the vast majority of cases. Pick bodies are the disease's defining feature, taking the form of protein tangles that appear as large bodies in neurons. In the last decade it was discovered that some of the patients thought to have Pick's disease actually had chromosome 17 frontotemporal dementia caused by tau gene mutations. During the early stages of the disease, they also exhibited behavioral deviations, such as disturbed social behavior, hyper- or hyposexuality, alcoholism, aggressive behavior, depression, and schizophrenic tendencies. New forms of dementia are still being identified.

And so it was, through damage, disease, and a process of backward reasoning, that the functions of the prefrontal cortex gradually came to light.

MORAL BEHAVIOR: THE HUMAN IN THE ANIMAL

> My object in this chapter is to shew that there is no fundamental difference between man and the higher mammals in their mental faculties.
>
> Charles Darwin, *The Descent of Man*

Adherents of the Intelligent Design movement believe that morality has no biological basis but is given to man through God's grace and that believers were at the front of the line when it was handed out. In *Schitterend ongeluk, of Sporen van ontwerp?* (Glorious Accident, or Traces of Design?) a book on Intelligent Design edited by the nanotechnologist Cees Dekker and others (2005), Henk Jochemsen, a molecular biologist and professor of neo-Calvinist philosophy, stated, "From the point of view of sociobiology and evolutionary ethics, truly altruistic behavior is biologically perverse and pathological, because it conflicts with human nature. Yet most cultures and great religions designate true altruism as a higher ideal."

Anyone who is familiar with the writings of Darwin or the Dutch primatologist Frans de Waal knows that Jochemsen's claim is nonsense. Over a century ago, Darwin was able to describe in detail how our moral awareness developed from social instincts that are important for the survival of the group. This type of behavior can be observed in all species whose members need to work together, like primates, elephants, and wolves.

Empathy—the capacity to recognize and share the feelings of others—provides the basis for all moral behavior. I remember being struck by the way in which our dog showed empathy for her playmate, our daughter's dog, after the latter's paw had been operated on. The two normally frolicked around wildly with one another, but after the operation our dog approached her friend very cautiously, sniffing as she did so, and remained motionless for a very long time, gazing at him and occasionally whimpering quietly to show her sympathy. She then went up to my daughter's dog and began very carefully to lick the paw that had been operated on. Similarly, when an elephant is hit by a bullet or a tranquilizer dart, other elephants trumpet loudly and try to help the victim back on his feet by pulling at him with their trunks or pushing him with their bodies, sometimes for hours on end.

Many moving instances have been described of truly moral behavior among animals. In one zoo, an old, sick bonobo ape was placed with a group of fellow apes. Other bonobos saw how confused he was by his new surroundings, took him by the hand, and led him to where he was supposed to be. When he then got lost and cried out in alarm, the others went to him, calmed him down, and brought him back to the group. You'd be lucky to encounter such civilized behavior on the streets of Amsterdam!

The fact that primates possess moral awareness is also evident from the empathy that they display toward other species. A bonobo's attempts to comfort an injured bird can be ascribed only to pure empathy. In 1996, a female gorilla named Binti Jua rescued a three-year-old boy who had fallen eighteen feet into the primate enclosure at a zoo near Chicago. Other species, too, can sacrifice themselves to save humans. A labrador in California gave new meaning to the phrase "man's best friend" when he jumped in front of his master to intercept a rattlesnake's bite. Dolphins are known to come to the aid of swimmers in addition to helping other dolphins caught in nets. Empathizing with and helping others may lie at the heart of human morality, but such behavior has a long evolutionary history and certainly isn't exclusive to humans.

These few examples alone show that Cees Dekker, a radical Christian who was formerly the most vocal proponent of Intelligent Design in the Netherlands, got it completely wrong when he claimed morality as an exclusively Christian trait. In an interview with the Dutch daily newspaper *De Volkskrant* (March 4, 2006), he stated, "Jesus said love God with all your heart, and love your neighbor as yourself. That's a moral precept, a law that's hard to fathom and that doesn't lend itself to investigation by scientific methods. Yet man can distinguish between good and evil." It would seem that adherents of Intelligent Design don't read the writings of those they criticize. As a result, they aren't forced to conclude that religion didn't invent moral precepts but simply took them over after behavior of this kind had evolved among social animals, including humans.

UNCONSCIOUS MORAL BEHAVIOR

> The greatest tragedy in mankind's entire history may be the hijacking of morality by religion.
>
> Arthur C. Clarke

Moral precepts serve to promote cooperation and support within social groups; they also act as a social contract, imposing restraints on the individual to benefit the community at large. As might be expected, Darwin's theory of moral psychology (1859) traced the emergence of ethical behavior not to selfish competition between individuals but to social solidarity within the group. As they evolved, humans developed altruistic behavior, based on the loving care shown by parents toward their offspring. Altruism was then extended to others of the same species according to the principle "Do as you would be done by." Such behavior, the product of millions of years of evolution, was ultimately made the cornerstone of human morality, a code of beliefs that was only recently incorporated into religions, a mere couple of thousand years ago. It's perhaps worth interjecting here (a touch cynically) that of all the stimuli that bind communities together, having a common enemy is the most powerful of all—a mechanism that many world leaders have exploited.

Inherent in the biological objective of morality—promoting cooperation—is that members of one's own group receive preferential treatment. Loyalty to the nuclear and extended families comes first, followed by loyalty to the community. Once your own survival and the health of those closest to you are assured, you can extend your circle of loyalty. "First grub, then morals," as Bertolt Brecht put it. These days we're doing so well that the circle of loyalty has been extended to the EU, the West, the Third World, animal welfare, and, since the Geneva Conventions of 1949, even to our enemies. However, the need for this approach was felt as far back as the third century B.C., when the Chinese philosopher Mo Zi, seeing the destruction caused by war, sighed, "When will there be a path to universal love and mutual benefit? When no one claims other countries as their own."

Although tests show no significant difference in the moral choices made by atheists and believers, the Intelligent Design movement claims that moral behavior is unique to man and derives from religion, especially Christianity. In his contribution to Cees Dekker's book on Intelligent Design (2005), Jitse van der Meer, who teaches biology at a Christian university, writes, "Humans are the only primates who are capable of moral thought." However, primatologist Frans de Waal has shown that people usually don't think at all about moral acts. Instead they act quickly and instinctively, on a biological impulse. It's only afterward that they think up reasons for what they did unconsciously in a flash.

Our moral values have evolved over the course of millions of years and are based on universal values of which we're unconscious. Moral behavior manifests itself at a very early stage of development. Since such behavior is also displayed by animals, it would seem to be hardwired. Before they are old enough to have acquired speech or reflect on moral issues, young children show an instinct to comfort family members who are in pain, just as primates comfort each other. In an experiment in which adults pretended to be sad, children aged between one and two responded by trying to comfort them. And this behavior wasn't confined to children. In the same experiment, pets also displayed a strong instinct to comfort. Just like eighteen-month-old human infants, chimpanzees can display altruistic behavior without the incentive of a short- or long-term reward. They will hand another chimpanzee a stick or a child a pencil simply because it's out of the other's reach. They will, moreover, do this repeatedly, again without the prospect of any reward. So the roots of our altruism go back a very long way. There's no foundation for Van der Meer's claim that "good behavior has no biological cause but must be inculcated, because it is not innate." Furthermore, it's inconceivable that the remarkable primate research carried out by

Frans de Waal and his team on the biological basis of social behavior should be dismissed by Henk Jochemsen in the above book as the "shriveling of life sciences and social sciences to biological specializations." It wouldn't hurt to put your unfounded ideas into perspective, adherents of Intelligent Design!

MORAL NETWORKS

> Moral decisions aren't just made in the prefrontal cortex; many other regions of the brain are involved.

We have a moral network in our brains made up of neurobiological building blocks that have evolved over time. First, brain cells known as mirror neurons come into play when we observe the actions of others. When you see someone move a hand, the same neurons fire in your brain as when you make that movement yourself. Mirror neurons help us to learn by imitation—a process that's largely automatic. Newborn babies can copy the mouth movements of adults before they are an hour old. The same neurons react to displays of emotion, enabling us to sense what others are experiencing and thus providing the basis for empathy. Mirror neurons have been discovered in the prefrontal cortex (PFC, fig. 15) and in other regions of the cerebral cortex.

Our large PFC contains important components of our moral network; it ensures that perceived emotions are linked to moral concepts, and it reacts to social signals and inhibits impulsive, selfish reactions. It's also crucial in deciding whether something is a fair deal. The importance of the PFC for our moral awareness is evident from studies of damage to that area from tumors, gunshot wounds, and other injuries, which has been shown to cause delinquent, psychopathic, and immoral behavior. A judge in the United States suffered damage to his PFC from grenade fragments. He completely lost the ability to relate to the defendants in court. Sadly (but perhaps fortunately for the local criminal community) he felt obliged to quit his job.

Damage to the PFC early in life impairs the ability to grasp moral concepts and can lead to psychopathic behavior. Men accused of murder often display malfunctions of the PFC. The first signs of frontotemporal dementia, a disorder that starts in the PFC, often take the form of antisocial, delinquent behavior, including sexual harassment, assault, robbery, burglary, hit-and-run crimes, and pedophilia. It usually takes a while for behavioral aberrations of this kind to be identified as the onset of such dementia. The PFC plays a central role when we face moral dilemmas, like whether to sacrifice the life of a single individual to save many lives. Most people find such decisions impossibly difficult, but individuals with damage to the PFC approach them very cold-bloodedly, being much more dispassionate and impersonal in their reasoning.

Besides the PFC, other cortical and subcortical areas of the brain play an important role in moral functioning, including the foremost part of the temporal lobe and the almond-shaped structure within it (the amygdala), the septum (the structure separating the ventricles of the brain, fig. 26), the reward circuitry (the ventral tegmental area/nucleus accumbens, fig. 16), and the hypothalamus (fig. 18) in the base of the brain. All of these areas are essential for the motivation and emotions that underlie moral behavior. The amygdala is also involved in calculating the social significance of facial expressions and the appropriate response. Along with abnormalities of the temporal lobe, malfunctions of the amygdala have been found in the brains of murderers and psychopaths, explaining why the latter are less responsive when their victims express grief and fear. Humans instinctively dislike acts that are inspired by evil motives. Malicious acts are condemned and in a criminal context result in stiffer sentences. But if you disable someone's moral circuitry (using transcranial magnetic stimulation to disrupt the right temporoparietal junction—the place where the temporal and parietal lobes meet) he no longer cares whether a person's motives are moral or not, since he can no longer understand the motivation behind an action.

So our moral networks aren't just confined to the neocortex, a comparatively new area of the brain. Much more ancient regions are equally crucial for our moral functioning. Typically moral emotions like guilt, pity, empathy, shame, pride, contempt, and gratitude, as well as disgust, respect, indignation, and anger, depend on interaction between the above-mentioned areas. Functional brain scans performed during tests in which people were confronted with terrible moral dilemmas, like suffocating a crying baby in order to save a number of lives, have revealed changes in patterns of activity in brain regions that we know (from the effects of damage or tumors) to be linked with moral functioning.

But we mustn't forget, when reflecting on our finer moral impulses, that while empathy enables us to understand others and share their emotions, it also allows us to imagine what others would go through if we deliberately hurt or tortured them, and these feelings, too, can be indulged in.

WHAT NATURE TEACHES US ABOUT A BETTER SOCIETY

> Man is a chimpanzee with ideas above his station.
>
> Bert Keizer, *Alzheimer Opera*

Frans de Waal is a world-famous Dutch primatologist who has worked in the United States since 1981 and has recently published his ninth book, a fascinating work optimistically titled *The Age of Empathy: Nature's Lessons for a Kinder Society*. In it he again draws parallels between human and animal behavior. The message of the book is that the age of empathy has dawned. The misguided belief of the Thatcher/Reagan era, that a free market economy would be self-regulating, culminated, under George W. Bush, in the nightmare of the financial crisis. It's time to call corporate greed to a halt, to curtail the covetous culture of CEOs and bankers. "Greed is out, empathy is in," claims De Waal. Humans aren't just the most aggressive primates but also the most empathetic, as shown by the aid response in the wake of Hurricane Katrina in 2005 and the Chinese earthquake of 2008. It's all a question of balance, something that has been conspicuously lacking in recent years. According to De Waal, empathy—the ability to relate to the feelings of others—needs to regain the upper hand. Empathy has a long evolutionary history, stretching back over 200 million years of mammalian development—which should provide a solid foundation for such a change. You might wonder whether De Waal is guilty of wishful thinking, but since the G20 summit agreed to curb the bonus culture in 2009, it's beginning to look as though he might just be right. De Waal is a die-hard Darwinist who demonstrates that, as far as emotions are concerned, the entire spectrum of behavior is already to be found in the animal world. Like Darwin, De Waal writes clearly and appealingly, illustrating each crucial step of his arguments with fascinating examples taken from the entire animal kingdom. He also proves his theories through a great many ingenious experiments. De Waal has one advantage over Darwin, however: a much better sense of humor.

The origins of empathy lie in a mother animal's caring behavior toward her offspring. It's an automatic response that involves not just the PFC, a comparatively recent brain area, but also much more ancient regions of our brains. Almost all humans—with the exception of psychopaths—show empathy. There's no doubting the importance of competition, whether in the monkey world or in human society, but cooperation and the pleasant feeling of sharing fairly and helping others are just as crucial. If you reward two apes equally for the same task, all is sweetness and light, but if you suddenly give one of them a nice grape instead of the piece of cucumber that his fellow ape is still getting, then the underpaid ape will stop cooperating and hurl the cucumber out of his cage in protest.

More than in his previous books, De Waal cites examples from neuroscience to explain behavioral mechanisms. (Indeed, so many new discoveries are being made in neuroscience that his next book could profitably explore the integration of behavior and neurobiology.) He of course mentions mirror neurons, which react to the emotions of others and thus form the basis of empathy. He also looks at gender differences. Women, it turns out, empathize with cheaters when they are punished, while men do not feel any empathy in such situations. Quite the opposite: Their reward circuitry (fig. 16) is activated, demonstrating that justifiable punishment generates actual pleasure.

I'm not yet quite convinced, however, by De Waal's claim that von Economo neurons (VEN cells or spindle cells) underpin self-consciousness, as expressed in the ability to recognize oneself in a mirror. To possess any marked degree of empathy, an animal must be able to distinguish itself from its surroundings. This ability has been studied using mirrors in the "mark test." In the test, a paint mark is made on an animal's forehead, after which the animal is shown its reflection in a mirror. If it realizes that it is seeing itself, it will touch the mark or try to rub it off. Dogs and cats fail this test; children over two, primates, dolphins, and—as De Waal himself showed with a particularly large mirror—elephants pass it.

Acceptance that animals, too, have emotions, is incidentally a fairly recent phenomenon in the Western world. When the first chimpanzee and orangutan were exhibited in the London Zoo in 1835, Queen Victoria pronounced them "frightful, and painfully and disagreeably human." The young Charles Darwin, however, declared that anybody who thought that man was superior to the apes should go and take a good look. De Waal attributes our difficulty in recognizing that animals have feelings to the religions on which our culture is based. According to the Judeo-Christian doctrine, only humans have souls, and man is the only intelligent creature, having been created in God's image. I don't find his theory convincing. The Chinese have been equally unreceptive to the idea of animal emotions until very recently. But increasing prosperity in China is being matched by increasing interest in and empathy with animals. More and more Chinese people have pets, animal abuse meets with public outrage, and a gigantic memorial to the rhesus monkeys that were sacrificed in the name of SARS research can be found at Wuhan University.

When a journalist from a religious newspaper asked De Waal what he would change in humans if he were God, he reflected for some time, being suspicious of movements that have sought to impose social change by diktat, like social Darwinism, Marxism, and radical feminism. De Waal pointed out that both sides of human nature, that of the friendly, empathetic, and sexy bonobo and that of the aggressive, dominant chimpanzee, are necessary for a stable society. He concluded that he wouldn't want God to change humans fundamentally but simply to increase their sense of "brotherhood" by endowing them with more empathy. I personally doubt whether this would rid the world of problems. Indeed, De Waal furnishes the other side of the argument. If you're open to everyone and trust everyone—a trait shown by people with Williams syndrome—others will perceive your behavior as abnormal and shut you out. Moreover, empathy has its dark side. The reason humans are so good at torture is precisely because they excel at imagining what others feel. In fact, the more empathetic they are, the crueler they can be. De Waal cites the example of Nazi guards who carried out acts of unimaginable brutality in concentration camps, yet who were loving husbands and fathers in their off-duty hours. We can possess plenty of empathy and at the same time use it very selectively. The millions who revered Hitler, Stalin, and Mao were no less empathetic than we are. So De Waal might do well to ask God if he could also curb our tendency to take our lead from charismatic alpha males. Not only might this prevent future genocides and cultural revolutions, it might also reduce the likelihood of another disaster caused by corporate greed.

14

Memory

KANDEL'S RESEARCH INTO MEMORY AND THE COLLECTIVE AMNESIA OF THE AUSTRIANS

> Mental activity stimulates the development of nerve cells and their axons in the part of the brain being used. In this way, existing connections between groups of cells can be reinforced by an increase in the number of terminal arborizations.
>
> Santiago Ramón y Cajal

The only thing I recall of the international committee on which I sat twenty-five years ago is the infectious laugh of Eric Kandel. It certainly wasn't a laugh that sprang from a happy childhood. He was born in Vienna in 1929 as Erich Kandel and was given a beautiful blue remote-controlled car for his ninth birthday. Two days later, during Kristallnacht, the Jewish Kandel family was ordered to leave their house by two Nazi policemen. When they were allowed to return home a few days later, they found that it had been ransacked. Everything was gone, including the blue car. After waiting a year for a visa, the family emigrated to the United States, where Erich changed his name to Eric. He trained as a psychiatrist, undoubtedly influenced by the fact that his first girlfriend's parents were renowned psychoanalysts who had worked with Sigmund Freud. He was so fascinated by psychoanalysis that in 1955, while working under the famous electrophysiologist Harry Grundfest at Columbia University, he enthusiastically announced that he wanted to find the biological foundation of Freud's theory of the psyche. Freud divided personality into three parts: the id, the unconscious, primitive component that is driven by the pleasure principle; the ego, which tries to balance the id's desires with reality; and the superego, which acts as conscience and moral guide. Freud himself had never given any thought to where these hypothetical elements were located in the brain. Grundfest, who introduced Kandel to neuroscience, listened patiently to his unfeasible research plans and gave him the most important advice of his career: "If you want to understand the mind, you will have to study the brain cell by cell." It was this approach, which involved studying first the cellular, then the molecular biology of memory, that ultimately won Kandel a Nobel Prize in 2000. Kandel fascinatingly describes this journey in his autobiography, *In Search of Memory*.

Memory is defined as the capacity to store and retrieve information. It provides us with conscious access to our past. Kandel initially focused his research on the hippocampus, a part of the brain that is crucial to memory. But the hippocampus proved too complex, and he went in search of a simpler organism to study, ultimately opting for the giant marine snail *Aplysia*. Kandel said that he had relied on his instincts when making the decision, just as when he took the plunge and decided to marry his girlfriend Denise. (It's perhaps worth noting that he described *Aplysia* as "large, proud, attractive, and obviously highly intelligent.") In primitive organisms like *Aplysia,* the various aspects of memory are present in simple reflex responses, initiated by a few extremely large neurons that make a relatively small number of synaptic connections. The simplicity of this circuitry made it comparatively easy to study aspects of the way in which neurons learn. Kandel showed that connections between neurons varied, becoming stronger or weaker in response to electrical stimuli. In other words, the nervous system didn't consist of fixed connections like an old-fashioned telephone exchange; instead its links proved to be plastic. There are circuits, formed during development, in which innate behavioral patterns are fixed. But nervous systems also contain components that can change through learning.

Learning proved to hinge on changes in the strength of synaptic contacts. These became stronger as the neurons learned from repeated stimuli, proof of "practice making perfect." This is the basis of memory. The various forms of learning, memory, forgetting, and thinking—and thus, in a sense, our minds themselves—are the result of synaptic contacts in different brain areas being affected by the many different chemical messengers contained in neurons. *Aplysia* has both short- and long-term memory, the latter of which, just like its human variant, requires repeated training interspersed with periods of rest. In the case of short-term memory (for instance briefly remembering a telephone number so that we can punch it in) only the strength of the existing synapses changes. In other words, a functional change takes place. The capacity of short-term memory is very limited: In the case of humans it is limited to fewer than twelve words or numbers, and if the information isn't repeated it will be retained for only a few minutes. Long-term memory requires the synthesis of new proteins, because it involves forming new connections between neurons. This amounts to a structural change for which glia cells produce the essential fuel, that is, lactate. Long-term memory is sometimes compared to a computer hard disk in which information is permanently stored. Short-term memory is likened to the working memory or random-access memory (RAM) of a computer, in which information changes every second depending on the tasks and programs being run.

Early on, memory storage can be disrupted by concussion, oxygen shortage after a heart attack, or electroshock treatment for depression. Factors like these can cause retrograde amnesia, wherein a person forgets everything that happened during the preceding period. Since memory can gradually be restored in such cases, it seems that the problem is caused by disrupted access to information rather than its storage in the brain. After a period of years, stored information is less vulnerable to disruptions of this kind. Ultimately, long-term memory contains an individual's entire knowledge and experience of the world and themselves.

Learning causes structural changes to the brain, as Ramón y Cajal observed back in 1894 (see the epigraph to this section). In professional violinists, for instance, the part of the cerebral cortex that directs the fingers of the left hand is five times as large as in people who don't play a stringed instrument. When I see the speed with which young children send text messages, I get the impression that the thumb areas in their cerebral cortex are considerably bigger than mine.

Kandel also unraveled the molecular processes that take place when synaptic strength alters and new synapses form, setting up a whole new research field in the process: the molecular neurobiology of cognition. He discovered the molecular mechanisms whereby information is transferred, through practice, from the short- to the long-term memory. The hippocampus (fig. 26) plays an important role in this process. At the same time, Kandel showed how a highly emotionally charged event bypasses the short-term memory and is immediately stored in the long-term memory. In those cases the amygdala (fig. 26) is crucial. Having found the likely molecular basis for normal memory deterioration as a result of aging, Kandel set up a company called Memory Pharmaceuticals. Unfortunately it hasn't yet come up with the perfect learning pill.

Shortly before being awarded the Nobel Prize, the seventy-eight-year-old Kandel was in Amsterdam to receive the Heineken Prize for Medicine. He still had the same old infectious laugh; it rang out during the lunch. After traveling to Stockholm to collect his Nobel Prize, he returned to Vienna, where he organized a symposium on Austria's enthusiastic response to National Socialism. It was his way of denouncing the collective denial of the role played by his homeland during the Nazi period. The man who had become famous for his work on memory was shocked to discover that schoolchildren in Austria knew nothing about Hitler or the Holocaust. While in Vienna, he was given a blue toy car exactly like the one the Nazis had stolen from him as a child. He responded laconically, saying that he was later glad that he'd had to leave the car behind in Vienna: "I went to the United States, where I had an absolutely wonderful life. And now I've got a Mercedes."

THE ANATOMY OF MEMORY

If memory is localized anywhere, it's everywhere.

Contacts between neurons change in response to neural activity. That's how memory is encoded, and it's a characteristic of every neuron. So you could argue that memory is located throughout the nervous system. But some brain structures are preeminently concerned with memory. Functional scans can show whether a brain area is *involved* in certain functions, but data on patients with local brain damage remains crucial in revealing whether or not those areas are actually *necessary* for a particular function. Valuable information on the role of parts of the cerebral cortex in memory has, for instance, been obtained through systematic studies of patients with brain disorders, bullet wounds, and other types of brain damage and of patients undergoing operations on the brain. Before operating on patients, the American-born Canadian neurosurgeon Wilder Penfield (1891–1976) stimulated their temporal lobes (fig. 1) with an electrical probe while they were still conscious so that he could more accurately target surgery. This sparked extremely detailed memories; some of the patients sang entire songs while lying on the operating table.

The importance of the temporal lobe for memory became clear

in 1953, when the American surgeon William Scoville removed large portions of the temporal lobe from a patient famously known as H.M., a man who had developed severe epilepsy after a bicycle accident. The operation cured his epilepsy but caused profound amnesia. He was unable to learn or retain information, although his short-term memory was intact. He could, for instance, briefly remember the number seven by constantly repeating it to himself, but if he was interrupted he completely lost track of what he'd just been trying to remember. In other words, the pathway from short- to long-term memory had been cut off. If Brenda Milner, the neuropsychologist who was treating him, reentered the room only a few minutes after speaking to him, he would invariably say, "It's been so long since I've seen you!" His personal history effectively ceased from the time of the operation. In his mind's eye, he remained about thirty years of age; as he grew older, he became unable to recognize recent photos of himself. Right up to his death in 2008, he was firmly convinced that Harry Truman was president. After moving to a new house, he invariably returned to his old home, and he eventually couldn't be trusted to go out alone.

The prefrontal cortex (fig. 15) has many functions and also coordinates the various parts of the brain that constitute the working or short-term memory. This is the memory that allows you to keep certain things briefly in mind, like the number you want to call, the plans you're making, and the problems you need to solve. The working memory is also crucial for processing language and is thought to be underdeveloped in children who suffer from dyslexia. The prefrontal cortex works closely with the hippocampus (fig. 26) by focusing attention and selecting stimuli. In memory tests, the words that cause heightened activity in both these areas of the brain are the ones remembered best. If we just want to retain a number long enough to make a phone call, our working memory will suffice. But if we repeat that number often enough, we can store it in our long-term memory. The working memory, a short-term storage space for general use, is crucial for carrying out complex tasks and for functional performance. It enabled H.M. to remember a few words or numbers, but he was then unable to transfer that information in the normal way, through repetition, to the long-term memory. The focus of H.M.'s epilepsy was in the region of the hippocampus, two-thirds of which had been removed. (The name *hippocampus,* meaning "seahorse," reflects the shape of this structure, with its ridges and curls.) He could still perfectly recall events that had taken place more than three years prior to the operation, which proved that the hippocampus wasn't the site of remote memory storage. It was H.M.'s complete inability to form new memories after the operation that gave clues to the hippocampus's function. Studies of neurological patients have since shown that even partial damage to the hippocampus can cause considerable, long-term impairment in the ability to create memories, or anterograde amnesia.

The hippocampus specializes in combining sensory information. The location of the restaurant you arranged to meet up at, what the person you're meeting looks like, the sounds and smells from the kitchen, and the position of the set table are all fused into a single coherent item of autobiographical memory, the chronicle of your life. And later, at least if the dinner was worth it, this information is transferred to the long-term memory. The hippocampus does all this in close partnership with an area located nearby on the underside of the cortex: the entorhinal cortex (also called the parahippocampal gyrus, fig. 26). Scoville also removed this latter structure from H.M.'s brain. But which of the two regions does information reach first? The answer was provided by studies of epileptic patients with electrodes in their brains who performed memory tests while the electrical activity in the two regions was selectively recorded. The entorhinal cortex proved to be activated first, followed by the hippocampus.

It's also in the entorhinal cortex that the first signs of Alzheimer's appear, and the memory problems in the onset of the disease indeed typically concern recent information. People with Alzheimer's may not know what happened an hour ago, but they can tell you detailed stories about a classmate at elementary school.

The hippocampus isn't only crucial to memory; we also need it for spatial orientation. Brain scans of taxi drivers who spent four years learning London's enormous and complex network of streets by heart showed a gradual increase in the volume of gray matter at the back of the hippocampus. Studies of people with damage to both sides of the hippocampus have shown that the hippocampus is necessary for imagining the future.

Fortunately, not all recent information is stored in the long-term memory. Who would want to retain every single detail of everything experienced during an entire lifetime, including every meal, every conversation, and every word in every book? That would make it incredibly difficult to locate and access really important information. There are people who are capable of remembering and reproducing enormous quantities of trivial information, like numerical series or entire phone books or train timetables. Yet this ability comes at the cost of other functions. These "savants" usually have a form of autism with severe impairments in areas such as social interaction or abstract thought (see chapter 9).

So what is normally sieved out for storage in the long-term memory? The deciding factors are the importance of the information and the emotional charge of a particular moment. Everyone knows where they were and what they were doing when they heard about the attack on the Twin Towers in New York on September 11, 2001. The amygdala (fig. 26), positioned just in front of the hippocampus in the temporal lobe, imprints memories that carry a strong emotional charge under the influence of the stress hormone cortisol. As a result, a traumatic experience is immediately stored for good in the long-term memory. And that explains why over 80 percent of our earliest memories have negative associations, as the psychologist Douwe Draaisma has shown. Remembering fear, shock, and sorrow is more important for survival than pleasant memories. However, this mechanism can cause problems. A woman with temporal lobe epilepsy, whose focus was the amygdala, kept having the same hallucinations during her seizures, in which she reexperienced a traumatic period of her youth, causing her terrible distress.

There's a clear evolutionary advantage in imprinting danger in the mind—for instance in wartime—so that when a similar situation occurs you're immediately on the alert. Sometimes this natural tendency becomes pathological, however, like when a soldier returns home from a war zone but can't shake off the feeling of being endangered. If he continues to feel fearful and under threat, if the images of war constantly replay in his brain, and if he immediately dives for cover when he hears a bang, then he's suffering from post-traumatic stress disorder (PTSD). During the First World War this was called "shell shock," and 306 British soldiers with the condition who refused to go back to the front were executed. PTSD is a sign that the amygdala has done its work too well, preventing the prefrontal cortex from signaling to the veteran that the danger is over. The amygdala is activated to respond to danger by the chemical messenger noradrenaline. So veterans with PTSD are treated with beta blockers (which have the opposite effect) to prevent dramatic experiences from being too strongly labeled by the amygdala and the individual from being overwhelmed by stressful memories. An exaggerated response by the amygdala to negative stimuli also underlies borderline personality disorder, whose symptoms include emotional instability and impulsiveness. In this disorder, negative emotions are linked to such a strong stress reaction that patients run an increased risk of retrograde and anterograde amnesia. H.M.'s amygdala had also been removed, along with other temporal lobe structures crucial to memory.

H.M.'s brain was sliced into wafer-thin sections at UC San Diego, a process that could be followed online. This was the first step in a process of intense microscopic research aimed at discovering exactly which brain structures were removed or damaged in the operation that he underwent fifty-five years earlier.

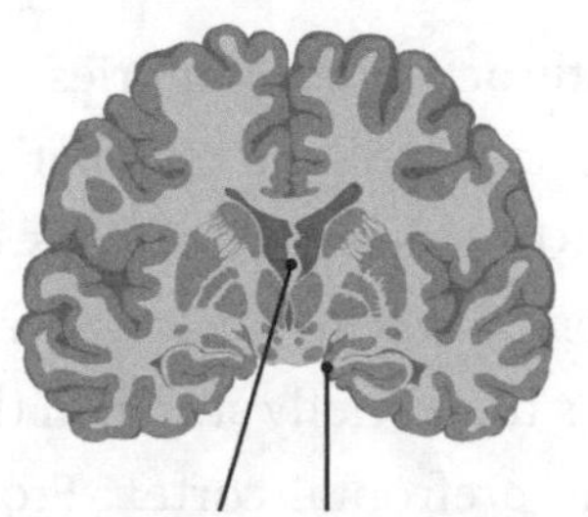
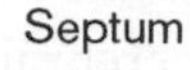

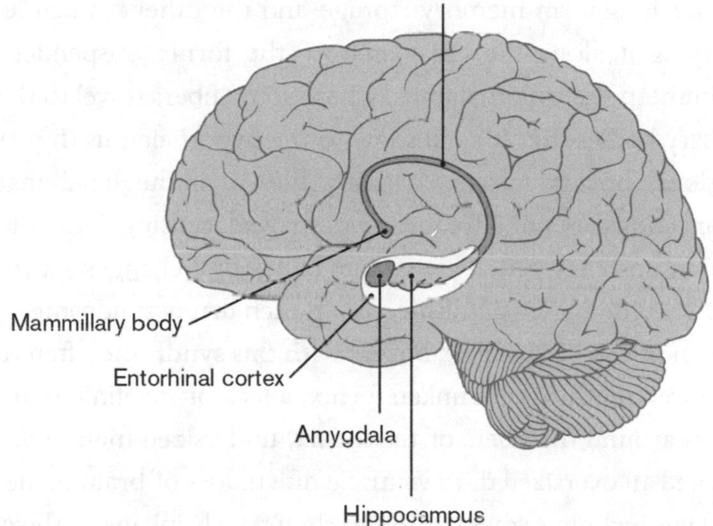

FIGURE 26. The route taken by information on its way to long-term memory starts in the entorhinal cortex, located deep in the brain in the parahippocampal gyrus. It's briefly stored in the hippocampus in a process directed by the prefrontal cortex. From there it follows two pathways: one taking it back to the cerebral cortex for long-term memory storage, and the other—much longer—carrying it along the great arch of the fornix, suspended in the septum, to the hypothalamus, where the fibers proceed to the mammillary bodies. The information then travels via the thalamus to various parts of the cortex. The amygdala, an almond-shaped structure positioned just in front of the hippocampus in the temporal cortex, imprints memories that carry a strong emotional charge.

THE PATH TO LONG-TERM MEMORY

Brain damage is seen in all contact sports, from boxing, kickboxing, and rugby to soccer.

While we sleep, the hippocampus constantly activates memories and transmits them to the cerebral cortex. The jury's still out on whether this mainly occurs during dream sleep (REM sleep) or periods of quiet sleep. The route information takes on its way to the long-term memory starts in the entorhinal cortex. It's then briefly stored in the hippocampus in a process directed by the prefrontal cortex. From there it follows two pathways, one taking it back to the cerebral cortex for long-term memory storage and the other—much longer—carrying it along the great arch of the fornix, suspended in the septum, to the hypothalamus, where some fibers travel to the mammillary bodies (fig. 26) and some to the hypothalamus (fig. 18). Professional boxers sustain so many blows to the head that these connections are not infrequently destroyed, causing dementia, tremors, an unsteady gait, and extreme behavioral changes—a condition known as dementia pugilistica, or "punch-drunk syndrome." Examination of the brains of ex-boxers with this syndrome often reveals a ruptured septum, a shrunken fornix, a lack of myelin (an insulating layer) around the fibers of the fornix, undersized mammillary bodies, and an oversized third ventricle due to loss of brain tissue. Other findings include Alzheimer's-type changes, shrinkage to the cerebral cortex, and cell death, mainly in the temporal region and the hippocampus (see chapter 12). Plenty of reason, in other words, for serious memory impairment and other malfunctions. Damage of this kind isn't confined to boxing but extends to all contact sports, from kickboxing and rugby to soccer and American football. Infarction or bleeding in the above-mentioned areas and pathways can also cause memory impairment or even dementia. In the case of Korsakoff's syndrome (caused by a combination of alcohol abuse and vitamin B_1 deficiency due to poor diet) small hemorrhages and scars are found in the mammillary bodies. People with Korsakoff's have memory impairments similar to those of patients with damaged temporal lobes. They fill in the gaps in their memory with made-up stories. The importance of the mammillary bodies to memory has emerged not just from problems associated with boxing, tumors, or operations (see chapter 5) but also from a bizarre accident that happened to a man during a game of billiards. His opponent's cue was accidentally forced up his nose, penetrating the underside of the brain and damaging the mammillary bodies, leaving the poor man with severe memory problems.

The mammillary bodies pass on information to the thalamus (fig. 2). Small infarctions in this area can lead to severe memory problems and even dementia. The information travels on from the thalamus to areas of the cerebral cortex from which memories of facts and events can be consciously recalled. This is known as the declarative or explicit memory.

SEPARATE MEMORY STORAGE

The case of the man who recognized his car but not his wife.

Different aspects of an event are stored in different sites in the brain. When we try to recall something that happened, the various elements have to be pieced back together again. Any missing bits are filled up by our brains, a process of which we're entirely unconscious. So the common comparison of memory to a computer hard disk that can reproduce everything perfectly isn't quite accurate. A better analogy would be the way in which an archeologist tries to reconstruct an entire skeleton from a few little bones—frequently getting it wrong. Our memory is notoriously unreliable, as is often shown in court cases.

That different types of information—music, images, and faces—are stored in different parts of the cortex emerged from cases of patients with very specific problems of recall. For instance, people who suffer damage to the temporal sulcus sometimes lose the ability to recognize faces, even of the person they're married to, despite there being nothing wrong with their eyesight. But they are able to recognize objects, like their cars, because those memories are stored in another place. Being able to identify your Ford Fiesta but not your wife must make for some interesting household scenes. This condition is known as prosopagnosia, or face blindness. Oliver Sacks described it in *The Mind's Eye* and *The Man Who Mistook His Wife for a Hat*. Dr. P., the man in question, was so severely afflicted that, instead of his hat, he tried to put on his wife's head. It's hard to conceive that he was meanwhile pursuing a distinguished career as a teacher at a music school. In its extreme form, the condition makes it difficult for people even to identify themselves in a mirror. A case is also known of a soldier who ran into his mother on the street without recognizing her when he came home on leave. Luckily it's not quite that bad in my case, but I've always had trouble recognizing faces, something that often leads to embarrassing situations. I occasionally introduce myself to a person who then gazes back at me in amazement and says, "Yes, I know who you are; we've been on the same committee for three years now." My father was troubled by this problem too, which does seem to run in families. It's one of the mutations that has been passed down to me. Yet defects in pattern recognition are clearly extremely selective, because I have excellent recall when it comes to microscope samples. More than once I've looked at a sample I haven't seen for several years and thought (rightly as it turned out), "Oh, that's Mr. X or Ms. Y." Yet if I'd met the individuals in question after a similar interval I would never be able to recognize them.

A study in which epilepsy patients with electrodes implanted in their temporal lobes were shown hundreds of pictures of different faces revealed the existence of neurons that fire only when the person sees a photo of a celebrity like Bill Clinton. So it's somewhere in that part of the brain that my facial recognition problem is located. In tests on monkeys, the neurons at the base of the temporal lobe that fired when they were shown a computer-generated face fired more strongly when they saw a face they knew. The strongest response came when they were shown images in which the most typical features of a familiar face had been caricatured—something that, given my own prosopagnosia, might explain my love of cartoons.

A recognition problem of an entirely different order occurs in Capgras syndrome. While being able to recognize a friend, partner, or close relative, the sufferer feels no emotional connection to them and is therefore convinced that they are impostors. This delusion that a loved one has been replaced by something else—a robot or extraterrestrial—leads to paranoid behavior. Capgras syndrome sometimes develops after brain damage or as a symptom of Alzheimer's.

That the various components of vision are processed in different parts of the brain can lead to very specific visual impairment. The psychologist Ed de Haan described the case of a patient who couldn't see movement. When cars were in motion, she couldn't see them, but when they stopped, they suddenly became visible. Some people can see but not recognize color or can see color but not shapes or have no perception of brightness and therefore have no idea whether they are switching a light on or off.

The safest storage place for information is our remote memory, where we keep language and music. It's the last part to be affected by Alzheimer's. Speech only disappears late in Stage 7 of the system devised by Dr. Barry Reisberg to chart the progress of the disease (see chapter 18). Alzheimer's sufferers can also retain musical skills much longer than other abilities. A professional pianist started to experience memory problems at the age of fifty-eight. By the time she was sixty-three, the dementia was so advanced that she could no longer retain anything that was said or written. But she was still able, on hearing a piece of music for the first time, to remember it and play it with musical feeling. Although her cognitive skills deteriorated sharply in the year that followed, she could still play the melodies she knew, an activity that gave her a great deal of pleasure. It seems that musical memory is regulated by a subsystem of the long-term memory located on the side of the brain (parietal cortex, fig. 1) and that it remains relatively intact. In the case of visual artists with Alzheimer's whose artistic skills remain unimpaired, the subsystem probably lies at the rear of the brain (visual cortex, fig. 1), an area that is less affected—and last affected—by the progression of the disease (see chapter 18).

THE IMPLICIT MEMORY IN THE CEREBELLUM

Someone who staggers around isn't necessarily drunk.

The cerebellum (figs. 1 and 2) is located at the back of the brain, under the large mass of the cerebral cortex. This relatively small structure (*cerebellum* is Latin for "little brain") contains 80 percent of our neurons and ensures that our movements and speech are flowing and coordinated. When you shake your head violently, for instance, it allows you to keep your eyes fixed on one point. It contains the memory of how to do things. It keeps track of motor learning during our development, from crawling to standing and walking, then cycling, swimming, playing the piano, and driving a car, and it constantly steers performance of these tasks. The program for these complex actions—our implicit memory—is stored and updated in this remarkable little computer, allowing us to perform them completely automatically. Practice makes perfect, even in the cerebellum. When we learn to drive, we initially have to think about every action ("I need to change gear, that means using the clutch, where was third gear again?"). This involves using explicit or declarative memory, the memory of facts and events, a time-consuming and highly inefficient process. By practicing the same tasks over and over, they become fully automatic and are transferred to the implicit or procedural memory in the cerebellum. When you have driven so often that you do it without thinking, it in fact becomes difficult to say (drawing on your explicit memory) exactly what actions are involved. H.M.'s implicit memory was intact, because he could learn new motor skills.

His ability to trace a star that he could see in a mirror improved as he practiced day after day, but he could remember nothing of these exercises. He no longer possessed that first, explicit stage in which his brain consciously trained, but his cerebellum was practicing and perfecting new tasks unconsciously.

The cerebellum also suppresses the impact that your own actions have on other parts of the brain. That's why you can't tickle yourself. Your brain wants to give priority to unexpected sensory input that might require an urgent response, and your attempts at tickling yourself (like your other actions) are expected, so the sensations they produce elsewhere are suppressed. Some people lose this mechanism after damage to the cerebellum and find that they can tickle themselves as a result.

Damage to the cerebellum doesn't cause paralysis, but it does make you unbelievably clumsy. Normally, if you shut your eyes, it shouldn't be at all difficult to touch the tip of your nose with your right or left index finger. If your cerebellum is damaged because of an infarct or hemorrhage, your finger will wave about from left to right and is just as likely to land in your eye. Damage of this kind also makes it hard to walk: You stagger about with your legs wide apart, trying not to fall. A colleague of mine once stumbled off an aircraft in this way, because a blood clot had shot into his cerebellum during the long flight. Alcohol and cannabis also impair the functioning of the cerebellum and have the same impact on walking ability.

The large neurons of the cerebellum, known as Purkinje cells, form while we're still in the womb. But the vast majority of the small neurons, called granule cells, form only after birth. So all developmental brain disorders, including autism and pedophilia, make their mark on the cerebellum. The many cerebellar deviations found in all cell types and chemical messengers in autism could explain certain impaired motor functions, like problems with movement coordination and speed and difficulty in learning how to tie shoelaces or ride a bike. But besides its crucial role in movement, it's becoming increasingly clear that the cerebellum is also involved in higher cognitive functions. Developmental disorders of the cerebellum, local damage, infarcts, or tumors can go hand in hand with a host of psychological problems, dyslexia, ADHD, impaired verbal intelligence, and learning disorders.

So the cerebellum is excellently designed for learning complex tasks and actions. But it also coordinates movements that take much less trouble to learn, like the involuntary muscle movements during orgasm. Gert Holstege, who teaches neuroanatomy at Groningen University, carried out brain scans of individuals experiencing orgasm, finding an incredible amount of activity in the cerebellum in both men and women. It makes you wonder what the world would be like if training the muscle movements involved in orgasm took as much time, patience, and effort as learning to play the piano. Problems like overpopulation, global warming, and environmental pollution would never arise!

15

Neurotheology: The Brain and Religion

> How so many absurd rules of conduct, as well as so many absurd religious beliefs, have originated, we do not know . . . but it is worthy of remark that a belief constantly inculcated during the early years of life, whilst the brain is impressible, appears to acquire almost the nature of an instinct; and the very essence of an instinct is that it is followed independently of reason.
>
> Charles Darwin, *The Descent of Man*

WHY ARE SO MANY PEOPLE RELIGIOUS?

> Whatever we cannot understand easily we call God; this saves wear and tear on the brain tissues.
>
> Edward Abbey

> Since it is obviously inconceivable that all religions can be right, the most reasonable conclusion is that they are all wrong.
>
> Christopher Hitchens

As far as I'm concerned, the most interesting question about religion isn't whether God exists but why so many people are religious. There are around ten thousand different religions, each of which is convinced that there's only one Truth and that they alone possess it. Hating people with a different faith seems to be part of belief. Around the year 1500, the church reformer Martin Luther described Jews as a "brood of vipers." Over the centuries the Christian hatred of the Jews led to pogroms and ultimately made the Holocaust possible. In 1947, over a million people were slaughtered when British India was partitioned into India for the Hindus and Pakistan for the Muslims. Nor has interfaith hatred diminished since then. Since the year 2000, 43 percent of civil wars have been of a religious nature.

Almost 64 percent of the world's population is Catholic, Protestant, Muslim, or Hindu. And faith is extremely tenacious. For many years, Communism was the only permitted belief in China and religion was banned, being regarded, in the tradition of Karl Marx, as the opium of the masses. But in 2007, one-third of Chinese people over the age of sixteen said that they were religious. Since that figure comes from a state-controlled newspaper, the *China Daily,* the true number of believers is likely at least that high. Around 95 percent of Americans say that they believe in God, 90 percent pray, 82 percent believe that God can perform miracles, and over 70 percent believe in life after death. It's striking that only 50 percent believe in hell, which shows a certain lack of consistency. In the Netherlands, a much more secular country, the percentages are lower. A study carried out in April 2007 showed that in the space of forty years, secularization had increased from 33 to 61 percent. Over half of the Dutch people doubt the existence of a higher power and are either agnostic or believe in an unspecified "something." Only 14 percent are atheists, the same percentage as Protestants. There are slightly more Catholics (16 percent).

In 2006, during a symposium in Istanbul, Herman van Praag, a professor of biological psychiatry, taking his lead from the 95 percent of believers in the United States, tried to convince me that atheism was an "anomaly." "That depends on who you compare yourself to," I replied. In 1996 a poll of American scientists revealed that only 39 percent were believers, a much smaller percentage than the national average. Only 7 percent of the country's top scientists (defined for this poll as the members of the National Academy of Sciences) professed a belief in God, while almost no Nobel laureates are religious. A mere 3 percent of the eminent scientists who are members of Britain's Royal Society are religious. Moreover, meta-analysis has shown a correlation among atheism, education, and IQ. So there are striking differences within populations, and it's clear that degree of atheism is linked to intelligence, education, academic achievement, and a positive interest in natural science. Scientists also differ per discipline: Biologists are less prone to believe in God and the hereafter than physicists. So it isn't surprising that the vast majority (78 percent) of eminent evolutionary biologists polled called themselves materialists (meaning that they believe physical matter to be the only reality). Almost three quarters (72 percent) of them regarded religion as a social phenomenon that had evolved along with *Homo sapiens*. They saw it as part of evolution, rather than conflicting with it.

It does indeed seem that religion must have afforded an evolutionary advantage. Receptiveness to religion is determined by spirituality, which is 50 percent genetically determined, as twin studies have shown. Spirituality is a characteristic that everyone has to a degree, even if they don't belong to a church. Religion is the local shape given to our spiritual feelings. The decision to be religious or not certainly isn't "free." The surroundings in which we grow up cause the parental religion to be imprinted in our brain circuitries during early development, in a similar way to our native language. Chemical messengers like serotonin affect the extent to which we are spiritual: The number of serotonin receptors in the brain corresponds to scores for spirituality. And substances that affect serotonin, like LSD, mescaline (from the peyote cactus), and psilocybin (from magic mushrooms) can generate mystical and spiritual experiences. Spiritual experiences can also be induced with substances that affect the brain's opiate system.

Dean Hamer believes that he has identified the gene that predisposes our level of spirituality, as he describes in *The God Gene* (2004). But since it will probably prove to be simply one of the many genes involved, he'd have done better to call his book *A God Gene*. The gene in question codes for VMAT2 (vesicular monoamine transporter 2), a protein that wraps chemical messengers (monoamines) in vesicles for transport through the nerve fibers and is crucial to many brain functions.

The religious programming of a child's brain starts after birth. The British evolutionary biologist Richard Dawkins is rightly incensed when reference is made to "Christian, Muslim, or Jewish children," because young children don't have any kind of faith of their own; faith is imprinted in them at a very impressionable stage by their Christian, Muslim, or Jewish parents. Dawkins rightly points out that society wouldn't tolerate the notion of atheist, humanist, or agnostic four-year-olds and that you shouldn't teach children *what* to think but *how* to think. Dawkins sees programmed belief as a byproduct of evolution. Children accept warnings and instructions issued by their parents and other authorities instantly and without argument, which protects them from danger. As a result, young children are credulous and therefore easy to indoctrinate. This might explain the universal tendency to retain the parental faith. Copying, the foundation of social learning, is an extremely efficient mechanism. We even have a separate system of mirror neurons for it. In this way, religious ideas like the belief that there's life after death, that if you die as a martyr you go to paradise and are given seventy-two virgins as a reward, that unbelievers should be persecuted, and that nothing is more important than belief in God are also passed on from generation to generation and imprinted in our brain circuitry. We all know from those around us how hard it is to shed ideas that have been instilled in early development.

THE EVOLUTIONARY ADVANTAGE OF RELIGION

> Religion is excellent stuff for keeping common people quiet.
>
> Napoleon Bonaparte

The evolution of modern man has given rise to five behavioral characteristics common to all cultures: language, toolmaking, music, art, and religion. Precursors of all these characteristics, with the exception of religion, can be found in the animal kingdom. However, the evolutionary advantage of religion to humankind is clear.

(1) First, religion binds groups. Jews have been kept together as a group by their faith, in spite of the Diaspora, the Inquisition, and the Holocaust. For leaders, belief is an excellent instrument. As Seneca said, "Religion is regarded by the common people as true, by the wise as false, and by rulers as useful." Religions use various mechanisms to keep the group together:

(a) One is the message that it's sinful to marry an unbeliever (that is, someone with a different belief). As an old Dutch proverb states, "When two faiths share a pillow, the devil sleeps in the middle." This principle is common to all religions, with attendant punishments and warnings. Segregating education according to faith makes it easier to reject others, because ignorance breeds contempt.

(b) Another is the imposition of numerous social rules on the individual in the name of God, sometimes accompanied by dire threats about the fate of those who don't keep them. One of the Ten Commandments, for instance, is lent force by the threat of a curse "unto the fourth generation." Blasphemy is severely punished in the Old Testament and is still a capital offense in Pakistan. Threats have also helped to make churches rich and powerful. In the Middle Ages, enormous sums were paid in return for "indulgences," shortening the time that someone would spend in purgatory. As Johann Tetzel, a preacher known for selling indulgences, is alleged to have put it, "As soon as a coin in the coffer rings, a soul from purgatory springs." In the beginning of the previous century, Catholic clerics were still automatically awarded indulgences based on the rank they held in the church. Threats and intimidation are effective even in this day and age. In Colorado, a pastor has introduced the idea of "Hell Houses," where fundamentalist Christian schools send children to frighten them about the punishments that await them in the afterlife if they stray from the straight and narrow.

(c) A further binding mechanism is being recognizable as a member of the group. This can take the form of distinguishing signs, like black clothing, a yarmulke, a cross, a headscarf, or a burka; or physical characteristics, like the circumcision of boys or girls; or knowledge of the holy scriptures, prayers, and rituals. You must be able to see who belongs to the group in order to obtain protection from fellow members. This mechanism is so strong that it seems senseless to try to ban people from wearing distinguishing accessories or items of clothing like headscarves. Social contacts within the group also bring with them considerable advantages and play an important role in American churches. The feeling of group kinship has been strengthened over the centuries by holy relics worshiped by the various faiths. It doesn't matter that there are wagonloads of Buddha's ashes in temples in China and Japan, nor that so many splinters of the True Cross have been preserved that, according to Erasmus, you could build a fleet of ships from them. The point is that such things keep the group together. The same applies to the twenty or so churches that claim to have Christ's original foreskin in their possession. (According to Jewish tradition, he was circumcised at the age of eight days.) Some theologians have argued that Christ's foreskin was restored on his ascension to heaven.

However, according to the seventeenth-century theologian Leo Allatius, the Holy Prepuce ascended to heaven separately, forming the ring around Saturn.

(d) Finally, most religions have rules that promote reproduction. This can entail a ban on contraception. The faith is spread by having children and then indoctrinating them, making the group bigger and therefore stronger.

(2) Traditionally, the commandments and prohibitions imposed by religions had a number of advantages. Besides the protection offered by the group, the social contacts and prescriptions (like kosher food) had some beneficial effects on health. Even today, various studies suggest that religious belief is associated with better mental health, as indicated by satisfaction with life, better mood, greater happiness, less depression, fewer suicidal inclinations, and less addiction. However, the causality of these correlations hasn't been demonstrated, and the links aren't conclusive. Moreover, the reduced incidence of depression applies only to women. Men who are regular churchgoers are in fact more likely to become depressed. An Israeli study showed that, in complete opposition to the researchers' hypothesis, a religious lifestyle was associated with a doubled risk of dementia thirty-five years later. Moreover, there are studies showing that praying is positively correlated with psychiatric problems.

(3) Having a religious faith is a source of comfort and help at difficult times, whereas atheists have to solve their difficulties without divine aid. Believers can also console themselves that God must have had a purpose in afflicting them. In other words, they see their problems as a test or punishment, that is, as having some meaning. "Because people have a sense of purpose, they assume that God, too, acts according to purpose," Spinoza said. He concluded that belief in a personal god came about because humans assumed that everything around them had been created for their use by a being who ruled over nature. So they viewed all calamities, like earthquakes, accidents, volcanic eruptions, epidemics, and floods, as a punishment by

that same being. According to Spinoza, religion emerged as a desperate attempt to ward off God's wrath.

(4) God has the answer to everything that we don't know or understand, and belief makes you optimistic (*"Yes, I'm singin' a happy song/With a Friend like Jesus I'll stand strong"*). Faith also gives you the assurance that even if times are hard now, things will be much better in the next life. Curiously, adherents of religion always claim that it adds "meaning" to their life, as if it were impossible to lead a meaningful life without divine intervention.

(5) Another advantage of religion, it would seem, is that it takes away the fear of death—all religions promise life after death. The belief in an afterlife goes back a hundred thousand years. We know this from all the items found in graves: food, water, tools, hunting weapons, and toys. Cro-Magnon people also buried their dead with large amounts of jewelry, as is still done in Asia today. You need to look good in the next life, too. Yet being religious doesn't invariably make people less afraid of dying. The moderately religious fear death more than fervent believers and those who are only very slightly religious, which is understandable when you see how often religion uses fear as a binding agent. Yet many appear to feel a little uncertain about the promised life after death. Richard Dawkins rightly wondered, "If they were truly sincere, shouldn't they all behave like the Abbot of Ampleforth? When Cardinal Basil Hume told him that he was dying, the abbot was delighted for him: 'Congratulations! That's brilliant news. I wish I was coming with you.'"

(6) A very important element of religion has always been that it sanctions killing other groups in the name of one's own god. The evolutionary advantage of the combination of aggression, a group distinguishable by its belief, and discrimination of others is clear. Over millions of years, humans have developed in an environment where there was just enough food for one's own group. Any other group encountered in the savanna posed a mortal threat and had to be destroyed. These evolutionary traits of aggression and tribalism

> can't be wiped out by a few generations of centrally heated life. That explains why xenophobia is still so widespread in our society. The whole world is full of conflicts between groups with different faiths. Since time immemorial the "peace of God" has been imposed on others by fire and sword. That's unlikely to change soon.

Though it comes at a price, belonging to a group brought with it many advantages. The protection it offered against other groups improved survival chances. But the harm caused by religions—largely to outsiders, but also to members of the group—is enormous. It seems as if this situation won't persist indefinitely, though. A study by the British politician Evan Luard showed that the nature of wars has been changing since the Middle Ages and that they are gradually becoming shorter and fewer in number. So we may perhaps be cautiously optimistic. Since the evolutionary advantage of religion as a binding agent and aggression as a means of eliminating outsiders will disappear in a globalized economy and information society, both traits will become less important over hundreds of thousands of years. In this way, freed from the straitjacket of outmoded religious rules, true freedom and humanity will be possible for all, no matter what their belief—or lack of it.

THE RELIGIOUS BRAIN

> Emotional excitement reaches men through tea, tobacco, opium, whisky, and religion.
>
> George Bernard Shaw (1856–1950)

Spiritual experiences cause changes in brain activity, which is logical and neither proves nor disproves the existence of God. After all, everything we do, think, and experience provokes such changes. Findings of this kind merely increase our understanding of the various brain structures and systems that play a role in both "normal" religious experiences and the type of religious experience that is a symptom of certain neurological or psychiatric disorders.

Functional scans of Japanese monks show that different types of meditation stimulate different areas of the brain, namely parts of the prefrontal cortex (fig. 15) and the parietal cortex (fig. 1). Religious belief is also associated with reduced reactivity of the anterior cingulate cortex (ACC, fig. 27), as is political conservatism. Although the causality of these correlations isn't clear, it's interesting that taking initiatives, by contrast, is associated with increased activity in the ACC. The EEGs of Carmelite nuns have shown marked changes during mystical experiences when they felt they were at one with God. In a state like this, individuals may also feel as if they have found the ultimate truth, lost all sense of time and space, are in harmony with mankind and the universe, and are filled with peace, joy, and unconditional love. Neuropharmacological studies show how crucial the activation of the dopamine reward system (fig. 16) is in such experiences. In this context, brain disorders are also instructive. Alzheimer's disease, for instance, is linked to the progressive loss of religious interest. The more slowly it progresses, the less religiousness and spirituality are affected. Conversely, hyperreligiosity is associated with frontotemporal dementia, mania, obsessive-compulsive behavior, schizophrenia, and temporal lobe epilepsy. A number of these disorders are known to make the dopamine reward system more active.

Carmelite nuns were asked to remember their most mystical Christian experience while undergoing functional scans. The scans showed a complex activation pattern of brain areas. Activation occurred in (1) the center of the temporal lobe, possibly relating to the feeling of being one with God (this region is also activated in temporal lobe epilepsy [fig. 28], sometimes causing intense religious experiences); (2) the caudate nucleus (an area in which emotions are processed, fig. 27), possibly relating to the feeling of joy and unconditional love; and (3) the brain stem (fig. 21), insular cortex (fig. 27), and prefrontal cortex (fig. 15), possibly relating to the bodily and autonomic reactions that go with these emotions and cortical consciousness of them. Finally, the parietal cortex was also activated, possibly relating to the feeling of changes in the body map similar to those in near-death experiences (fig. 28).

It's sometimes hard to draw a line between spiritual experiences and pathological symptoms. The former can get out of hand, leading to mental illness. Intense religious experiences occasionally spark

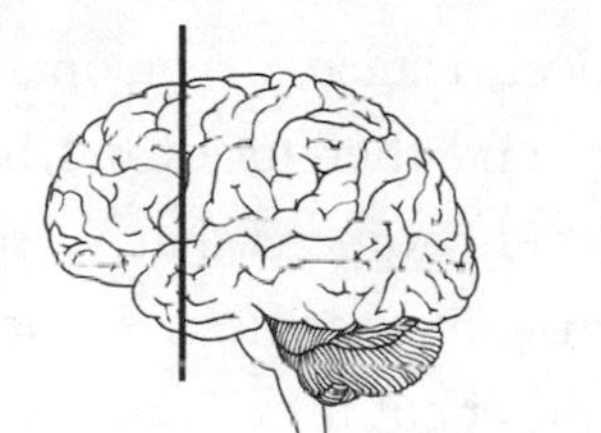

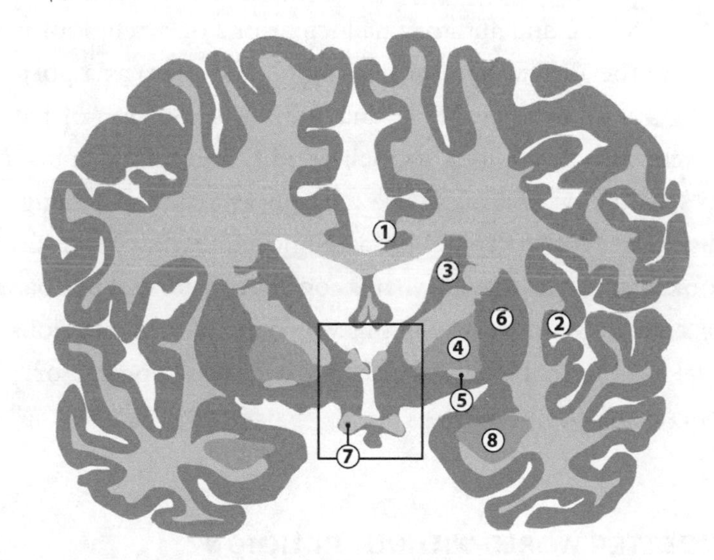

FIGURE 27. Some brain systems involved in emotions. (1) The cingulate cortex, the brain's alarm center. (2) The insular cortex, which is active in emotional experiences and coordinates bodily, autonomic reactions. (3) The caudate nucleus, active in the motor system and emotions. (4) The globus pallidus, active in the motor system. (5) The ventral pallidum/nucleus accumbens, active in reward. (6) The putamen, active in the motor system. (7) The optic chiasm, or crossing of the optic nerves. (8) The amygdala, active in fear, aggression, and sexual behavior. The hypothalamus is indicated by the box.

brief episodes of psychosis. Paul Verspeek, hosting a local Dutch radio show on Boxing Day 2005, asked psychiatrists how they would recognize Jesus Christ if he returned to Earth. How would they distinguish between him and mentally ill patients who claimed to be Christ? The psychiatrists were stumped for an answer. During the 1960s, when meditation and drug use were popular, many people developed psychiatric problems. They were unable to control their spiritual experiences, which derailed their psychological, social, and professional functioning. In some cultures and religions, however, voluntary engagement in meditative practices, trance, depersonalization, and derealization are quite normal and therefore can't be seen as symptoms of a psychiatric disorder. Phenomena that Western culture classifies as chicanery or nonsense, like magic arts, voodoo, and sorcery, are considered normal in other cultures. Some also regard visual and auditory hallucinations of a religious nature (like seeing the Virgin Mary or hearing God's voice) as a normal part of religious experiences. That said, a high proportion of patients with psychoses are religious, as their condition often prompts an interest in spirituality. And many use religion as a way of coping with their disorder. So problems with a religious bearing always need to be looked at in the light of what is considered normal in a particular era or cultural setting. Only in this way can "purely" religious and spiritual problems be distinguished from neurological or psychiatric ones.

A BETTER WORLD WITHOUT RELIGION?

> We shall cast terror into the hearts of those who disbelieve, because they ascribe unto Allah partners, for which no warrant hath been revealed. Their habitation is the Fire, and hapless the abode of the wrongdoers.
>
> Qur'an 3:152

> If a man would follow, today, the teachings of the Old Testament, he would be a criminal. If he would follow strictly the teachings of the New, he would be insane.
>
> Robert G. Ingersoll

> Well, I believe that there's somebody out there who watches over us. . . . Unfortunately, it's the government.
>
> Miles Monroe (Woody Allen) in *Sleeper*

Like all faiths, the Christian belief has always presented itself as a religion of freedom and humanity. Protestants and other Christians did indeed make heroic efforts to protect Jews living in hiding during the Second World War, and they are still high on the list when it comes to taking in foster children. But humanity, persistence, and courage certainly aren't exclusive to believers; socialists, Communists, and atheists also possess those qualities. Moreover, the good intentions inspired by faith unfortunately often don't turn out at all as envisaged.

Would people be better off without religion? I think they would. Let me give a few examples. Throughout history, countless people have been imprisoned and killed in the name of Christianity and of other religions. The Old Testament is awash with murders, and that can have a stimulating effect. Experimental psychological studies show that reading a Bible text in which God sanctions killing clearly raises levels of aggression—though only among believers. Nor is the New Testament all about love and peace. When Pilate washes his hands of the decision to have Christ crucified, Matthew 27:25 states that the people answered, "His blood is on us and on our children!" This has been used to justify Christian anti-Semitism and has resulted in the discrimination, persecution, and murder of countless Jews. Moreover, passages like "I did not come to bring peace, but a sword" (Matthew 10:34) don't sound very peace-loving.

Pope John Paul II apologized (albeit reluctantly) for the Crusades and the persecution of the Jews. So far, however, the Catholic Church has yet to publicly condemn Pope Pius XII's silence about the Holocaust during the Second World War, despite being very well aware of it. Nor has it yet apologized for the Inquisition or the Church's contribution to the slave trade (which made the Netherlands so rich) or for discriminating against women, homosexuals, and transsexuals or for banning contraception, thus condemning millions in South America to a life of poverty and millions in Africa to infection with AIDS. In 2005, three million people died of AIDS, and five million people were infected with HIV. And what does the Catholic Church do? It opposes the use of condoms. The president of the Pontifical Council for the Family, Cardinal Alfonso López Trujillo, who one assumes was hardly able to speak from practical experience, claimed that condoms were permeable "in 15% or even up to 20% of cases" and that their use implied "sexually immoral conduct." In 2009 Pope Benedict maintained, contrary to all the statistics, that the use of condoms would make the African AIDS crisis worse. In recent years the extensive and systematic sexual abuse of children by Catholic priests throughout the world has come to light; the Church pretended that it had never been aware of it. In a tasteless echo of the excuse made by many Germans about the Holocaust after the war, the Dutch cardinal Adrianus Simonis responded with the German words, "*Wir haben es nicht gewusst*" (We did not know about it). It had of course long been common knowledge. It's said that Alfred Hitchcock once saw a priest talking to a child when he was visiting Switzerland. The priest had his hand on the boy's shoulder. Hitchcock leaned out of his car window and shouted, "Run, little boy! Run for your life!" There seems little point in waiting for the pope to save the situation by inveighing against sexual abuse within the church. As George Bernard Shaw famously said, "Why should we take advice on sex from the pope? If he knows anything about it, he shouldn't!"

Blame certainly shouldn't be confined to a single religion. Almost every religion has fundamentalist, outdated ideas that are proclaimed as the "truth" and imposed on others, sometimes at all costs. Nor is religious extremist aggression confined to a particular faith, as witness the 169 deaths caused by the right-wing Christian extremist Timothy McVeigh (the "Oklahoma City Bomber") when he blew up a government building; the massacre of 29 Muslims in the Cave of the Patriarchs in Hebron by Baruch Goldstein, a Zionist extremist and racist; and the destruction of the Twin Towers on 9/11 in 2001. The list could go on forever.

Sacrificing children to the gods has taken place throughout the ages. Mexican history has many terrible examples. In 2007, a grave was found near Mexico City containing the skeletons of twenty-four children between the ages of five and fifteen, all carefully laid on their sides facing east. They died between A.D. 950 and 1150 after their throats were slit by the Toltecs in a mass sacrifice to the rain god Tlaloc. Child sacrifices are still made to this day in the Netherlands by fundamentalist Christians, Bible in hand, who oppose inoculation against polio, German measles, mumps, and meningitis. The Bible is in fact silent on the subject of inoculation, but they see it as conflicting with God's providence. Likewise, Jehovah's Witnesses aren't allowed by their church to sanction a blood transfusion for a sick child. If the child develops some terrible complication, then that must have been the will of God. When she retired, the Dutch judge Anita Leeser-Gassan said how grateful parents were when the courts jointly decided with doctors that children should be given a blood transfusion in such cases. Jehovah's Witnesses base this prohibition on transfusions on a passage from the New Testament, which states, "It seemed good to the Holy Spirit and to us not to burden you with anything beyond the following requirements. You are to abstain from food sacrificed to idols, from blood, from the meat of strangled animals and from sexual immorality" (Acts 15:28–29).

How can they interpret a reference made two thousand years ago to "blood" as meaning "blood transfusion"? This ban means that women from this sect are six times as likely to die in childbirth. Isn't it appalling that the interpretation placed on that one little sentence prevents a lifesaving intervention from taking place?

As far as Islam is concerned, we can cite honor crimes, the killing of innocent people by suicide bombers, the hacking off of right hands, and the decapitation of hostages and apostates (people who convert to another religion) as a few examples of violent actions blessed by religion. In Iran in July 2007, a man was stoned to death for adultery. The local judge was the one who threw the first stone. And then there's violence against women, including female circumcision, a mutilation that still causes the deaths of large numbers of young girls every year and ruins the lives of countless women. In Sudan, almost 90 percent of girls under ten are circumcised, and a WHO report published in 2006 revealed that 100 million girls and women around the world have undergone this mutilation. Female circumcision isn't prescribed by the Qur'an, and many Christian women in Egypt are also circumcised. But the practice is confined to the Islamic world and is strongly endorsed by reactionary clerics, who give reasons for their stance. The Egyptian scholar Yousuf Al-Badri believes that female circumcision would solve many problems in the Western world. "Western women are not circumcised and behold the result: a licentious society. Women always want sex. Over 70% of the children are illegitimate. A large proportion of the Egyptian women have a clitoris of over 3 cm. They need to be circumcised, in order for them to be able to control their emotions and sexual desires. Otherwise they will be continually excited and frustrated, because they do not get satisfied." The consequences of female circumcision are appalling; women who have undergone it often suffer the most excruciating pain during urination, menstruation, and sexual intercourse. In Africa, almost 50 percent of babies born to circumcised women die during or shortly after birth. The mothers themselves often have severe hemorrhages during childbirth.

Piety, alas, isn't often linked with an ability to see things in perspective or with a sense of humor. There isn't a single joke in the Bible. Islamic governments whip up popular frenzies at the slightest reason. In September 2006, twelve cartoons making fun of Islamic extremism were published in the Danish newspaper *Jyllands-Posten:* One showed suicide bombers being told at the heavenly gates that they needed to wait because the supply of promised virgins was running low. The cartoons enraged the Danish Muslim community, which subsequently mobilized Muslims in the Middle East. In Jordan and other Middle Eastern countries, Danish products were taken off supermarket shelves. The Muslim Brotherhood, Syria, the Islamic Jihad Union, the interior ministers of Arab countries, and the Organisation of the Islamic Conference all behaved as if they were themselves models of tolerance toward other religions and demanded apologies. The paper's editor in chief apologized to any Muslims who felt offended by the cartoons, but that didn't appease feelings. Mobs took to the streets in many places, and deaths resulted. In 2006, during an address at the University of Regensburg, Pope Benedict linked Islam and violence. In order to demonstrate their peaceableness, Islamic fundamentalists responded by burning down Christian churches in West Jordan and killing the Italian nun Leonella Sgorbati in Somalia. It seems that the Islamic world isn't yet ready to respond to perceived slurs with intellectual debate.

Extremist organizations like the Taliban in Afghanistan, Hamas in the Palestinian territories, and Hezbollah in Lebanon are rapidly gaining in popularity and strength. And once again, this isn't a specifically Muslim problem. Under the Bush administration, fundamentalist Christians in the United States frequently stirred up public opinion with their fanatical pro-life campaign, their anti-Darwinist ideas, and their homophobia. Jewish right-wing extremists have been similarly active in Israel. For the time being, religions around the world will continue to take their meaningless toll. It's a shame, because there's no need to indoctrinate children with religion. Their spirituality can be put to excellent use in art, science, and the environmental sphere or simply to make the lives of the less privileged happier.

UNCLEAN MUSSELS AND WOMEN

> Some religious precepts have a rational basis. We just don't know which ones.

Some apparently bizarre religious precepts have a rational basis. The ban on Jews and Muslims eating pork was probably very sensible in an age before meat had to be passed by a health inspector. The concept that menstruating women are "unclean," expressed in both the Bible and the Qur'an, is less easy to fathom. Leviticus does not allow any room for doubt on the subject, making it clear that anything a menstruating woman "lies on . . . will be unclean and anything she sits on will be unclean. Anyone who touches her bed will be unclean. . . . Anyone who touches anything she sits on will be unclean . . . till evening. . . . If a man has sexual relations with her . . . he will be unclean for seven days." After each menstruation, women are obliged to make a sacrifice and "purify" themselves by taking a ritual bath, a mikveh. I can see no hygienic reason for this precept, but it does benefit reproduction. According to the law, a woman who has menstruated (usually for around five days) has to wait seven days in order to "purify" herself on the eighth day, which falls on the thirteenth day of her cycle—right around the most fertile period. Clearly, ending periods of sexual abstinence precisely when fertilization is most likely promotes the survival of the group. Might that be the clever reasoning behind this misogynistic rule? Whatever the case, the idea that you need to steer clear of menstrual blood is common to many cultures. Before Mao's day, menstruating women in China not only were held to be unclean, they were also deployed in battle on account of their magical powers. Lined along the city walls, they waved their sanitary towels in a bid to deflect the enemy's cannon fire.

It seems that the fear of menstrual blood has not greatly diminished since Leviticus. According to Vincent de Beauvais (1478), menstrual blood could prevent wheat from germinating, turn grapes sour, kill herbs, render fruit trees barren, make iron rust, tarnish bronze, and cause rabies. And these ideas didn't die out in the Middle Ages. When my mother-in-law was menstruating, she wasn't allowed to enter her grandmother's kitchen when fruit from the garden was being canned. Surinamese women still aren't allowed into the kitchen when they are menstruating. According to certain traditions, a menstruating woman can cause bread to grow moldy, meat to rot, and plants to die—simply at a touch or even a glance. During the Cultural Revolution in China, however, the traditional Chinese rule that only non-menstruating women were allowed to prepare food for the ancestors was done away with, along with many other traditions.

Other religious precepts have a more rational basis. Shellfish like mussels aren't "unclean" just according to Jewish dietary laws; North American Indians were also prohibited from eating them. There seems to be a good reason for this. In 1987, around a hundred people suddenly fell seriously ill within a day of eating mussels from Prince Edward Island in eastern Canada. Besides nausea and vomiting, the victims also had serious neurological symptoms including disorientation, headache, and paralysis. Four died, and seven fell into a coma. A year later some of the victims were still experiencing severe memory problems. They could no longer recall events that would normally be unforgettable, like a daughter's wedding. Postmortems were carried out on the brains of four people who died in this incident. It turned out that the hippocampus and the amygdala, two structures essential to memory, had been severely damaged. That summer, extreme weather conditions in Canada had led to an explosive growth of algae (*Nitzschia pungens*). The shellfish filtered and ingested high concentrations of the algae, which proved to contain domoic acid, a substance that's toxic to the nervous system, destroying brain cells by overactivation. This effect isn't confined to humans. In 1961, bizarre behavior was seen in a colony of sooty shearwaters in Santa Cruz County, California. The birds were flying into windshields and lampposts and pecking and vomiting on people. There were many dead birds in the streets. Alfred Hitchcock requested the local papers to send him reports of the strange behavior. Two years later he made the film *The Birds,* probably inspired by these accounts and of course by Daphne du Maurier's short story of the same name. In 1991, in a similar epidemic in the same area, in which cormorants and brown pelicans suddenly started behaving strangely, the corpses were indeed found to contain high concentrations of domoic acid. So we should definitely take account of certain biblical precepts—it's just a pity that we don't know which ones. Playing safe by keeping all the laws in Leviticus is no longer an option in this day and age. So read its punishments and tremble!

PRAYING FOR ANOTHER: A PLACEBO FOR YOURSELF

> Then I saw all that God has done. No one can comprehend what goes on under the sun. Despite all their efforts to search it out, no one can discover its meaning. Even if the wise claim they know, they cannot really comprehend it.
>
> Ecclesiastes 8:17

> I have never made but one prayer to God, a very short one: "O Lord make my enemies ridiculous." And God granted it.
>
> Voltaire

Sir Francis Galton (1822–1911), a cousin of Darwin's, was the first person to investigate the efficacy of prayer statistically. He concluded that public prayers for the health and longevity of kings and queens were ineffective, since their lives were not on average any longer than those of other classes and professions. Equally, the many prayers said for the safety of voyages undertaken by missionaries and pilgrims didn't result in fewer shipwrecks than average.

Many recent studies have shown that prayer has no effect on patients with leukemia or rheumatism or those dependent on kidney dialysis. Praying for patients by means of a headset while they were anesthetized for open heart surgery proved equally ineffectual. Certain publications make claims for the efficacy of prayer, but their experimental methods are fundamentally flawed. For instance, in a trial involving prayer for patients in a heart-monitoring ward, it emerged that the secretary who divided the patients into groups was also registering the results (so it was not a blind trial) and that the group being prayed for was already in better health than the control group at the start of the trial. A combination of fourteen good studies led to the conclusion in 2006 that praying for others didn't have a curative effect. However, one large, well-controlled study showed that it had an injurious effect on heart patients. It looked at 1,802 patients who were undergoing coronary bypass operations. The patients were divided into three groups. The first two groups were told they might or might not be prayed for. Group 1 was prayed for, while Group 2 wasn't prayed for. Prayers were also said for Group 3, who knew that they would be prayed for. To everyone's surprise, this last group had the highest percentage of complications. Perhaps it was because being told they were being prayed for made them fear that they must be in a bad way. Another study showed that the more psychiatric symptoms patients had, the more they prayed. But you can't conclude from that that praying causes psychiatric problems; these patients were probably desperately seeking aid for their mental health difficulties. Whatever the case, no universally convincing evidence has been presented that praying for others is effective. No one, for instance, has ever seen an amputated limb grow back after prayer.

Despite the negative literature on the efficacy of prayer, the vast majority have no doubt whatsoever: 82 percent of Americans believe that prayer can cure serious diseases, 73 percent believe that praying for others can cure disease, and 64 percent want their doctors to pray for them. Why does the majority believe in the efficacy of prayer, when research doesn't support it? I think it's because people who habitually pray feel good when they do so. It provokes a relaxation response, reducing the level of the stress hormone cortisol. So praying for others is largely something you do to relax yourself. It's not a new idea. Spinoza didn't see the point of petitionary prayer, because he didn't believe in a personal God who responded to appeals. He did, however, regard prayer as a means of concentration and meditation. You can achieve the same result with yoga, meditation, or listening to your favorite music. Doing yoga also reduces cortisol and increases production of the sleep hormone melatonin (as does meditation); the sympathetic autonomic nervous system is also less active after yoga exercises. So it has a de-stressing effect.

Incidentally, studies on the efficacy of prayer are beset by unique methodological obstacles:

- Sometimes only the first name of the person being prayed for was mentioned, and in some cases only the photo was shown. Would that be enough for God to be able to identify the individual in question?
- How do you prevent people in the control group from being prayed for? Many people in hospitals might well be receiving the benefit of prayers from partners, friends, or acquaintances.
- You might also wonder, even as a believer, whether God listens to everyone who prays and whether he is prepared and able to intervene in the affairs of humankind.
- Believers could also ask themselves whether God's ways are a legitimate object of study and whether God would allow himself to be tested ("Do not put the Lord your God to the test," Deuteronomy 6:16).

In the face of all these methodological problems, the only way to establish whether praying for someone else is effective would be through a well-controlled experimental study with animals. As yet I haven't come across one.

RELIGIOUS MANIA

> When one person has a delusion, they are considered crazy. When millions of people have the same delusion, they call it religion.
>
> Richard Dawkins

Certain neurological and psychiatric disorders can give rise to religious mania, at least if religion has been programmed into the brain during an individual's youth. After an epileptic seizure patients can lose contact with reality, and a quarter of these psychoses take a religious form. Religious delusions can also result from mania, depression, or schizophrenia or constitute the first symptom of frontotemporal dementia. The murder, in 2003, of the Swedish foreign minister Anna Lindh was, for instance, committed at the "command of Jesus" by the twenty-five-year-old schizophrenia sufferer Mijailo Mijailović, who had stopped taking his medication. He thought he'd been chosen by Jesus and couldn't resist the voices that were instructing him to kill. John Nash, who in 1994 won a Nobel Prize in Economic Sciences, had been diagnosed at the age of twenty-nine with paranoid schizophrenia. Some of his delusions were religious in nature; he saw himself as a secret messianic figure and the biblical Esau. Near-death experiences can also have a religious slant. One woman with a lung embolism claimed that Jesus himself had sent her back from heaven to earth to look after her children.

Ger Klein, a former Dutch state secretary for education and science, wrote a vivid description of his own religious mania. I met him in the summer of 1975, when he was forced to cut 200 million guilders from his budget in the space of a single week. With one stroke of the pen he abolished the Netherlands Institute for Brain Research (NIN), which was waiting for a new director; the Institute for War Documentation (RIOD), headed by the historian Loe de Jong; the

Dutch astronomical satellite program; and several other important bodies. I was a thirty-one-year-old researcher at the time, and although I had no administrative experience whatsoever, I took it upon myself to try, with the rest of the NIN's staff, to get the government decision reversed. After talks with all the parliamentary factions and a hard-fought campaign, we ultimately succeeded; the House of Representatives voted unanimously to preserve the institute. In the negotiations that followed, Klein and I got along well, despite our very different interests, backgrounds, and characters.

In 1978, Loe de Jong revealed at a press conference that Willem Aantjes, the leader of a new party called the Christian Democratic Appeal, had signed up with the SS during World War II. Klein, who only shortly before had been responsible for the RIOD in his capacity of state secretary, was dismayed. According to Klein, De Jong was completely out of order in subjecting Aantjes to what was in effect a summary execution. Klein thought that De Jong, with whom he'd clashed in the past, should have left the government to deal with this issue. Getting more and more agitated, he started preparing for the upcoming debate in the House like a man possessed. By four o'clock in the morning he'd already drunk three liters of strong coffee. But the Labor Party elected someone else as spokesman for the debate, and to make matters worse, Klein was torn to shreds by the education minister, Arie Pais. When he drove back home after the debate on November 17, 1978, he suddenly felt as if he had been given an almighty punch in the forehead. That marked the start of his manic phase, which he compellingly described in his 1994 book *Over de Rooie* (Seeing Red). He was under the impression that he'd undergone a brain operation and was being controlled by external forces. A booming voice said to him, "You are not just God, no, you are the God of Gods." He went and stood outside a supermarket, announcing to all passersby that a humanist salvation awaited them. He wasn't at all surprised when people hurried away, rather than stopping to listen, because that meant they had taken the urgency of his message seriously. Later he took off all his clothes and ran in circles around his house in the middle of winter. This manic period made way for a terrible depression a few months later.

After reading Klein's enthralling book, I wrote to him to ask if he still remembered me and to tell him that now, nineteen years on, we both had a common interest: manic depression. By way of a subtle hint, I enclosed some of our publications on postmortem studies of the brain tissue of patients with the disorder. I received a long and kind letter back, which I still preserve between the pages of his book. "Of course I remember the consultations between your delegation and the ministry about the closure of the Institute. . . . The decision to ax it following the budgetary cuts that were forced on me almost cost me my political career, but I'm sure you must be satisfied with what was achieved. . . . The research into manic depression naturally interests me very much. As a layman I know little about it: Perhaps we could meet in the near future so that you could explain the medical advances in this field to me. Would this be a possibility?" Of course I immediately invited him to visit the institute. However, he died two years later, in December 1998, without having taken up my invitation.

TEMPORAL LOBE EPILEPSY: MESSAGES FROM GOD

> I don't want to go to heaven; I wouldn't know anyone there.
>
> Harm Edens, *HP/De Tijd*

> There shall not be found among you anyone . . . who uses divination, . . . or a spiritist, or one who calls up the dead. . . . For whoever does these things is detestable to the LORD.
>
> Deuteronomy 18:10–12

Patients with temporal lobe epilepsy sometimes have ecstatic experiences, making them think that they are in direct contact with God and receiving orders from him. One man had visions of a bright light and a figure resembling Jesus. He turned out to have a tumor in the temporal lobe that was causing epileptic activity. After it had been removed, the ecstatic seizures disappeared for good. The divine visions in attacks of this kind are usually very brief—between thirty seconds and a couple of minutes—but they can permanently affect a person's personality, transforming them emotionally or inducing hyperreligiosity. Between attacks, these individuals often develop Geschwind syndrome, whose symptoms include obsessive writing, loss of interest in sex, and extreme religiousness. This rare form of epilepsy may well explain the behavior of various historical figures.

When the apostle Paul was still using his Hebrew name Saul and was on the way to Damascus to track down and imprison Christians, he had an ecstatic experience: "As he neared Damascus on his journey, suddenly a light from heaven flashed around him. He fell to the ground and heard a voice say to him, 'Saul, Saul, why do you persecute me?' 'Who are you, Lord?' Saul asked. 'I am Jesus, whom you are persecuting,' he replied. . . . For three days [Saul] was blind" (Acts 9:1–9). It seems distinctly possible that the apostle had temporal lobe epilepsy, as there are other known cases in which it has provoked ecstatic experiences culminating in temporary cortical blindness and conversion to Christianity. The text of Corinthians 12:1–9 and the visual hallucinations reported by Luke, Paul's chronicler, lend weight to this diagnosis. In one of the hallucinations Jesus spoke to him encouragingly; in another he fell into a trance while praying in Jerusalem and saw Jesus.

The Prophet Muhammad, founder of Islam, had a history of epileptic seizures linked to religious experiences from the age of six. He had his first visions in A.D. 610. While asleep in a remote place in the hills near Mecca, he heard a voice that he later ascribed to the Archangel Gabriel, who commanded him, "Read" (*iqra*). He answered, "I cannot read." The voice repeated, "Read, in the name of Allah who created!" Terrified that something was wrong with him, Muhammad considered throwing himself down the mountain. But he subsequently heard a voice saying, "O Muhammad! You are the Messenger of Allah, and I am Gabriel." From the first time in the cave of Hira he continued to receive revelations from Gabriel up to the time of his death. These were subsequently written down and collected as the suras of the Qur'an.

Joan of Arc was born in 1412 to a farmer in the French village of Domrémy and was burned to death at the stake in Rouen on May 30, 1431, at the age of nineteen. Her life history, including her epileptic attacks, was minutely documented by the Inquisition and the Catholic Church. She was thirteen when she first heard the voice of God. It came from the right and was usually preceded by a bright light on the same side. Not long after the voice, saints appeared to give her daily advice during her campaigns. Her epileptic attacks were sometimes provoked by church bells, the sound of which affected her deeply and caused her to drop to her knees and pray, even on the battlefield. The ecstatic seizures were accompanied by a feeling of bliss, making her cry when they were over. Between seizures she displayed all eighteen characteristics of Geschwind syndrome, including emotionality, euphoria, a conviction of dedication, a lack of humor, modesty, a strong moral sense, asexuality, impatience, aggression, depression, suicidal tendencies, and extreme piety.

In 1889, Vincent van Gogh committed himself to the hospital in the French town of Saint-Rémy-de-Provence. He was suffering from epilepsy, along with many other health problems. During bouts of psychosis he had visual and auditory hallucinations as well as bizarre religious and paranoid delusions. During one such attack, he cut off a piece of his ear and sent it as a present to a local prostitute named Rachel. In between attacks he displayed Geschwind syndrome characteristics. His hypergraphia manifested itself not just in the over six hundred letters he wrote to his brother but also in his enormous productivity as an artist, turning out an oil painting every other day. He'd become increasingly religious from the age of twenty and reread the Bible obsessively. He wanted to become a pastor but was rejected on the grounds of his personality. In 1887 he spent his time translating the Bible into French, German, and English. On Sundays he went to four different churches, and on the wall of his house in Arles he wrote, "I am the Holy Ghost."

In 1849, the Russian writer Fyodor Dostoyevsky was arrested for being a member of a liberal group of intellectuals and sentenced to death. After a mock execution in which he and fellow prisoners stood waiting to be shot by a firing squad, he heard that his sentence had been commuted to four years of exile with hard labor in Siberia. He suffered hundreds of epileptic seizures, and his novel *The Idiot* contains lyrical descriptions of his religious experiences during the ecstatic periods just before the seizures. He wouldn't have missed them for the world: "You all, healthy people, have no idea what joy that joy is which we epileptics experience the second before a seizure. Mahomet, in his Qur'an, said he had seen Paradise and had gone into it. All these stupid clever men are quite sure that he was a liar and a charlatan. But no, he did not lie, he really had been in Paradise during an attack of epilepsy; he was a victim of this disease as I am. I do not know whether this joy lasts for seconds or hours or months, but believe me, I would not exchange it for all the delights of this world." This account shows how the ecstatic periods, which last only a few minutes at most, can seem much longer. Dostoyevsky also wrote about the religious visions he experienced, describing how heaven came down to earth and absorbed him, how he felt the presence of God and was filled with it, and cried, "Yes, God exists." After that, he remembered nothing, which suggests that he subsequently had a generalized epileptic seizure. His seizures were frequent, coming once a week or every three days, and are also described in his book *Demons* (*The Possessed*): "There are seconds—they come five or six at a time—when you suddenly feel the presence of the eternal harmony perfectly attained. It's something not earthly. . . . If it lasted more than five seconds, the soul could not endure it and must perish. In those five seconds I live through a lifetime, and I'd give my whole life for them, because they are worth it."

Many may be disappointed to hear that people from non-Western cultures with this syndrome have never reported seeing Jesus or a

Western image of God during a seizure. In Haiti, temporal lobe epilepsy is interpreted as possession by the spirits of the dead and a voodoo curse. It seems that the divine image imprinted in our brains during early development reemerges during epileptic seizures, along with artistic, literary, political, or religious creations and our mental store of thoughts and convictions.

PUBLIC REACTIONS TO MY VIEWS ON RELIGION

> Forgive, O Lord, my little jokes on Thee, and I'll forgive Thy great big joke on me.
>
> Robert Frost

It all started amicably. Around nine years ago, the Dutch newspaper *Trouw* published an article (September 30, 2000) about a lecture I'd given on the brain and religion, using the title of my lecture as a headline: "We Are Our Brains." Not long afterward, Monsignor Everard de Jong, the auxiliary bishop of Roermond, wrote a long, eloquent letter to the newspaper setting out his criticism (which boiled down to us being more than our brains) and ending with the question, "Surely Professor Swaab's wife doesn't love him solely—or primarily—for his perishable brain?" A little later he came up to me in the intermission of another debate and introduced himself as the author of the letter.

"I'm delighted to hear that," I said, "because I think I can answer your question. My wife said that if my brain was transplanted into the body of Steve McQueen, she wouldn't object at all." The bishop, completely perplexed, responded with a glassy look. Apparently he'd never heard of Steve McQueen. After Cees Dekker presented a copy of his book *Looking Up in Flatland*, about science and faith, to the education minister Ronald Plasterk, I was invited to join a debate with Dekker—as was Monsignor de Jong. As soon as I saw him I asked him if he now knew who Steve McQueen was. He had to admit that he still didn't! The bishop subsequently made a sympathetic attempt to put me back on the straight and narrow by sending me a copy of *The Spiritual Brain: A Neuroscientist's Case for the Existence of the Soul,* by Mario Beauregard and Denyse O'Leary. The book hasn't shaken my unbelief, though.

In 2005, two television producers, Rob Muntz and Paul Jan van de Wint, asked me if I wanted to take part in a television program about the brain and religion. At the time their names meant nothing to me, and I wasn't aware of their wild reputation, but we got along very well. Van de Wint planned to interview five believers in their homes and five atheists in church, while in between Muntz asked passersby for their views. The program was to be broadcast by an educational organization called RVU. That all sounded fine to me. We had a nice conversation, and I agreed to take part. Later, the interviews with the believers were dropped on the grounds that they were too boring.

The interview with me, the first in the series, took place in St. Nicholas Church in Amsterdam. I talked about the ecstatic experiences of Joan of Arc, the apostle Paul, and the Prophet Muhammad, as well as about manic patients who think that they are God and schizophrenic patients who receive instructions from God. I also talked about the way in which you can induce out-of-body experiences (like the ones associated with near-death experiences) through electrical stimulation of the cortex. We discussed the aspects of our behavior that are fixed at a very early stage of development, including aggressive behavior and what that implies about our moral accountability for our actions. I also spoke about my personal views on religion, heaven, and life after death.

At an advance screening just before the start of the series, which turned out to be called *God Doesn't Exist,* I was dismayed to see for the first time the absurd clips that had been edited into the interviews (images of a black woman being crucified, for instance), and I realized that we were in for trouble. But it was too late to do anything about it. I was then asked to take the stage to give my comments. What did I think of the program? I hid my concern and said, "Great—just a shame about all that chattering in between the clips." It was a lively evening, but my family left feeling worried about the broadcasting of the interview—rightly so, as it turned out.

On June 4, St. Nicholas Church instituted summary proceedings in an attempt to ban the broadcast. But the RVU had kept to its agreements with the church and paid the fee of €50 an hour for using the premises. The RVU's offer to display a message both before and after the broadcast in which the church distanced itself from the program was accepted by the court, and the application to ban the program was turned down. Meanwhile, Muntz and Van de Wint got hate mail from thousands of Christians. The Dutch Roman Catholic and Protestant churches protested jointly before the broadcast but couldn't stop it going ahead. It was rescheduled to a time when the fewest people would be watching (a few minutes before midnight), and the Sunday rerun was scrapped. Those around me were encouraging about the interview, but a lot of people were alienated by the film clips. On June 9, the Reformed Political Party and the Christian Union blocs in the House of Representatives requested that the "downright blasphemous program be banned." Their written questions were sent to the prime minister and the ministers responsible for justice and the media. According to the main Dutch press agency ANP, the blocs believed that the broadcast "mocked God and the Christian faith deliberately and in the most damaging way possible." I never heard any more of those formal complaints, nor of the criminal complaint filed on June 23, 2005, by the League Against Blasphemy and Swearing against the RVU program on the grounds that it was "blasphemous and insulting." So much for the famous Dutch tolerance.

16

There Isn't More Between Heaven and Earth...

SOUL VERSUS MIND

So far, the everyday reality of the production of mind by the brain is something that no one's yet managed to formulate in a way that doesn't cause unholy confusion.

Bert Keizer, *Inexplicably Inhabited*, 2010

As Freud already said, the idea of the continued existence of an immaterial "something" after death is common to all cultures and all religions. That "something" is usually called the soul. It's presumed to remain in the neighborhood of the body for a short time after death and then to depart permanently for another place. I've been assisted in postmortems several times by people of Surinamese origin; they would knock on the door three times before coming in, so as to warn the soul. In the aboriginal community in Australia, the name of the deceased may not be spoken or written for a period of time (determined by the family) in order to leave the soul in peace. If an aboriginal dies unexpectedly as the result of an accident or crime, a warning is broadcast on the news, and the period of silence starts. In an ancient Chinese tradition, the deceased was buried with a beautifully decorated container for the soul. So far, however, these have always been found to be empty. The great Jewish scholar and physician Maimonides (1135–1204) wrote about the immortality of the soul. The Qur'an allows for no doubt on the subject, stating that perfect souls will immediately be admitted to paradise.

Over the centuries, people have argued about the moment at which a newborn baby acquires a soul. The religious views on the subject still resonate in the political debate over issues like abortion, stem cell research, and embryo selection. Talmudic scholars hold that the embryo acquires a soul after forty days of pregnancy, perhaps because that is when a fetus becomes recognizable. Before forty days, the Talmud describes a fetus as "water." This distinction has led to the sanctioning of human embryo stem cell research in Israel. According to the ancient Greeks, the moment at which a baby acquired a soul depended on its sex. Hippocrates (460–370 B.C.) believed that male fetuses acquired a soul on the thirtieth day of pregnancy and female fetuses on the forty-second day. According to Aristotle (384–322 B.C.), the difference was somewhat greater: forty days for male fetuses, while female fetuses had to wait for eighty days. The prejudice on which this difference was based was finally explained by the Italian theologian and philosopher Thomas Aquinas (1225–1274). A woman was thought to be a *mas occasionatus,* a defective male, and therefore needed more time to develop a soul (*Summa Theologiae* 1:92).

In 1906, the American doctor Duncan McDougall weighed dying patients, placing them (along with their beds) on scales with a seesaw mechanism. After death he weighed them again, and he found that the head end of the bed had lost an average of twenty-one grams. McDougall concluded from the difference in weights that he had weighed the "soul." It seems a strange experiment: Since the soul is said to be immaterial, it surely wouldn't weigh anything. The loss of weight when the heart stops beating is more likely to be attributable to the redistribution of blood between the various organs. But the precise figure of twenty-one grams caught the public imagination and even became the title of a film. McDougall's subsequent finding, after another experiment, that dogs didn't become lighter at the moment of death tied in with Descartes's claim in 1662 that animals were "soulless machines." In the 1930s, however, a Los Angeles schoolteacher by the name of H. Laverne Twining carried out more precise experiments, which showed that all animals lost a few milligrams or grams of weight at the moment of death, from which he concluded they had some sort of soul.

So throughout history, all cultures have postulated the existence of a soul. Nowadays, there is a field of study that you'd think was all about the soul: psychology. Yet psychologists don't study the soul, only behavior and, more recently, the brain. A "psychon" doesn't exist; a neuron does. When you die, you don't give up the ghost, your brain merely stops working. I've yet to hear a good argument against my simple conclusion that the "mind" is the product of the activity of our hundred billion brain cells and the "soul" merely a misconception. The universality of the notion of a soul seems merely to spring from mankind's fear of death, the longing to see the dear departed once again, and the misplaced, arrogant idea that we're so important that something must remain of us after death.

HEART AND SOUL

> A man or a woman who is a medium or a necromancer shall surely be put to death. They shall be stoned with stones; their blood shall be upon them.
>
> Leviticus 20:27

Some people still cling to the idea that the heart plays a special role when it comes to feelings, emotions, character, love, and even the soul. The editors of the Dutch newspaper *NRC Handelsblad* once forwarded me a reader's letter, which read, "The professor keeps going on about the brain, but the heart, as the seat of the emotions, is the brain's exact counterpart." I don't deny that we sometimes feel our hearts beating with excitement, but that's in response to a command from the brain, which prepares our bodies, through the autonomic nervous system, to fight, flee, or make love.

The mythical role of the heart is reinforced by anecdotes "proving" that donor qualities are transplanted into the bodies of people who receive heart transplants. In 2008, the Dutch newspaper *De Telegraaf* published the story of an American man who had committed suicide by shooting himself through the head. His name was Sonny Graham, and he'd received a heart transplant twelve years previously. The donor heart had come from Terry Cottle, who had committed suicide at the age of thirty-three. Sonny Graham had been so grateful for his new life that he'd started to correspond with Terry's widow, Cheryl. One thing led to another. "I felt like I had known her for years," Graham told a local paper in 2006. "I couldn't keep my eyes off her. I just stared." In 2004 Fox News reported that the widow married the man who'd been given her first husband's heart. Her new husband went on to kill himself in the same way as the heart's first "owner." At the age of thirty-nine, Cheryl was widowed for the second time. *De Telegraaf* didn't conclude from this that it might not have been easy to live with Cheryl. No, instead it asserted, "And this breathes new life into the story that when you transplant an organ like the heart, the soul of the deceased is transplanted with it." In fact *De Telegraaf* is rather partial to stories of this kind. The heading of one of its weekend supplements ran, "Does your soul live in your heart? Claire Sylvia (47) was given a boy's heart. She now whistles at girls and drinks beer." Sylvia, who published a book about her experience in 1997, was convinced that she had inherited the characteristics of the young motorcyclist who was the donor of her heart and lung transplant.

The *Journal of Near-Death Studies,* a periodical that until recently I hadn't heard of, publishes many stories of heart transplant patients whose tastes or abilities changed to match those of their donors. Some stories tell of patients whose musical preferences changed. One tells of a man who, after being given a woman's heart, suddenly became mad about pink, a color that he'd loathed before the operation. Another patient, a woman who'd been given the heart of a chess player, claimed that she'd now mastered the game. And a man who received the heart of someone who had been murdered said that he'd seen the killer's face in a dream. The problem with accounts like these is that the heart recipients knew a lot about the donors, ranging from their sex, age, and cause of death to many other details about their lives. To take such anecdotes seriously, we need well-controlled studies in which heart recipients know nothing at all about their donors. A heart transplant is an extraordinarily serious, stressful, and life-threatening operation that can drastically affect personality for many years. Heart recipients often become more spiritual, feel guilty about the deceased donor, and feel that the latter lives on in their body. Their behavior is also affected by the heavy medication needed to counteract rejection of the transplanted organ. So people have every reason to feel different after a heart transplant. And how could a transplanted heart, which has no nerve connections with the recipient's brain, transmit complex information about the donor to the brain of the recipient in such a way as to influence their behavior?

Until well-controlled studies have proven otherwise, we must assume from the available clinical and experimental literature that all of our characteristics are located exclusively in our brains and that the heart is merely a pump that can be replaced without donor characteristics, good or bad, being transplanted along with it.

PSEUDOSCIENTIFIC EXPLANATIONS OF NEAR-DEATH EXPERIENCES

> People who leave their bodies shouldn't sneakily take their five senses with them.
>
> Bert Keizer, *Inexplicably Inhabited,* 2010

One of my PhD students gave me an analytical account of his two near-death experiences (NDEs), describing them with scientific interest:

> The first time I had an NDE I was eleven years old and had pneumonia with pleuritis. My temperature went up to 42.3°C and I was drenched in sweat. But our family doctor thought I was just being a crybaby. Then I seemed to glide into a tunnel, with a light at the other end. I had the pleasant sensation of being absolutely at peace. I didn't consciously hear background music, but I did feel like you do when you hear the kind of music that gives you goose pimples. So I can well imagine why some people think they hear heavenly choirs during an NDE. The light surrounds you like a warm bath; it's very bright, but not painfully so.
>
> The second NDE, which happened when I was thirty-four, was more interesting because I "saw" myself lying on the ground. It later turned out that this was due to a heart rhythm disorder. I was having a meal and stood up to get something. I grew dizzy and collapsed on the floor, for all intents and purposes unconscious. However, I was very conscious of myself and "saw" myself lying on the floor. My wife rushed to my side. She was obviously panic-stricken and kept calling my name. I wanted to tell her that it wasn't serious and that I was all right, but I couldn't speak. You'd expect to be worried or frightened in such a situation, but I found that if you're floating around the room in a relaxed state this isn't the case. However, at the same time I was perfectly aware that this

wasn't real and that I was imagining that I was floating. For one thing, I could hear my wife's voice next to me, rather than below me. I remember analyzing my situation: I'm lying here on the ground, the window's there, the door's there, the sofa's ten feet away; I'm lying here in front of the door and can still hear, but can't see, can't respond. So the floating sensation must at least in part be a visual projection. It's very strange, but I felt no panic or worry at all. After a while I sensed that I was regaining control of my body, the floating sensation disappeared and I could see normally. I don't exactly know how long it took, but it must have been somewhere between thirty seconds and a minute. Yet it seemed an eternity to me: I'd lost all sense of time. And I can normally accurately guess the time to within five minutes if I don't have a watch. But I think that sense of timelessness contributes to the feeling of euphoria during an NDE. After all, we spend our lives in a perpetual hurry, always wanting to do as much as possible. At a moment like that you're completely freed from such concerns. In neither case did I "meet" anyone, nor did I hear voices.

These days, most people have heard of NDEs. Familiarity with the phenomenon increased after the Dutch cardiologist Pim van Lommel published his bestselling book *Consciousness Beyond Life: The Science of the Near-Death Experience* in 2007. NDEs can be caused by oxygen deprivation, extreme fear, high fever, or exposure to chemical substances. Twenty percent of people who suffer a heart attack say afterward that they felt a sense of tranquility, that the pain stopped, and that they thought they had died. Some have the sensation of leaving their bodies and seeing themselves lying down. Others feel as if they're traveling at high speed from a dark place into a tunnel with a bright light at the other end, find themselves in a beautiful landscape, or hear music. Some are reunited with deceased friends and relatives or experience divine apparitions, while others see their entire lives flash before their eyes. And all of this happens in less than a minute. The compromised brain responds to the situation by retrieving an incredibly rapid stream of memories, thoughts, images, and ideas. Christians see Jesus, while Hindus see messengers of Yama, the god of the dead, coming to take them away. Memories appear to be recalled at much greater speed than normal, and people have visions of the future. At a certain point a border is reached, after which people return to their bodies and the NDE ends. Some people report that Jesus himself sent them back because their children needed them.

Van Lommel's achievement was to describe patients' near-death experiences in detail in *The Lancet* in 2001, sparking debate on the phenomenon in the medical world. He related how greatly changed his patients often were after an NDE, to the extent that their marriages frequently failed. People who have had NDEs lose their fear of death, become more religious or spiritual, and believe more in the paranormal. NDEs have such an overwhelming impact on many people that they really don't want to have them explained in neurological terms. They believe that they have been granted a glimpse of the hereafter; their lives take on a more spiritual and religious focus. My PhD student, an exception to this rule, remained a critical scientist even during and after his NDE.

Throwing Out Four Nobel Prizes

Unfortunately, Van Lommel doesn't take the same rational approach, allowing himself to be carried away by his patients' belief in paranormal explanations for NDEs. His pseudoscientific interpretations must be popular, because his book is selling like hotcakes. He categorically rejects any neurobiological explanation of NDEs, putting forward his own theory that allegedly explains in a single stroke not only NDEs but all spiritual and paranormal phenomena, including predictive dreams, reincarnation, seeing things at a great distance, and moving objects with the power of thought. According to him, consciousness isn't produced by the brain as we "shortsighted, materialist, reductionist brain researchers" think. No, according to him it's present "everywhere in the universe," simply being received by the brain in the way that "programs are received by radio or television." According to Van Lommel, thoughts don't have a material basis either. Apparently he's unaware of recent experiments proving the contrary. Take the case of someone whose arm had been amputated who was able to control a computer mouse and a prosthetic arm by the power of thought, using equipment that registered the electrical activity of neurons. In other words, Van Lommel has gotten things backward: The presumed "radio" (the brain) makes its own programs.

Van Lommel claims that his spiritual theory is necessary because the brain lacks sufficient storage capacity for long-term memory. This is nonsense; he appears to be unaware that in 2000, Eric Kandel won the Nobel Prize for describing how short- and long-term memory are formed at the molecular level. Van Lommel also claims that our organisms lack the information necessary for our embryonic development and immune responses; he believes all this information to be stored in the universe. Once again, he appears to be unaware that in 1995 a Nobel Prize was awarded for the discovery of genes involved in early embryonic development, while in 1987, Susumu Tonegawa received the Nobel Prize for his discovery of the genetic mechanism responsible for antibody diversity. As a crowning touch, Van Lommel claims that DNA isn't the carrier of genetic information but merely receives this information from the universal consciousness. Can anyone seriously believe that Watson and Crick didn't deserve the Nobel Prize they won in 1962 for determining the structure of DNA? By dismissing four Nobel Prizes without a single scientific argument, Van Lommel shoots down his own book—or at least its scientific pretensions.

Generating Near-Death Experiences

NDEs can be caused by anything that impairs brain function, inducing a state between consciousness, dream sleep, and unconsciousness. NDEs have been reported in cases of severe blood loss, septic or anaphylactic shock, electrocution, coma due to brain damage or cerebrovascular accident, suicide, near-drowning (especially in the case of children), and depression. NDEs can also be caused by excessive levels of CO_2, hyperventilation, LSD, psilocybin, or mescaline and have been experienced by pilots accelerating too rapidly in fighter planes. They've also been reported in connection with ketamine, a drug used as an anesthetic.

Gert van Dijk, a Dutch neurologist at Leiden University Medical Center, carries out weekly trials that involve making patients faint. They regularly experience NDEs in which they hear voices, have pleasant sensations, or feel that they're in another world entirely. All of these symptoms are caused by well-documented temporary impairment to blood circulation in the brain; Van Dijk measures the point at which the EEG flatlines (though the brain stem continues to function, because the patients are still breathing). Yet Van Lommel repeatedly maintains in his book that NDEs can't be caused by lack of oxygen to the brain, because then everyone who had a heart attack would experience them. He forgets that longer periods of oxygen deprivation impair memory so much that you can't remember an NDE, even though he mentions this himself in a study published in *The Lancet*. Van Lommel's study also shows that some people are more prone than others to experience or recall NDEs. This has to do with the extent to which dream sleep (REM sleep) can surface during consciousness, as the American neurologist Kevin Nelson has shown. When REM sleep intrudes during an NDE, people's muscles relax as they normally do in dreams, preventing them from moving or speaking, just as in narcolepsy (see chapter 5). This REM paralysis contributes to the feeling of being dead. Van Lommel also rejects the notion that NDEs are caused by oxygen shortage on the grounds that severe stress has also been shown to induce NDEs. However, that can easily be attributed to the brain's response to cortisol and other stress signals. When pain or stress becomes unbearable, when escape from a life-threatening situation is impossible, the brain resigns itself and REM sleep intrudes. The stress system is switched off and the dream stage of the NDE begins, along with pleasant sensations. How can Van Lommel actually be so sure that all brain activity stops during unconsciousness? Gert van Dijk disproves this on a regular basis. (His flatlining EEGs don't show that all activity has ceased—an EEG only measures activity in the upper part of the cerebral cortex.) Moreover, when someone has a heart attack, the interval between normal brain function and the onset of unconsciousness is long enough for them to have an NDE. The same applies to the interval between unconsciousness and recovery, which is when Dostoyevsky would have his seemingly endless experiences during epileptic seizures. The possibility of NDEs occurring when consciousness returns is underpinned by the fact that some people feel themselves flopping back into their bodies at the moment of successful reanimation.

There are no scientific grounds for Van Lommel's peculiar theory, whereas brain research perfectly explains all aspects of NDEs. An out-of-body experience can be triggered by stimulating the place where the temporal lobe and the parietal lobe meet. If the processing of information from the muscles, the organ of balance, and the visual system is disrupted at the spot where the temporal lobe and the parietal lobe meet—the angular gyrus (fig. 28)—you get the sensation of leaving your body and floating around. Similar experiences have been generated by cannabis use, which influences a great many chemical messengers in the brain. In one patient's case, electrical stimulation at the rear of the hypothalamus near the fornix (fig. 26) had the side effect of activating the medial temporal lobe, causing him to reexperience events that had taken place thirty years earlier (see chapter 11) and creating the NDE-like effect of his life flashing past. This area is involved in storing the episodic autobiographical memories that form the chronicle of our lives. It is, moreover, extremely sensitive to lack of oxygen and therefore can be easily activated. Stimulating the hippocampus provokes extremely clear, highly detailed autobiographical memories, including memories of people who have died. When one's life is in acute danger it appears that all of these memories are retrieved not one after the other but virtually simultaneously, leading to what is called "panoramic memory." As we know from temporal lobe epilepsy and other situations in which the temporal lobe is stimulated (see chapter 15), this can go hand in hand with strong spiritual or religious feelings. People feel that they are one with the universe, the world, or God, or they think that they have gone to heaven or the afterlife and are in direct contact with God, Jesus, or some other religious figure. The feeling of tranquility and the absence of pain in NDEs is ascribed to the release of opiates or stimulation of the brain's reward system. The vision of a tunnel is caused by reduced blood circulation in the eyeball, starting on the periphery of the field of vision. The periphery grows dark while the center of the visual field remains clear, creating the impression of a tunnel with light at the far end. Fighter pilots trained in a giant centrifuge that impaired blood circulation in their eyes also saw such tunnels. The bright, attractive colors and the beautiful light at the end of the tunnel are caused by stimulation of the visual cortex, just as when we dream. And just as in a dream, a person experiencing an NDE finds themselves taking part in a bizarre story.

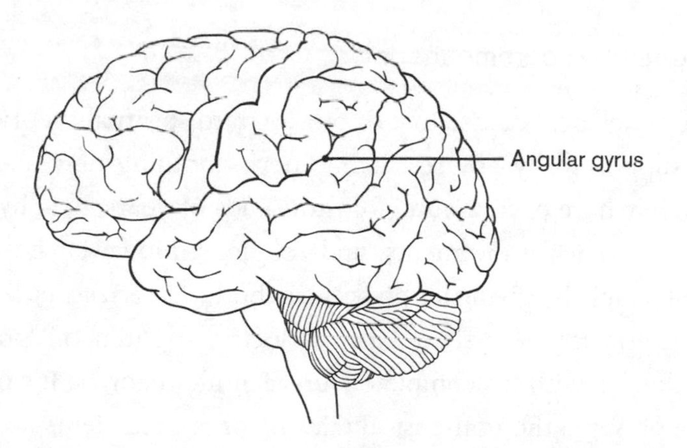

FIGURE 28. An out-of-body experience can be induced by stimulating the place where the temporal lobe (which is very susceptible to oxygen deficiency) and the parietal lobe meet: the angular gyrus. If the processing of information from the muscles, the organ of balance, and the visual system is disrupted at this location, you get the sensation of leaving your body and floating about.

Van Lommel's theory boils down to our brains and our DNA being receivers for "waves of consciousness," a term that he doesn't go on to define. In his explanations he frequently uses terms like *entanglement, nonlocality,* and other terms borrowed from quantum mechanics. The theoretical physicist Robbert Dijkgraaf, president of the Royal Netherlands Academy of Arts and Sciences, better qualified than I to pronounce on questions of physics, is clear in his opinion. "When faced with the inexplicable, people like to believe that quantum mechanics must provide the answer. But unfortunately all the characteristics of quantum systems vanish with incredible speed if you put more than a few particles together. Entanglement and nonlocality (involving particles being connected with each other and able to influence each other at a distance) occur only under exceptional circumstances: at a temperature of one billionth of a degree above absolute zero and in extremely well-isolated surroundings. The quantum world is incompatible with a warm, complex system like the human brain or the world around it. That can be demonstrated in five minutes on the back of an envelope."

Irresponsible Scaremongering

Van Lommel is of course free to put forward spiritual theories that aren't supported by any research. There's nothing new about his ideas. They have been around for thousands of years, held by many cultures, mystical movements, and religions. However, he should not fool people by giving his book the subtitle *The Science of the Near-Death Experience*. Nor should he, a doctor, frighten off potential organ donors with his completely unscientific theories. It's baffling that he presents the nonsensical tales of organ recipients acquiring donor characteristics (see earlier in this chapter) as if they were truths. While claiming not to be against organ transplantation, he is wilfully scaring potential donors and their families unnecessarily.

Various hospitals are currently attempting to collect evidence of out-of-body experiences during an NDE. Researchers have placed playing cards on top of cabinets in the rooms of patients in order to test whether people see the cards when they are ostensibly floating above their bodies. As you might expect, patients who felt they were leaving their bodies haven't been able to identify the cards. So all in all, there's no reason whatsoever to regard NDEs as proof of observations outside the brain, or as evidence of life after death. Those patients never got as far as the hereafter.

In short, near-death isn't the same as death, just as near-pregnant isn't the same as pregnant.

EFFECTIVE PLACEBOS

> The art of medicine consists in amusing the patient while nature cures the disease.
>
> Voltaire (1694–1778)

The discovery that the most commonly used antidepressants aren't significantly more clinically effective than a placebo was met with general amazement. It seems that physicians are strangely reluctant to sing the praises of the placebo effect, which occurs when patients show improvement after being given an essentially ineffective compound or treatment. As a rule, ineffective pills that are red, yellow, or orange are perceived to have a stimulating effect, while blue and green pills are deemed calming. Placebos can also have side effects, like nausea or stomachache. You can even get addicted to placebos, experiencing withdrawal symptoms when treatment stops. So the effectiveness and neurobiological mechanisms of placebos make a very interesting subject for investigation.

The placebo effect results from unconscious changes in brain function that reduce the symptoms of a disease. It is caused by the patient's own expectations of treatment. While the substances in placebos are pharmacologically ineffective, their effect can be quite considerable. The placebo effect isn't confined to pills; it also extends to psychotherapy, surgical interventions, and other therapies. For many years, psychiatric patients were advised to breathe into a paper bag if they felt a panic attack coming on, an approach that proved very successful. It was based on the theory that when you hyperventilate, you exhale too much carbon dioxide, ultimately causing a panic attack. But it later turned out that hyperventilation wasn't the cause but the consequence of a panic attack and that breathing in extra carbon dioxide from a paper bag should theoretically induce a panic attack rather than relieve it. But because people believed in it, it worked.

Placebos can help to relieve the symptoms of Parkinson's disease (caused by a lack of the chemical messenger dopamine) by making the brain release more dopamine. A similar effect can also be obtained by using electrodes implanted in the brain to inhibit the subthalamic nucleus. If a doctor tells a patient that he's switching the electrode's simulator on or off but doesn't in fact do so, the patient's symptoms nevertheless improve or worsen accordingly. During an operation to implant depth electrodes, an ineffective substance was injected into the intravenous line of Parkinson's patients, who were told that it was a new anti-Parkinson's drug. Electrical activity in that same brain area diminished as a result, reducing symptoms in over half of the cases. It seems that the brains of the patients responding to the placebo "know" in which area a change in activity is needed to alleviate symptoms.

Depressive patients who were treated with a placebo showed the same improvements after six weeks as patients who'd been given real antidepressants. Brain scans showed that changes in activity patterns were very similar in both categories of patient. So the brain can be prompted by a placebo to bring about the exact functional changes that are needed to reduce the symptoms of depression: increased activity in the prefrontal cortex and reduced activity in the hypothalamus.

If a patient is in pain and is given a placebo, the brain "knows" how to suppress the pain by releasing more endorphins (morphine-like substances) and cannabis-like compounds and altering activity patterns in certain areas of the brain and spinal cord. An expensive placebo proves more effective than a cheap placebo. By contrast, because of their dementia, Alzheimer's patients don't expect painkillers to help. As a result, they don't work as well, and such patients need to be given higher doses to achieve the same effect. The placebo effect is the result of the brain's own unconscious self-healing potential. That mechanism can contribute little or nothing to the treatment of cancer but has been proven effective in a number of brain disorders. Studies of the mechanism of placebo effects and of why some people are more receptive to them than others (including whether a predisposition to spirituality plays a role) could have important clinical consequences. In the meantime we certainly mustn't underestimate the effectiveness of the doctor figure, whose traditional power to inspire confidence makes him a walking placebo.

TRADITIONAL CHINESE MEDICINE: SOMETIMES MORE THAN A PLACEBO

> Acupuncture outperforms placebos.

Since time immemorial, traditional Chinese medicine (TCM) has held that an incredible number of substances and foods are good for your health. In fact, the Chinese like to claim that everything that's tasty is good for you and will prolong life. But there are also serious studies suggesting that the widely publicized benefits of unfermented green tea are well founded: It appears to reduce the risk of cardiovascular diseases and certain forms of cancer. Based on seventeen national screening trials, it's estimated that you're 10 percent less likely to have a heart attack if you drink three large cups of green tea a day. Green tea is thought not only to counteract high blood pressure and obesity but also to protect the brain. Various traditional Chinese herbal preparations are believed to diminish dementia symptoms, and modern techniques are being used to determine their active ingredients and functional mechanisms. Cat's claw herb (*Uncaria rhynchophylla*) combats beta-amyloid plaques and so might be effective against Alzheimer's. However, claims that green tea can combat Parkinson's and Alzheimer's still need further evidence. In TCM, millipedes, beetles, and worms are traditionally held to counteract dementia. Extracts from these creatures have indeed been shown to inhibit acetylcholine esterase activity, just as the Western medicine prescribed for Alzheimer's does, which does help certain patients to a degree. It's by no means impossible that Chinese research into traditional medicines will bring to light entirely new active substances.

Acupuncture, that exotic therapy with its impressive rituals, creates high expectations in patients and certainly has a placebo effect. The question is whether that entirely accounts for its effectiveness or whether the ancient Chinese notion of meridians and classic acupuncture points is meaningful. But studies show that determining the source of acupuncture's effectiveness can be very complex. Here are a few examples.

One study, testing the efficacy of acupuncture for migraine, arbitrarily divided patients into three groups. The first group was given genuine acupuncture, in which the needles were inserted at the classic points and the doctors had to elicit "Qi," a radiating sensation regarded as a sign of the needle's effectiveness. The second group underwent fake acupuncture, in which the needles were inserted at predetermined non-acupuncture sites. The third group was placed on a waiting list. The treatment of the real acupuncture group didn't prove more effective than that of the fake acupuncture group, but both groups benefited more than the group on the waiting list. From this one could conclude that acupuncture is effective but that the significance of classical acupuncture points, at least in the case of migraine treatment, is debatable. However, it's impossible to say whether the benefits were achieved by a physiological mechanism or a very strong placebo effect. The same result was achieved in a similar study of three groups of patients with tension headaches.

However, in a similar experiment with patients suffering from osteoarthritis of the knee, acupuncture proved more effective. This study compared "real" acupuncture treatment with minimal acupuncture (in which needles were superficially inserted in non-acupuncture points) and a waiting list group. After eight weeks of treatment, the patients who had undergone real acupuncture showed significant improvements in terms of pain and knee function compared to the group receiving minimal treatment. Although the difference between the two groups diminished over time, a clinically relevant treatment effect was established. In the case of chronic mechanical neck pain, acupuncture proved statistically effective but clinically ineffective. In that study, electroacupuncture was compared with fake electroacupuncture (in which the treatment was identical but the needles weren't hooked up to a power supply). In the absence of a control group, the effect of the acupuncture needle itself couldn't be determined.

A brain imaging study looked at differences in brain response to acupuncture treatment, specifically at the expectations that patients suffering from painful osteoarthritis had of the treatment. In a single blind randomized crossover trial, three interventions were compared: real acupuncture, placebo acupuncture, and overt placebo acupuncture. The placebo group was treated using a Streitberger needle—a special kind of needle that, when pressed against the skin, moves back into the handle, giving the impression that it has pierced the skin. The overt placebo group knew that their treatment wasn't therapeutic, having been told that the needle wouldn't pierce the skin. None of the three treatments had the effect of reducing pain. However, scans showed that the ipsilateral insular region, which coordinates bodily autonomic reactions, was activated more by the real acupuncture needle than the Streitberger needle, although both treatments created the same expectations in patients. The two interventions caused greater activity in the prefrontal cortex, anterior cingulate cortex, and midbrain than the overt placebo treatment, of which patients had no therapeutic expectations. This experiment shows that acupuncture needles can have a specific physiological effect and that a patient's expectations of treatment stimulate brain areas associated with reward. So acupuncture can do more than just achieve a placebo effect caused by a patient's expectations of therapy. But in order to establish acupuncture as an evidence-based form of medicine, one would need to perform similar experiments for every disorder. Animal trials can play an important role here. Painkilling in rats by means of electroacupuncture was shown to be linked to higher concentrations of vasopressin in the paraventricular nucleus, while levels of oxytocin and opioid peptides remained the same. Vasopressin can be measured in the blood, which may prove useful in evaluating the effectiveness of acupuncture and its functional mechanisms in humans.

HERBAL THERAPY

Herbs can contain active substances but also very toxic ones.

Herbal medicine is an immensely popular form of alternative therapy. Around thirty thousand herbal products are offered for sale in the United States, and around $4 billion a year is spent on them. If you have a chronic illness and the doctors can't really help you, there comes a moment when most of us feel the need to try alternative therapy. Everyone knows someone who knows someone who was suddenly cured by it. (Curiously, no one ever mentions that diseases sometimes go away by themselves.)

An important contributing factor to the perceived success of alternative therapy is probably that alternative doctors are much nicer and make much more time for their patients than regular doctors. Belief in the effectiveness of alternative therapy, both by the practitioner and the patient, is often the best form of placebo.

People trying out herbal therapies often rationalize it by saying "There's no harm in trying." After all, the thinking goes, these are "natural" substances and therefore can't hurt you. This is a misconception that I would like to set right. Herbs can be completely ineffective against the maladies they allegedly cure and yet be extremely dangerous. What's more, most of the toxins we know of are also "natural" substances. That makes sense, because when a chemical substance affects our organism, it's usually through a specific protein receptor, and we have receptors only for natural substances or chemical substances resembling them.

Reading the medical literature on the possible toxic effects of "safe" herbs is enough to make your hair stand on end. A whole host of neurological and psychiatric disorders has been shown to result from the use of herbal medicine, ranging from vascular infections in the brain, swelling of the brain, delirium, coma, disorientation, hallucinations, cerebral bleeding, motor disorders, depression, muscle weakness, and tingling to epileptic seizures. Ginseng can cause insomnia, vaginal bleeding, and mania. Valerian can make you nauseous and give you a hangover. Thorn apple (*Datura stramonium*) can cause disorientation, while passionflowers can generate hallucinations. Taking kava-kava (*Piper methysticum*), sold as an antistress cure, can lead to life-threatening liver inflammation and cirrhosis of the liver, while ma huang (*Ephedra sinica*) can induce psychosis. Preparations of ma huang contain ephedra alkaloids, substances contained in diet products, pep pills, and "smart" drugs, and are also used to enhance performance in sports. These preparations have rightly been banned in the Netherlands. The ginkgo tree, which is common in China, has fan-shaped leaves that inspired Art Nouveau designs in Europe at the beginning of the twentieth century. Ginkgo is prescribed as a remedy for memory problems and dementia (it does have some effect, but even less so than Western drugs for those conditions, which are themselves not very effective) but can cause headache and dizziness. Eucalyptus can induce delirium. St. John's wort (*Hypericum perforatum*) is taken for depression and has been shown to improve mood, but it can also cause anxiety and fatigue.

Some herbs, especially Asian ones, are, moreover, polluted with heavy metals. And don't let yourself be fooled: The argument that herbal mixtures have been used for centuries, for instance in traditional Chinese medicine, gives no guarantee whatsoever, either of effectiveness or of the absence of toxic effects. It's also important to know that herbs can interact unexpectedly and dangerously with conventional medicines. St. John's wort, for instance, can counteract oral contraceptives and disrupt the effect of antiretroviral drugs and Prozac.

Armed with this basic knowledge and a critical attitude, let us now go online. Type in "herbal therapy" on Google, and you'll get over 20 million hits for herbs against every kind of disease, along with all the rubbish spouted by the con artists who sell them. Read and tremble! Particular caution is advised when they try to sell you something that they claim has no side effects. Any medicine that works has side effects. If someone claims that a product has no side effects, there are three explanations: (1) It doesn't work, (2) its side effects have never been tested, or (3) (the most likely) both (1) and (2) apply. The only one who will certainly benefit from the herbs touted online is the herb supplier himself.

This doesn't mean that herbs can't contain chemical components that have a therapeutic effect. A great deal of work is being done in China to support traditional medicine with scientific research. Efforts are being made to identify the active chemical substances in herbs that have been used for centuries, but since TCM is based on the notion that drugs are most efficacious when there's a mixture of active substances, Chinese scientists face a daunting task. But these substances are now being isolated and their effectiveness is being researched along Western lines by means of cell cultures and animal experiments. There's considerable pressure within China itself to give TCM a scientific basis, and physicians who simply go on prescribing TCM according to ancient Chinese traditions are sharply criticized.

Sometimes the isolation of chemical substances from herbs provides some familiar results. For instance, plants that are traditionally used against aging disorders turn out to contain a lot of melatonin. In the West, too, there are claims that melatonin is an antioxidant that inhibits the aging process, but there's no hard clinical evidence to back the claims up. However, we do know that melatonin as a purely chemical substance is effective in restoring the sleep-wake rhythm in dementia patients, reducing nocturnal restlessness, and achieving modest improvements in memory. As far as I know, no controlled trials have yet been carried out to establish the same effect using herbs containing high concentrations of melatonin. In TCM, ginseng is prescribed for problems relating to sexual dysfunction. Animal experiments in the United States have shown that ginseng indeed increases libido, facilitates erections, and stimulates sexual behavior. Now these findings only need to be confirmed in a clinical setting.

The effectiveness of traditional plant extracts is increasingly being tested in controlled clinical trials, just as in Western drug research. Sometimes the results contradict one another, as always happens in drug research. Some studies show that ginkgo leaf extract indeed causes slight improvements in older patients with memory disorders and dementia sufferers, while other studies show no improvement in memory function. A comparative analysis has been made between ginkgo and Western anti-dementia drugs (acetylcholine esterase inhibitors), which aren't very effective and have many side effects. Nevertheless, the Western drugs scored slightly better than ginkgo, so it isn't a magic bullet against dementia.

Time will reveal the reasons for the discrepancies between the studies and show who is right. The important point is that TCM is now being investigated using controlled, Western methods, so that we will eventually know whether it can indeed be used to develop effective drugs that don't have too many side effects and pose no risk of toxicity. The latter danger is far from imaginary. In 2006, samples of a TCM aloe preparation sold in Britain were found to contain 11,700 times the permitted level of mercury. It's findings like these that are currently placing extra pressure on China to modernize TCM.

17

Free Will, a Pleasant Illusion

> Perhaps the conscious mental representations are afterthoughts—ideas thought after the deed to provide us with the illusion of power and control.
>
> Irvin D. Yalom, *When Nietzsche Wept*

FREE WILL VERSUS CHOICE

> Here I stand, I can do no other.
>
> Attributed to Martin Luther when he appeared at the Diet of Worms in 1521

It's often claimed that human beings have "free will" because we make choices. This is faulty reasoning. Every organism constantly makes choices. The point of contention is whether those choices are entirely free. The American researcher Joseph L. Price has defined free will as the ability to choose to act or refrain from action without extrinsic or intrinsic constraints. Applying this definition, can we ever be said to make a decision freely? Back in 1838 Darwin wrote that free will was a "delusion," arguing that people rarely analyzed their motives and usually acted instinctively. Indeed, free will is such a complex issue that philosophers have yet to agree on what it actually is, though it's often said to have three components. First, an action is only free if you could also have *abstained* from it (you must have alternative options). Second, an action must be carried out for a *reason*. Third, you should feel that you're truly carrying out the action of your own *volition*. But feelings are, of course, entirely subjective.

No one who has experienced the suddenness and intensity of falling passionately in love will classify partner choice as a "free choice" or a "well-considered decision." Plato was of exactly the same mind regarding the autonomy of this process. He regarded the sexual impulse as a fourth species of soul, located below the navel, describing it as "rebellious and masterful, like an animal disobedient to reason." Spinoza was no believer in free will either. He demonstrated that in *Ethics III*, proposition 2, where he states, "An infant believes that of its own free will it desires milk, a hot-headed youth believes he freely desires vengeance, a coward believes he freely desires to run away; a delirious man, a garrulous woman, a child . . . believe that they speak from the free decision of their mind, when they are in reality unable to restrain their impulse to talk." Spinoza shows that characteristics are innate. You can't change them.

Our current knowledge of neurobiology makes it clear that there's no such thing as *absolute* freedom. Many genetic factors and environmental influences in early development, through their effects on our brain development, determine the structure and therefore the function of our brains for the rest of our lives. As a result, we start life not only with a host of possibilities and talents but also many limitations, like a congenital tendency to addiction, a set level of aggression, a predetermined gender identity and sexual orientation, and a predisposition for ADHD, borderline personality disorder, depression, or schizophrenia. Our behavior is determined from birth. This view—the polar opposite of the belief in social engineering that held sway in the 1960s—has been referred to as "neurocalvinism," alluding to the doctrine of predestination that shaped Calvinist thinking. To this day, adherents of strict Protestant sects believe that God has predetermined the course of everyone's life from the moment of birth, including whether you will go to heaven or hell.

That a great deal is determined during our early development applies not just to psychiatric disorders but also to our functioning in everyday life. We may have a theoretical choice between a heterosexual and a homosexual relationship, but our sexual orientation, already programmed in the womb, doesn't allow us to choose freely between these theoretical possibilities. We're born into a linguistic environment that shapes our brain structure and function without us being free to choose our mother tongue. The religious environment in which we end up after birth also determines how we shape our spirituality (its level being genetically predetermined)—that is, whether our focus will be belief, materialism, or environmental concerns. In other words, our genetic backgrounds and all the factors that permanently affected our early brain development saddle us with a host of internal limitations; we are not free to decide to change our gender identity, sexual orientation, aggression level, character, religion, or native language. Nor can we decide to have a certain talent, or to abstain from thought. As Nietzsche wrote, "A thought comes when 'it' wants, not when 'I' want." Our influence on our moral choices is also limited. We approve of things or reject them, not because we have thought about the matter so deeply but because we cannot do otherwise. Ethics are a product of our ancient social instinct to do what is good for the group, a finding that goes back to Darwin. So we're left with the paradox that the only individuals who are still free to a degree (apart from their genetic limitations) are fetuses in the early stages of gestation. But they can't exploit this limited freedom because their nerve systems are still too immature. By the time we're adults, the capacity of our brains to be modified has become very limited and, along with it, the potential for our behavior to change. By then we have been issued with a certain "character." And by then our last little bit of freedom is further curtailed by the obligations and prohibitions that society imposes on us.

THE BRAIN AS A GIANT, UNCONSCIOUS COMPUTER

> When making less important decisions I have found it useful to weigh up all the pros and cons. Yet in the case of truly significant matters the decision needs to come from the subconscious, from somewhere within ourselves.
>
> Sigmund Freud

We make a great many decisions "in a fraction of a second" or "instinctively" or on the basis of our "intuition," without thinking about them consciously. We "choose" a partner by falling in love at first sight, and an accused man will tell the court in all sincerity that he killed the victim before he knew it. In his book *Blink*, the science journalist Malcolm Gladwell paints a fascinating picture of the important and complex decisions made by the unconscious brain in a couple of seconds. Yet this happens only after its internal computer has carried out a gigantic number of calculations. Just as today's planes can fly and land on automatic pilot, without the assistance of a flesh and blood captain, our brains can to a very great extent function excellently without conscious thought. But they have to be trained to do so. It's only by feeding the unconscious brain a huge amount of data over a long period of time that an art expert is able to "sense" that he's looking at a forgery, and it's only by seeing a great many patients that a medical specialist develops the "clinical glance" allowing her to make a diagnosis almost as soon as a patient enters the room. Functional scanning has shown that conscious reasoning involves different brain circuits from those used for intuitive decisions. Only in the latter case are the insular cortex and the anterior cingulate cortex activated; these areas are important for autonomic regulation. They also play a role in our gastrointestinal system, so it's rather appropriate to say we make a decision based on "gut feelings."

Our brains *have* to work on automatic pilot to a very great extent. We're continually bombarded with an enormous amount of information and unconsciously use selective attention to extract what is important to us. Even when photographs of naked people are flashed before an individual too briefly to register the images consciously, a heterosexual man's attention will still be caught more by naked women than naked men. Homosexual men and heterosexual women will focus more on the images of nude men, while the response of lesbian and bisexual women falls between that of heterosexual men and women.

Emotions also play an important role in unconscious processes. In the case of moral judgments, emotions are decisive. An area in the brain's frontal lobe, the ventromedial prefrontal cortex, is crucial to solving moral dilemmas, like whether to sacrifice the life of a single individual to save many lives. Most of us find these extremely emotional decisions well-nigh impossible, but individuals with a damaged prefrontal cortex weigh them in a clinical, highly detached way. They don't experience emotions like empathy or sympathy when faced with dilemmas of this kind.

Decisions that involve social norms and values are apparently made on an emotional basis even when it's possible to weigh them rationally. The products of conscious reasoning processes are by no means always superior to unconscious decisions. In fact, conscious reasoning can even get in the way of good decisions. According to psychologist Ed de Haan, it's sometimes better to make important financial decisions, like buying a house, on the basis of intuition—that is, without conscious reasoning. Just like Dustin Hoffman in *Rain Man,* the autistic savant Daniel Tammet tried to win money at blackjack by counting cards in a Las Vegas casino. He lost very badly until he decided to use intuition instead, whereupon he started winning again (see chapter 9). When you drive to work in busy traffic you make hundreds of decisions in complex, potentially life-threatening situations completely automatically. You can also, as it were, "park" a problem at the back of your mind for a while, giving it no conscious thought, and then all of a sudden, while you're doing something completely different, the solution pops up. In other words—in perpetual homage to Sigmund Freud—our behavior is for a very great part steered by unconscious processes. A hundred years later we have returned to the subconscious, but this time without the Freudian vision of repressed, infantile, sexual, and aggressive urges and other dubious claims.

Physical factors, like temperature and light, can also greatly affect our actions. Outbursts of aggression can be triggered by long hot summers. A study of the 2,131 major conflicts of the last 3,500 years has shown that in both the northern and southern hemispheres the decision to go to war has tended to be made in the summer, while in countries near the equator such decisions are unaffected by the seasons. In other words, it isn't military strategy or "reason" or "free will" but the temperature that appears to be decisive when taking the momentous step of declaring war.

Of course, making so many unconscious decisions also has drawbacks. The racist and sexist views that we unconsciously hold are often unexpectedly influential, for instance in job interviews. But on the whole our brains have to function as efficient, unconscious computers that nevertheless make rational decisions. Unconscious, "implicit" associations enable us to make countless complex decisions quickly and effectively, something that would be impossible if we were to consciously weigh up all the pros and cons in every instance—it would simply be too time-consuming.

Yet all these unconscious decisions leave no room for a purely conscious free will. This has far-reaching implications, because when we hold somebody responsible for their actions, we're assuming the existence of free will, which—at least as far as most of our actions are concerned—simply doesn't exist.

THE UNCONSCIOUS WILL

> We must accept the fact that it is possible we know something without knowing why we know it.
>
> Malcolm Gladwell, 2005

Because our overburdened brain constantly makes decisions using unconscious processes, Harvard psychologist Daniel Wegner speaks of an unconscious will rather than a free will. The unconscious will makes split-second decisions on the basis of events in our surroundings, a process that's determined by the way our brains formed during development and by what we have learned since. The complex, ever-changing environment in which we live means that our lives can never be predictable, and the way in which our brains have developed means that there can be no such thing as complete free will. Yet we believe that we're constantly making free choices, and we call this "free will." According to Wegner, this is an illusion.

Wegner has carried out an experiment that supports his theory. Person A stands in front of a mirror with his arms tucked out of sight. Person B stands behind him and stick his arms out under A's armpits, where A's arms would normally be. When B's arms carry out commands that are given aloud (like "scratch your nose," "wave your right hand") A sees the movements in the mirror and feels as though he is controlling them with his will. Wegner's work clearly shows that both actions themselves and the "conscious" idea of initiating an action are prompted by unconscious processes in the brain. You can't oversee these processes, but you can interpret the resultant action. The "conscious picture" that our brains register when we carry out an action gives us the feeling that we have knowingly performed that action. But that feeling doesn't constitute proof of a conscious, causal chain of events leading to the action. According to the Amsterdam psychologist Victor Lamme, the illusion of conscious will only occurs belatedly, when the information about the action being performed is transmitted back to the cerebral cortex. Wegner believes that the illusion of free will is necessary in order to give an action personal legitimacy. It's like a rubber stamp saying "I did this!"

In his famous experiments, Benjamin Libet showed that when we initiate actions, it takes half a second before our brains consciously register the action. His conclusion that "conscious" experiences are preceded by half a second of unconscious brain activity ("readiness potential") raised serious doubts as to the possibility of acting from free will. Although Libet's observations have been hotly debated, recent fMRI scans have shown that there are areas of the cerebral cortex in which motor actions are prepared for as much as seven to ten seconds before they are consciously perceived. And the time lag between action and consciousness has been demonstrated in numerous ways. In one experiment, people were given the task of quickly touching a spot that lit up on a computer screen. Their visual cortex worked with great speed. One-tenth of a second after the light appeared, it fired off a message to the motor cortex to initiate the movement to touch the light (fig. 22). If the processing in the visual cortex was interrupted by a magnetic pulse, the action was carried out, but the person wasn't conscious of the screen lighting up. All of these observations support the idea that the notion of acting from free will is indeed illusory. Whether it's possible, as Libet believed, that we do at least have the power of veto over an action as soon as we become aware of it ("free won't") remains to be seen. It's of course equally possible that vetoing an action, too, is preceded by unconscious brain activity.

But even if consciousness is somewhat slow on the uptake, it remains useful. We plan consciously (see chapter 13) and learn to drive consciously (a process that can be carried out automatically after a lot of training; see chapter 14). If you weren't conscious of the pain caused by a wound or infection, you'd be unlikely to do something about it, and your chances of survival would be slim. What's more, consciousness ensures that you try to avoid similar hazards in the future. The fact that many of our actions occur unconsciously doesn't mean that we can't act consciously when we pay attention. Driving a car on autopilot is fine until something unexpected happens that requires your attention. Then slow, conscious action takes over—with all its own inherent dangers.

WHAT FREE WILL *ISN'T*

Every brain is unique, so it's not so unique to be unique.

While Francis Crick had doubts about the existence of free will, he theorized that part of the prefrontal cortex, the anterior cingulate cortex, might provide a neural basis for will. (He meant "will" in the sense of "taking initiative," certainly not free will as defined by Price—that is, the ability to decide to act or refrain from action without intrinsic or extrinsic constraints.) Antonio Damasio, too, believes the anterior cingulate cortex and the medial prefrontal cortex to be the source of both external action (movement) and internal action (thought and reasoning). In the case of Alzheimer's patients, a strong correlation is indeed found between the degree of apathy and the shrinking of the anterior cingulate cortex. But that doesn't constitute proof of the localization of "free will."

Many examples have been presented to brain researchers in attempts to undermine the theory that free will is illusory. Decisions to perform deeds of resistance, for instance, are cited as evidence for the existence of free will. One wonders whether that is such a good example, given that religious extremists who have been indoctrinated from an early age would see their deeds in the same light. And while we're at it, the famous words allegedly spoken by Martin Luther when he appeared before the Diet of Worms in 1521, "Here I stand, I can do no other," don't sound like the expression of a completely free decision either.

Science and art have often been hauled out to "prove" the existence of free will. The Australian electrophysiologist and Nobel Prize winner John Eccles invoked the creativity of the scientist as overriding proof of the existence of free will. Indeed, our brains are unique and can therefore produce unique poems or paintings or devise unique experiments. But this doesn't constitute proof of free will. It's not for nothing that researchers working independently and in different parts of the world often make the same "unique" discovery simultaneously. And this has always been the case. Darwin was forced to publish his theory of evolution much against his will, because Alfred Russel Wallace had arrived, entirely independently, at the same idea. The theory was made public in the form of a joint paper read out at a meeting of the Linnean Society in London on July 1, 1858. Darwin wasn't present, as he and his wife were burying their son that day. Wallace wasn't there either, because he was in the Far East. The joint paper didn't prompt further debate. Even more remarkably, a Scottish gardener had already come up with the notion of natural selection just as Darwin was setting sail on the *Beagle*. However, as his idea was published in a book, no one took any notice of it. (This is still very much the case today. Any scientific article published as a chapter in a book is lost as a publication.) What this shows is that the time was apparently right for this entirely novel concept. If these men hadn't come up with the idea, someone else soon would have. That doesn't detract from the brilliance with which Darwin set out this theory, illustrating each step of his argument with countless examples in works that are not only scientifically significant but also highly readable.

This phenomenon of independent but simultaneous discovery is also to be found in art. The precursors of artistic development go back about 164,000 years, to our common roots in Africa, but mankind appears to have "discovered" art about 30,000 years ago, more or less simultaneously in France, Australia, and Africa. The world's oldest sculpture, a female figurine carved out of a mammoth's tusk found in Germany, also dates from this period. These "unique" expressions of human creativity apparently depended on the stage reached by the brain in its evolutionary development. Similarly, the most influential factor in the "unique" experiments carried out by researchers appears to be the progress of scientific thought and the development of new technologies and instruments, enabling the next logical step to be taken.

The uniqueness of scientific or artistic discoveries is therefore insufficient proof of the existence of free will. Better arguments are needed.

FREE WILL AND BRAIN DISORDERS

Free will is illusory, particularly in the case of mental disorders.

Free will entails being able to predict the consequences of one's actions. A person suffering from a brain disorder may be unable to do so. This can have legal implications, because under Dutch criminal law, a person who commits an offense for which they can't be held accountable due to impaired development or a mental disorder isn't criminally liable. In 2003, a resident of a nursing home, an eighty-one-year-old woman with severe dementia, murdered her eighty-year-old roommate, who also suffered from dementia. The Public Prosecution Service of course refrained from prosecution. Schizophrenia sufferers occasionally commit crimes of violence. In 1981 John Hinckley Jr. tried to assassinate President Ronald Reagan. In 2003, Mijailo Mijailović murdered the Swedish foreign minister, Anna Lindh, believing that Jesus had commanded him to do so.

Very few people will claim that these acts were committed out of free will. When a polite, nicely dressed lady suffering from Tourette's syndrome, sitting with her handbag on her lap in her doctor's consulting room, suddenly pours out a stream of tic-induced obscenities, she doesn't do so of her own free will. But how far can this exception be taken? Can you hold a pedophile morally responsible for his sexual orientation if you know that it results from a combination of genetic factors and atypical brain development? His pedophilia certainly isn't a free choice. How responsible is someone who, as a result of his genetic background and his mother's smoking during pregnancy, develops ADHD and commits criminal acts? We know that malnutrition during pregnancy increases the risk of a child going on to develop antisocial behavior. How much free will can be attributed to him if he gets into trouble with the police? Can you hold a teenager fully accountable for committing an offense while he's still getting used to a brain that has been completely reconfigured by sex hormones?

The complexity of the concept of free will is also illustrated by alien hand syndrome, a rare neurological condition that occurs when communication between the two sides of the brain breaks down. This can happen when the link between the two hemispheres, the corpus callosum, gets damaged. As a result, the activity initiated by the individual sides of the brain can no longer be coordinated, and the patient loses their sense of control over one of their hands. Their "alien hand" carries out involuntary actions that may be completely at odds with those of the normal hand. One hand will be putting on a pair of trousers, while the other makes strenuous attempts to take them off. How would free will apply in such cases? One person with the syndrome described how she woke up several times in the night to find that her left hand was trying to strangle her. It was a detail seized upon by the makers of the film *Dr. Strangelove,* in which one of Peter Sellers's hands is constantly wrestling to prevent the other from choking him to death. When the above-mentioned patient was awake, her left hand would try to undo the buttons of her dress against "the will" of her right hand. Her left hand also fought with the right in trying to pick up the phone. The feeling of having lost control of your own limbs and the lack of a sense that you're initiating actions are terrifying; they create the illusion that someone else or something else is controlling those actions. Indeed, the patient in question, seeking to account for the fact that she apparently wasn't in charge of her own hand, thought that it was being controlled "from the moon." It seems that if we're aware of what we do but lack a sense of autonomy (free will), our body feels like an alien object. It has accordingly been suggested that the illusion of acting out of free will might well be the price we have to pay for consciousness. In the case of alien hand syndrome there appear to be two wills in a single brain, each wanting something different. So the illusion of possessing free will also depends on effective links between the right and left hemispheres of the brain.

The notion that we're free to choose how to act isn't just mistaken, it has also caused a great deal of misery. It used, for instance, to be generally accepted that our sexual orientation, that is, heterosexuality, homosexuality, or bisexuality, was a matter of choice, and since homosexuality is regarded as a wrong choice by all religions, it was until recently criminalized and regarded as an aberration. Indeed, it was removed from the ICD-10 (International Classification of Diseases) only in 1992. Before that time, all kinds of attempts were made to "cure" homosexuals of their supposed disorder by imprisoning them and subjecting them to all manner of terrible interventions and operations, none of which had the slightest effect. I'm curious to see how long it will take before society starts to think differently about other actions and behaviors that are now thought to be subject to free will, like aggressive and delinquent behavior, pedophilia, kleptomania, and stalking. Society's acceptance of those behaviors as innate would have far-reaching consequences in the treatment of offenders.

It is also sometimes claimed that one's free will can combat physical ailments. Some argue that a positive attitude to certain neurological disorders like MS promotes healing. Not only is there no evidence whatsoever for this view, but one of its consequences is that if the disease worsens, the poor patient is told that he or she didn't "try" hard enough to fight their illness!

Wouldn't it be better for us to accept that complete freedom of will is illusory? It isn't exactly a new idea; as Spinoza said in his *Ethics II,* proposition 48, "In the Mind there is no absolute or free will."

18

Alzheimer's Disease

AGING OF THE BRAIN, ALZHEIMER'S DISEASE, AND OTHER FORMS OF DEMENTIA

> The prospect that you'll finally forget that you've forgotten everything and that this loss will no longer trouble you isn't a consoling thought, because it signifies your ultimate erasure as an individual.
>
> Douwe Draaisma

Life is sometimes allegorically portrayed as a staircase that we climb step by step, ascending as we develop and grow until we reach our prime and then start to descend around the age of fifty, as the aging process sets in. But when it comes to the aging of the brain, a better metaphor might be a film of our life being wound back to the start. Alzheimer's reverses the stages of development so that sufferers gradually lose their personality and their faculties, culminating in a state of complete dependence that leaves them curled up in a fetal position, demented and effectively brain-dead (see later in this chapter).

"Normal" brain aging and Alzheimer's have a lot in common. First, age is the main risk factor for Alzheimer's: The likelihood of developing the condition increases exponentially as you get older. Second, all the alterations found in the brains of Alzheimer's patients are also present in old but dementia-free "control brains," albeit to a much lesser degree and starting at a later age. Alzheimer's can in many ways be seen as a premature, accelerated, and severe process of brain aging.

Every living organism ages. Why is that? Probably because it takes less energy for nature to replace individuals every now and then with younger specimens than to keep the original organisms intact forever by constantly repairing damaged cells. This has led to the theory of the disposable individual, a notion that is, alas, reflected in the way society views the elderly, as a problem to be dealt with as cheaply as possible.

Many Forms of Dementia

Alzheimer's is the most common form of dementia. Given demographic aging and the fact that age is the main risk factor for the disease, the number of Alzheimer's patients is expected to double over the next thirty years.

The only way of finding out whether someone with dementia was correctly diagnosed with Alzheimer's is by carrying out a postmortem examination of their brain under the microscope to look for the telltale changes involved (fig. 29). That is because there are many other forms of dementia that can be distinguished from Alzheimer's only through analysis of brain tissue. Brain infarcts and hemorrhages can cause multi-infarct dementia, very often combined with Alzheimer's-type changes. Parkinson's disease can also cause dementia; when it spreads to the cortex it takes the form of Lewy body dementia. The various types of dementia in the prefrontal cortex used to be collectively referred to as Pick's disease. Nowadays a distinction is made between different frontotemporal forms of dementia caused by tau gene mutations linked to chromosome 17 (fig. 30). The first sign of such dementia isn't usually memory loss but bizarre and inappropriate behavior. Alcohol abuse can lead to a form of dementia called Korsakoff's syndrome, in which patients fill in the gaps in their memories with made-up stories that they firmly believe to be true. That's not to say that they can't remember anything at all. I was about to introduce myself to a Korsakoff's patient when the person in question said, "I know you, aren't you Dick Swaab?" It was rather chastening to discover that my memory of faces is worse than that of someone with Korsakoff's! In the early stages of the AIDS epidemic, dementia used to be a common symptom, resulting from the brain damage caused by the many different infections. Thanks to the new multitherapy approach, dementia in AIDS patients is now a thing of the past.

I once gave a lecture at an Alzheimer's café—an informal gathering for Alzheimer's patients that has become popular in the Netherlands—and was approached during the break by a man of about forty-five who said that he was in the early stages of dementia. I responded that I could see no outward sign of it. He told me that he'd already had a few small cerebral hemorrhages that he knew would persist, ultimately resulting in dementia. I asked him if he had relatives in the little Dutch fishing village of Katwijk. "Yes," he answered. "Your diagnosis is spot-on, Professor!" The reason I asked was that there's a family in that village with a rare mutation that causes amyloid buildup in the blood vessels, leading to cerebral hemorrhage and dementia. Its members know exactly what fate has in store for them, because they have seen so many relatives deteriorate in this way. Yet it's a very rare form of dementia, just like Creutzfeldt-Jakob disease, which is caused by infections in abnormal proteins. It can have a genetic cause but was previously also transmitted in brain operations, before surgeons knew that their instruments needed to be sterilized in a certain way. People have also been infected through corneal transplants and the pituitary extracts formerly given to children lacking growth hormones. These hormonal extracts, which probably come from Russia, are sometimes still found in gyms, where they are used by bodybuilders who want to increase muscle mass. It's a highly risky business: You need only one Creutzfeldt-Jakob carrier to make the entire batch deadly.

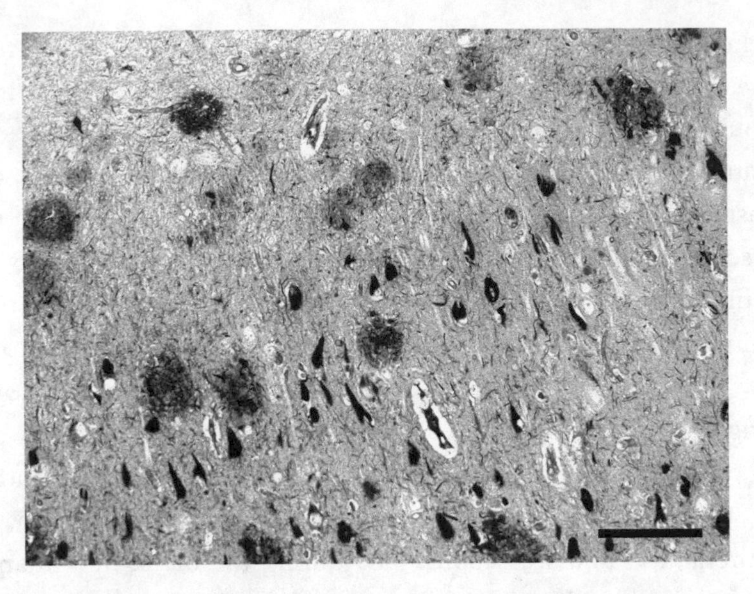

FIGURE 29. A Gallyas silver stain of brain tissue (cortex) from an eighty-five-year-old, showing the two types of lesions associated with Alzheimer's: the large, round, amyloid-containing plaques between the neurons and the black neurofibrillary tangles in the neurons. The bar is a size marker (100 micrometers). Courtesy of Unga Unmehopa.

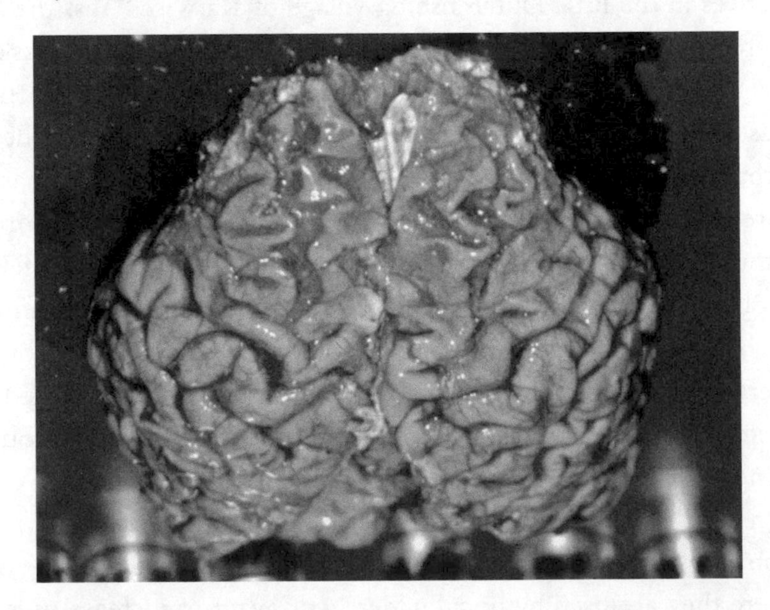

FIGURE 30. In frontotemporal dementia, the front of the brain (shown at top center) shrinks dramatically, while the rest of the brain remains intact. Courtesy of the Netherlands Brain Bank.

A variant of Creutzfeldt-Jakob disease is mad cow disease, which originated when infected protein from the brains of cows ended up with other offal in hamburgers. Huntington's disease is an inherited form of dementia. People in whose families it runs are familiar with the symptoms, having seen them in relatives: jerky movements and lack of coordination. When they start displaying those symptoms they know that they are on the road to developing dementia.

So there are many forms of dementia, but most result from Alzheimer's. If you knew nothing about all these different types and diagnosed all patients with dementia as having Alzheimer's, you would ultimately be proved right in most cases; microscope samples would reveal either Alzheimer's or a mixture of vascular changes of which Alzheimer's was a part.

What Causes Alzheimer's?

> Alzheimer's can be seen as a premature, accelerated, and severe process of brain aging.

In recent decades, research into Alzheimer's has devoted considerable attention to a couple of rare genetic forms of the disease. In Belgium, there are two families in which members develop Alzheimer's at the age of thirty-five. Most die between the ages of forty and fifty. In families like these, mutations have been found in the genes for beta-amyloid precursor protein (βapp) and presenilin 1 and 2. However, we need to bear in mind that these mutations account for less than 1 percent of all Alzheimer's patients. Age and a variant form of a gene called apolipoprotein E-ε4 (ApoE-ε4) are by far the main risk factors for the form of Alzheimer's that occurs in 94 percent of Alzheimer's patients over sixty-five. The ApoE-ε4 gene is thought to be responsible for around 17 percent of all cases of Alzheimer's. But unlike the three above mutations, simply having that form of the gene doesn't mean that you will definitely develop the disease, just that you're more likely to do so. After learning how to identify the ApoE-ε4 gene, our students wanted to find out whether they had it. But we forbade them from testing themselves for the gene. Knowing that you have the ApoE-ε4 gene can lead only to worry. You may never develop Alzheimer's—but you'll be tormented by the knowledge that if you do get it, there's no cure. Molecular genetic research of archived samples taken from the brain of the first person diagnosed with Alzheimer's a hundred years ago, a fifty-one-year-old woman named Auguste D., revealed neither any of the known mutations nor ApoE-ε4. So this was a case of someone who developed Alzheimer's at a very early age without having any of the genes that are mostly responsible for the condition.

Clearly, extremely complex interactions between genetic background and the environment play a role in determining whether someone will develop Alzheimer's. But how do all those different factors lead to the same form of dementia? The most popular hypothesis is that risk factors result in a buildup of toxic amyloid beta peptides (β A4) in the form of plaques, which are thought to alter transport proteins and make them stick together (the tangles), disrupting cell function and causing the neuron to die. Like deadly relay runners, infected neurons then pass on toxins, spreading the disease through the brain in a six-stage pattern described by Braak and Braak. Indeed, Alzheimer's does seem to follow a set neuroanatomical route, starting in the same brain structure (the entorhinal cortex, fig. 26), traveling on to the limbic system and finally to the cerebral cortex. Although there's much to be said for this theory, known as the amyloid cascade hypothesis, in the case of the rare families that have βapp mutations, there are at least as many arguments against it in the case of the most common form of Alzheimer's, which is non-inherited. So far, studies of transgenic mice haven't shown that amyloid cascades are responsible for the creation of tangles in the sporadic type of Alzheimer's. I lean toward the theory that Alzheimer's is simply an accelerated form of brain aging. Every active neuron sustains damage, just as a car engine does, through wear and tear. Unlike car engines, neurons can repair themselves—but only to a certain extent. Tiny flaws remain and mount up over the years, causing the degeneration that is the aging process. In the case of individuals whose brains aren't good at repairing themselves or who incur a lot of brain damage, like professional boxers (see chapter 12), this degeneration is more serious and occurs faster, resulting in plaques and tangles, which lead to the onset of Alzheimer's. If this theory is correct, the only way of preventing Alzheimer's would be to halt brain aging. And that's bad news, because we're a long way from being able to do that.

ALZHEIMER'S: THE STAGES OF DETERIORATION

> Be nice to your kids, they'll choose your nursing home.
>
> Text on a mug my daughter gave me

A third of those who suffer from Alzheimer's are unaware that they have the disease. (Lack of awareness of illness is a medical condition in its own right, known as anosognosia.) They deny that there's anything wrong with them and have to be dragged to a doctor by their partner. Someone I knew had taken his wife, who was developing dementia, to a symposium of mine at which there was much discussion of Alzheimer's. A concerned friend asked her, "Wasn't that a bit confrontational for you?" to which she replied, "No, but it must have been for anyone with Alzheimer's." Others, though, realize early on that something is wrong. When Harold Wilson was reelected as British prime minister in 1974, he noticed that he was beginning to lose his power of perfect recall. In 1976, to universal surprise, he decided to resign. Two years later, he experienced the first symptoms of Alzheimer's.

The onset of Alzheimer's can be insidious and its progress extremely protracted. When Ronald Reagan became president of the United States in 1981 at the age of nearly seventy, he solemnly declared that he would resign if he got Alzheimer's. Looking back, there are indications that he developed the disease in 1984. Analysis of his performance in debates shows that he was starting to misuse articles, prepositions, and pronouns. He also paused five times more frequently and spoke 9 percent more slowly than before. In 1992, the condition manifested itself more plainly, and in 1994, ten years after those first changes in his speech patterns, Reagan wrote to his fellow countrymen to announce that he was one of the million Americans with Alzheimer's. He died a decade later, twenty years after the onset of the disease.

Alzheimer's travels through our brains by a fixed route. When we look at a brain sample under a microscope, we can see the first telltale signs of the disease, the tangles in the cerebral cortex of the temporal lobe (the entorhinal cortex, fig. 26). The next sign is abnormalities in the hippocampus. These changes appear before any symptoms do; in fact, the person who gave us permission to use their brain as a "control" in our studies wasn't aware that they were already ill. At present, the very first signs of the condition can't be identified while the sufferer is alive. But once it has progressed, severely damaging the temporal cortex (fig. 1) and the hippocampus (fig. 26), the first memory problems appear. The sufferer is unable to remember recent events, yet can still recall minute details of events in the distant past, like a party at elementary school. When the disease attacks the remaining areas of the cerebral cortex, dementia ensues. The rear part of the brain, the visual cortex (fig. 22), is the last to be damaged. Some painters with Alzheimer's can have full dementia and yet at the same time retain their creative and artistic powers. Artists have been able to make excellent portraits in this state while being incapable of determining their value or negotiating a price for them. Their visual cortex functions right up to the end.

In Alzheimer's, not only do microscopic changes follow a set pattern, but functional losses do as well. We lose abilities in almost exactly the reverse order in which we acquire them. Dr. Barry Reisberg of New York has identified seven stages of Alzheimer's. In Stage 1 you still function normally. In Stage 2 you start to lose things and find it hard to carry out your job but can still maintain a semblance of normality. In Stage 3, your co-workers notice that you can no longer handle difficult situations at work. In Stage 4 you have trouble with complex tasks, like handling finances. You then start to need help choosing what to wear (5). After that you need help getting dressed (6a) and getting washed (6b), you can no longer go to the toilet unaided (6c), and you develop urinary incontinence (6d) and fecal incontinence (6e). By Stage 7a you can only speak about half a dozen intelligible words, after which you lose the power of speech entirely (7b). You can no longer walk (7c) or sit unaided (7d). You then lose the ability to smile (7e)—a skill that made everyone so happy when you were a baby—and the ability to hold your head up (7f). The patient ends up in bed, curled up in a fetal position (fig. 31); if you insert a finger in his mouth, he will show a sucking reflex, having at that point fully regressed to the condition of a newborn baby.

Language and music are stored in a part of the memory that's

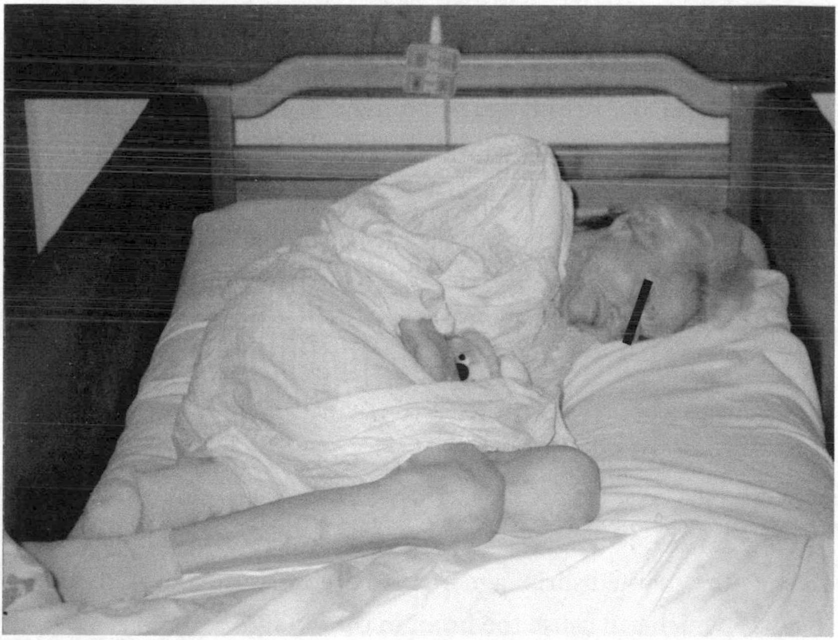

FIGURE 31. In the final stage of Alzheimer's, the patient lies curled up in bed in a fetal position. Courtesy of Professor E.J.A. Scherder of the Clinical Neuropsychology Department of VU University in Amsterdam.

only affected at a late stage of Alzheimer's. The ability to speak only disappears in Stage 7. Musical skills can be retained for a very long time. A professional pianist with Alzheimer's could no longer comprehend anything that was said or written, including sheet music. Yet she could retain pieces of music that she heard for the first time and reproduce them with musical feeling. At a later stage she could still play the melodies with which she was familiar, a pastime that gave her great satisfaction. A case has also been described of a violinist with Alzheimer's who retained his musical skills. The inverse of the brain's retention of musical skills is true too; music is one of the earliest influences in an infant's development. Indeed, the effects of music on brain function can be seen very early on. Premature babies in incubators become calmer, have better oxygen values, and are able to leave incubators earlier if music is played to them. Newborn babies are much more interested when a mother sings than when she speaks, and they already have a sense of rhythm. So, rather like businesses undergoing reorganization, Alzheimer's functions along the lines of "last in, first out," with the most senior members of staff being allowed to stay put. But of course the brain isn't being reorganized in Alzheimer's; it's being demolished.

"USE IT OR LOSE IT": REACTIVATING NEURONS IN ALZHEIMER'S DISEASE

> As long as a brain with Alzheimer's still has neurons—even if they are shrunken and no longer function—they can in principle be reactivated.

Despite the marked shrinkage of the cerebral cortex in Alzheimer's (fig. 32), which can cause the brain in the skull to resemble a walnut in its shell, it retains all its neurons. In contrast to what's generally thought, brain cells don't die en masse as a result of Alzheimer's. Cell death is limited to regions like the entorhinal cortex, part of the

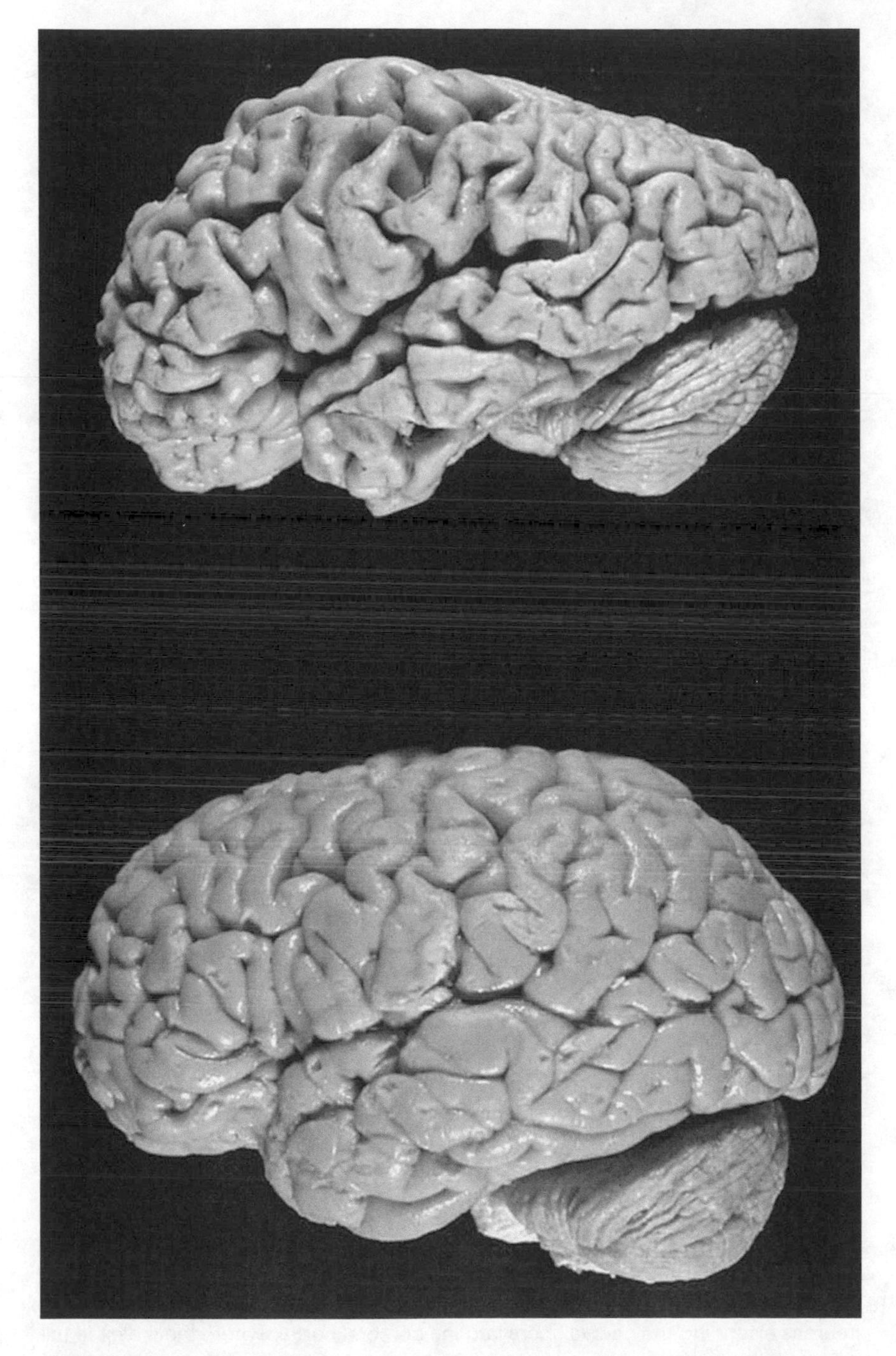

FIGURE 32. A characteristic symptom of Alzheimer's is marked shrinkage (atrophy) of the entire cerebral cortex, which can cause the brain to look like a walnut (a normal brain is shown underneath). Courtesy of the Netherlands Brain Bank.

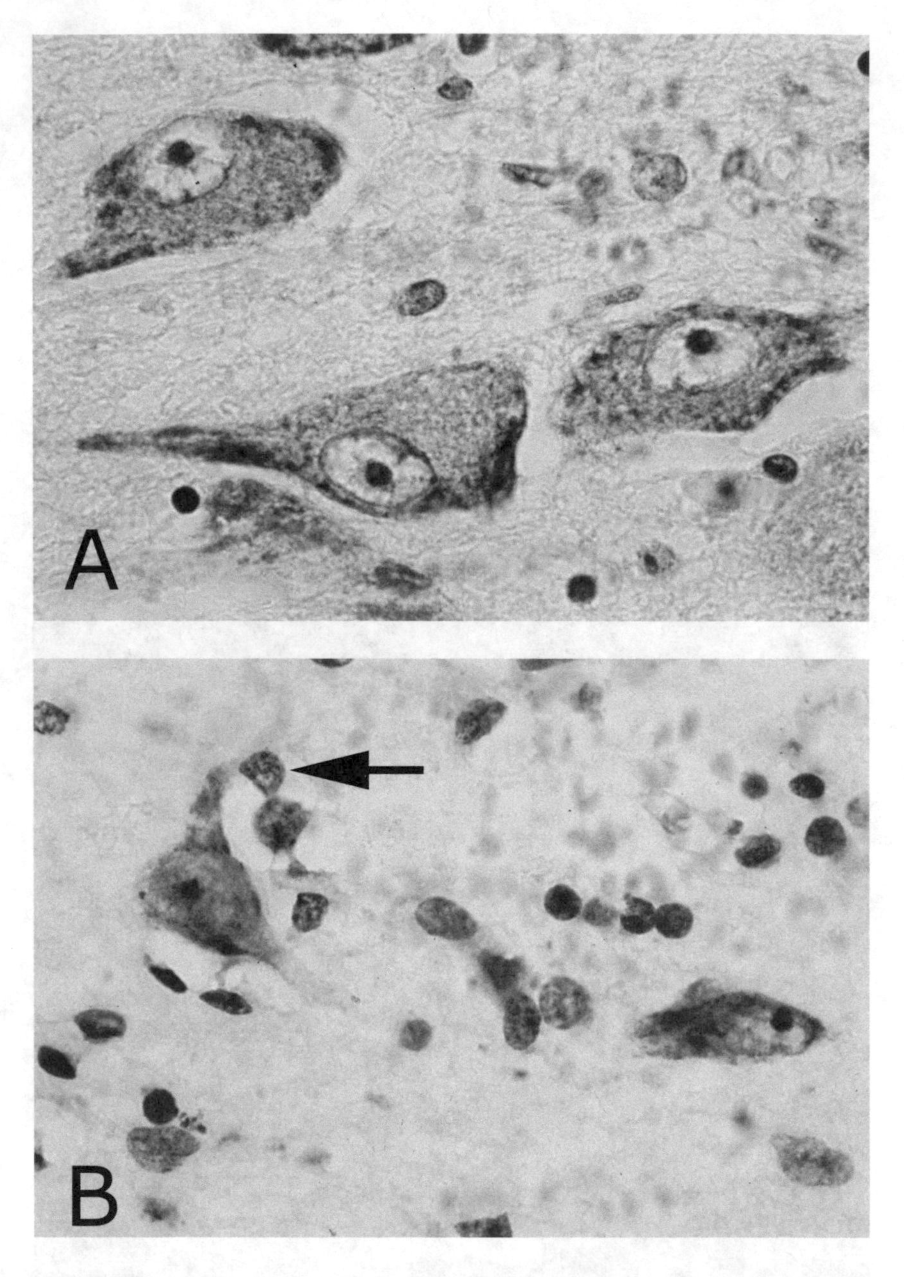

FIGURE 33. Microscope slides showing atrophied neurons in the nucleus basalis of Meynert. A is a control sample taken from a healthy patient, showing the large neurons extending their nerve fibers into the cerebral cortex, where they release the neurotransmitter acetylcholine (see also fig. 25). B shows how Alzheimer's causes these cells to shrink (the arrow points to a group of three dramatically shrunken neurons). Courtesy of Dr. Ronald Verwer.

hippocampus, and the locus coeruleus and only occurs at an advanced stage of the disease. Conversely, reduced activity, leading to neuronal shrinkage (fig. 33), affects the entire brain from an early stage. This also explains why symptoms can fluctuate so strongly at the beginning of the disease. Someone can show marked signs of senility one moment but be able to carry on an intelligent conversation the next. If the memory disorders in the early stages of Alzheimer's were indeed due to cell death, such fluctuations wouldn't occur, as cell death isn't reversible.

Activation Versus Alzheimer's

While patients are of course indifferent to whether their dementia is due to loss of neurons or reduced neural activity, the difference is crucial for developing therapeutic strategies. If neurons are still there, albeit atrophied and nonfunctioning, it should, in principle, be possible to reactivate these cells, which is a focus of our research.

Growing up bilingual, having a good education and a challenging job, and remaining active in old age reduce the likelihood of Alzheimer's. This suggests that maximizing brain activity has a preventive effect. There are, moreover, areas of the brain in which neurons aren't affected by the disease. We have found such areas to be extremely active, sometimes even becoming more active as aging progresses. In 1991, I paraphrased the hypothesis that activating neurons appeared to provide some protection against aging and Alzheimer's as "Use it or lose it."

Studies also show that activation can reduce Alzheimer's pathology. Transgenic mice with considerable buildup in their brains of the toxic protein amyloid (a symptom of Alzheimer's) were placed in an enriched environment—a large cage in which they could play with one another and in which they were regularly treated to new toys. When mice remained in this environment, their amyloid levels decreased (and went down even further if they also got more exercise than usual). Sadly, the research team headed by Erik Scherder at the

VU University in Amsterdam couldn't link extra physical exercise to improved function in Alzheimer's patients. Yet he did find, in an earlier study, that general stimulation of the group's brains (through transcutaneous electrical nerve stimulation) had a beneficial effect on cognition and mood. The team headed by Mark Tuszynski in San Diego is using gene therapy to stimulate the nucleus basalis of Meynert (fig. 25), a brain structure that's important in memory, in Alzheimer's patients with promising results (see chapter 11).

Stimulating the Biological Clock with Light

To test the effectiveness of activating neurons affected by Alzheimer's, we opted to stimulate the circadian system. The study also had clinical importance, because nighttime restlessness is the main reason that people with dementia are institutionalized. It makes them putter around late at night, turning on the gas and doing other potentially dangerous things, or go out and wander the streets. Sooner or later their partner can no longer cope with the Herculean task of caring for them and keeping an eye on them day and night. The circadian system (from the Latin *circa,* meaning "around," and *diem,* meaning "day"), which is responsible for all our day and night rhythms, is affected very early on in Alzheimer's. Patients no longer get their nightly peak of melatonin, the sleep hormone secreted by the pineal gland. We established that these early changes are caused by the biological clock, the suprachiasmatic nucleus, a structure that can easily be stimulated by means of light therapy. As expected, light therapy improved circadian rhythms and reduced restlessness in Alzheimer's patients. It didn't work in patients with impaired vision—a nice control demonstrating the effectiveness of light. A three-and-a-half-year follow-up study by Eus van Someren and his team showed that more light doesn't just stabilize rhythms but also improves mood and even slows the deterioration of memory. The combination of more light during the day and melatonin supplements before sleep proved even more effective in some respects.

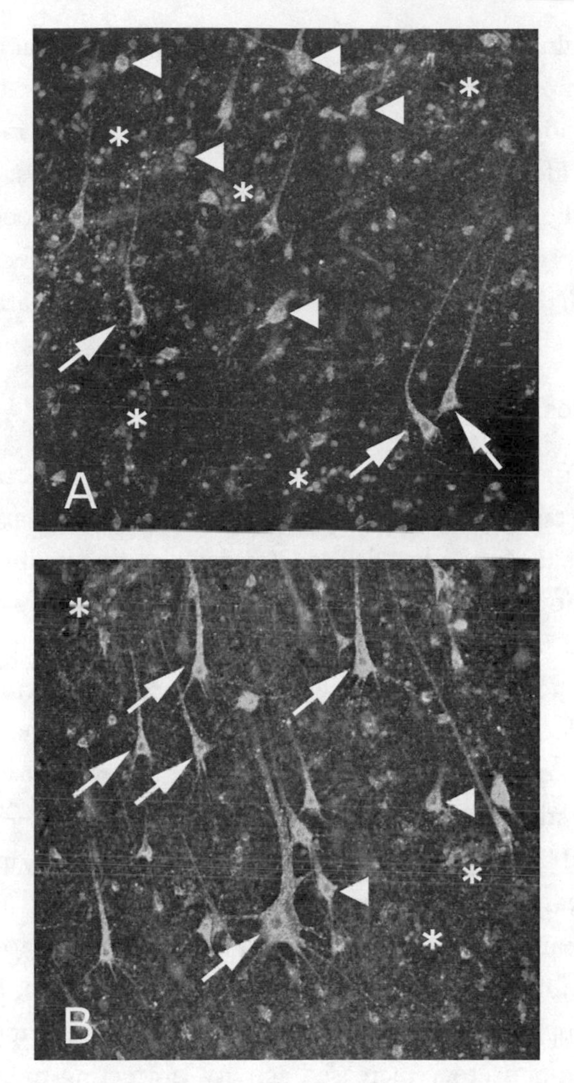

FIGURE 34. Thin slices of tissue removed from the brain of a patient with Alzheimer's within ten hours of death. At this stage, neurons can be cultured for many weeks. In this model, stem cells proved to secrete an unknown substance that improved the survival of the cultured neurons. After forty-eight days in normal culture conditions (A), only a few neurons are still active and intact (indicated by arrows). These are outnumbered by neurons with leaky membranes (triangles), evident from their colored nuclei. Many nuclei of dead cells can also be seen (small dots, some of which are marked with an asterisk by way of example). B shows how, after a slice of this kind is cultured with stem cells, there are many more active, intact neurons (arrows) and fewer leaky (triangles) and dead (asterisks) cells. Courtesy of Dr. Ronald Verwer.

The results of this simple intervention are fully equal to those of current anti-Alzheimer's drugs and do not have any of the side effects. Although stimulating the biological clock can improve the quality of life of Alzheimer's patients and their carers, it isn't, of course, a therapy for the disease itself. However, it does prove an important principle, namely that even if neurons are affected by Alzheimer's, their function can be restored through stimulation.

Current Research

The Netherlands Institute for Neuroscience (NIN) is currently looking at substances that can activate neurons in other areas of the brain. Ronald Verwer has devised a procedure that enables neurons in thin sections of brain tissue obtained within ten hours of death to be cultured for several weeks. This allows us to test the effects of potentially activating substances without inconveniencing patients. In this model, stem cells proved to secrete a substance that promoted the survival of the cultured neurons (fig. 34). But as yet we have absolutely no idea what the nature of this compound is.

A second line of research is based on the finding that, in the initial stages at least, the brain appears to resist Alzheimer's successfully. Our NIN team with Koen Bossers and Joost Verhaagen found this to be the case in the very earliest stage of the condition, before any memory impairment occurs. A number of brain areas appear to engage in compensatory increased activity, protecting memory for a time. Our team found that hundreds of genes are activated in the prefrontal cortex before typical Alzheimer's-type changes affect that region. The study's conclusion is that the prefrontal cortex (fig. 15) does its utmost to continue to function normally when Alzheimer's first strikes. However, this compensatory mechanism eventually fails, metabolism decreases, and memory impairment begins its relentless progress. We hope that by studying the brain's initial defense mechanisms against Alzheimer's we can develop new medication. It's just a pity that the pace of research is often so frustratingly slow.

PAIN AND DEMENTIA

Dementia is a humiliating condition that tends to go hand in hand with depression and, especially in its early stages, with fear. It's a reason why many of us decide that we don't want to undergo this process. A committee of the Dutch Association for Voluntary Euthanasia (NVVE) on which I sat concluded that the current euthanasia legislation provides scope for people with dementia to opt for euthanasia, providing that they do so in time (see later in this chapter). The suffering associated with dementia can indeed be considerable, and it isn't just caused by the fear of deterioration. Neuropsychologist Erik Scherder is one of the few to have pointed out that the very brain disorder that underlies dementia also makes it much harder to diagnose and treat pain. Inadequate pain treatment for patients with dementia is a common, extremely worrying problem that increases with the severity of dementia. Sometimes, as in the case of vascular dementia, the disease generates its own "central" pain. In addition, many elderly people suffer chronic pain already, for instance because of arthritis, and since dementia is an aging disorder, it stands to reason that many elderly patients with the condition suffer from chronic pain as well. Yet if you look at painkiller use, a strange picture emerges. A patient with dementia who, say, breaks their hip, is prescribed fewer painkillers than a patient without dementia. This isn't because people with dementia don't suffer pain. Inadequate pain treatment has more to do with the difficulty that doctors experience in estimating the degree of pain suffered by this category of patients. People whose brains are intact can tell you how much pain they are experiencing. Their blood pressure and heart rate also go up in response to pain—an automatic reaction of the nervous system. In Alzheimer's, however, this system is impaired. As a result, moderate pain doesn't affect patients' blood pressure and heart rate; by the time an increase is perceptible, they are in severe pain. But there are methods for estimating pain levels, not just for patients in the early stages of dementia, who can still communicate, but also for patients with severe dementia who have lost all power of communication. For patients in the early stages of dementia, pain scales to indicate pain intensity have been devised. For patients with severe dementia, pain assessment has to be done on an observational basis, just as in the case of very young children.

Pain stimuli follow two routes. In the lateral system they travel up the side of the spinal cord to the part of the cerebral cortex where sensory input is processed. Since this part of the brain is left largely intact by Alzheimer's, pain stimuli are received and processed as normal, and people with the condition have normal pain thresholds. The second route transports pain stimuli up the center of the spinal cord to the cingulate cortex, an alarm center and an area much damaged by Alzheimer's. This is the medial pain system, which processes the emotional aspects of pain. Since the lateral pain system functions properly, people with Alzheimer's experience pain, but because their medial system is damaged, they don't understand what is going on. So they respond in ways that we don't associate with pain. They frown or appear fearful or agitated.

The extent to which patients with dementia suffer pain, moreover, depends on the cause of their dementia. Patients with vascular dementia experience more pain through disruption of fiber systems in the brain, while patients with frontotemporal dementia lose the ability to process pain emotionally. It wouldn't be my own choice, but there are people with dementia who decide to carry on to the bitter end. In their case it's crucial to ensure professional diagnosis and treatment of their pain, because no scientific evidence has yet been found to support the notion that suffering is ennobling.

ALZHEIMER'S AND THE RIGHT MOMENT FOR CHOOSING TO END ONE'S LIFE

> We have let him go. Respecting his courageous decision to be a step ahead of the disease that affected his memory, we take our leave of a resolute man and loving partner, father, father-in-law, grandfather, friend, and father of my children.
>
> Death announcement in *Het Parool,* March 20, 2010

On the evening of Tuesday, November 11, 2008, Nan Rosens's impressive film *Before I Forget* was the subject of debate in the Rode Hoed cultural center in Amsterdam. The building was packed. In the documentary, Paul van Eerde explains that he doesn't want to experience the humiliation and loss of dignity that Alzheimer's entails. His wife and children support him in his difficult decision, and the family enjoys their remaining time together. However, the one person who doesn't endorse Paul's own choice is his family doctor. Paul isn't the only one to find out belatedly what his doctor's stance is. Whereas the vast majority of the Dutch population favors euthanasia, assisted dying, or the concept of a suicide pill, 91 percent of the population don't know their own family doctor's views on the matter. But that doesn't have to be the case. I know an eighty-year-old entrepreneur who moved to a new city and found a new doctor. The first thing he said to the doctor was, "I have two questions I'd like to ask you; one is more urgent than the other: How do you feel about abortion and how do you feel about euthanasia?" Unfortunately not everyone is comfortable being so assertive.

The second lesson of the film was that the doctor who didn't want to cooperate also didn't refer Paul to a colleague who would be prepared to help. We should start tackling this issue by training medical students properly and making sure that doctors get on-the-job training to deal with the difficult issue of euthanasia. It's essential for a patient to build up a good long-term relationship with their doctor so that both parties can anticipate the right moment to end life. You can't start such preparation early enough, and making a living will while you're still in the peak of health is a good way to establish your doctor's views on euthanasia and to start to build the necessary relationship—or to determine that you need a different doctor.

In its earliest stage, dementia can only be reliably diagnosed by a memory clinic. Get a referral to such a clinic if you or your partner is worried about your memory. If you're diagnosed with early dementia, you must start to anticipate "the right moment." It's natural to want to enjoy life as long as possible, but if you wait too long you will no longer be able to confirm your wish for euthanasia, and your doctor will be unable to help you. In the early stages of Alzheimer's, people are still mentally competent. After that they have lucid moments in which they are aware of their situation, but there comes a time when those, too, disappear. The former health minister Els Borst-Eilers said that she would want to end her life when she was unable to recognize her children and grandchildren. That's so late in the progression of the disease that it may create problems for a doctor. The right moment to end one's life is different for everyone and must be decided in close consultation with one's physician. Bear in mind that doctors, too, find this far from easy. The pioneer in the field of euthanasia, Sytske van der Meer, has said that she prefers the lethal substance to be given as a drink, because the patient has to drink it of his own volition, showing that he has stuck to his decision right up to the end. Others prefer a drip, because death then follows very quickly. This choice, too, must be properly discussed with the doctor. The debaters in the Rode Hoed agreed that the current euthanasia legislation provides more scope for assisted dying in early dementia than is commonly thought. To date, an assessment committee has judged thirty-five cases of euthanasia or assisted dying to have been conducted with all due care. This shows that the law gives doctors the protection they need. And awareness of this fact is fortunately increasing.

19

Death

Die, my dear doctor! That's the last thing I shall do!

Last words of Lord Palmerston, British prime minister

Death is peculiar. First you make a fantastic organism, then you throw it away again fifty years later. It's a rotten trick, and if God existed I'd like to meet him in a dark alley to have a little chat about it.

Midas Dekkers, *De Volkskrant*

THE MAGIC OF LIFE AND DEATH

Being dead will be no different from being unborn.

Mark Twain

Life is a sexually transmitted disease that is invariably fatal.

Life and death are hard to define. Life must meet certain criteria, like movement, metabolism, growth, independent reproduction (for which information-carrying molecules like DNA or RNA are needed), integration, and regulation. Although the last two characteristics can also be found in unicellular organisms, they only really came into their own when neurons evolved. In isolation, these characteristics aren't proof of life. Flowing water moves, iron undergoes metabolic changes that we call rusting, a crystal can grow, and these days an increasing number of young people believe that life has more to offer *without* reproduction. Integration and regulation are characteristics that can also be programmed into a computer. A combination of all of these criteria must be present for life to be said to exist.

For centuries, doctors have based their pronouncement of death on the absence of a heartbeat and breathing and the assumption that those functions won't resume. After a few tense minutes, a doctor can be increasingly certain of his diagnosis. As a one-line Dutch poem entitled "Death" puts it, "Nothing you can do about it." We were always taught that neurons are extremely sensitive to oxygen deprivation. After four or five minutes without oxygen, irreparable brain damage was said to result. And that's true, but it's the capillary cells, not the neurons, that are particularly sensitive to lack of oxygen. It makes them swell so much that even if the heart starts to beat again and breathing resumes within four to five minutes, the red blood cells can no longer get through the capillaries in the brain to give off oxygen. Moreover, in the stage that follows, toxic substances are released and the neurons ultimately die.

In "The Roller Coaster" (1953), a charming story about transplanting beautiful memories, the Dutch writer and doctor Belcampo predicted that by the year 2000 it would be possible to culture human neurons. Indeed, if the Netherlands Brain Bank receives postmortem tissue within ten hours of the donor's death, we can culture neurons in thin slices of brain tissue for weeks (fig. 34). In 2002, Ronald Verwer discovered that the cells in these slices are still able to produce proteins and to transport substances. They can also be electrically active. In fact, glial cells, which provide the support structure for the neural system, can even be cultivated from brain tissue a full eighteen hours after death.

Culturing slices of postmortem brain tissue shows that brain cells can survive ten hours of oxygen deficiency and that the death of an individual isn't the same thing as the death of his neurons. The question of what life and death actually are becomes even more intriguing when one considers that these living cells are constructed from dead molecules like DNA, RNA, proteins, and fats. Would it be possible to create life from dead molecules? In 2003, Craig Venter made the first step in this direction by synthesizing a virus (Ph-X174) from dead material. But to reproduce, a virus needs the entire molecular machinery of the cell that it infects. Since a virus can't reproduce independently, occupying a space somewhere between living and dead material, Venter's experiment doesn't count as synthesizing life.

Full reincarnation can be said to exist at the level of molecular building blocks. Atoms have such a long life that each of our own atoms existed in many millions of organisms before being built into our bodies. So there's a good chance that your body houses atoms that were once part of a famous historical figure. Cells also contain water molecules, and these too are recycled. We drink water that once came from rivers. It leaves our bodies as urine, is purified, travels to the sea, evaporates, and then returns to our taps via rain and rivers. The biologist Lewis Wolpert calculated that the number of water molecules in a glass of water is so great (outnumbering the number of glasses of water in the sea) that there's a real likelihood of us drinking one that has been through, say, Napoleon's bladder. So our molecules have been constructed from hand-me-down atoms and are surrounded by water that has passed through countless bodies.

In principle, the molecular building blocks of life can be synthesized. The theory goes that if all the necessary molecules are brought together in the right way, life will appear as an emergent, new characteristic. That this is the case can only be proved by synthesizing, say, a living bacteria from dead material. In early 2008 Craig Venter synthesized the complete genome of the bacteria *Mycoplasma genitalium,* which involved over half a million building blocks. In 2010 he succeeded in getting the bacteria to replicate. After thirty cell divisions the original proteins were diluted out, so that the proteins of the cells were all produced by the synthetic genome. At the time

Venter estimated that the project to synthesize an entire bacteria would be complete by the end of that year. While he failed to meet this deadline, he certainly came closer to his objective. If he does eventually succeed, he can't be sure of a Nobel Prize, because creationists will surely claim that his experiment has been done before, citing a previous magical experiment described in Genesis 2:7, "Then the Lord God formed man of dust from the ground, and breathed into his nostrils the breath of life, and the man became a living being."

DR. DEIJMAN AND BLACK JAN

> "Those who did wrong during their life become useful after their death."

Brain research has been conducted in Amsterdam since as early as the seventeenth century, although the context back then was rather different. Criminals who were condemned to death were hanged in the north of Amsterdam or on Dam Square. After execution, their bodies could be donated to the guild of surgeons for public dissection. The city authorities allowed such dissections to take place once a year (in winter, because they took three to five days, and the stench would have been unbearable in summer). Initially they were carried out in St. Margaret's cloister on Nes, where the Flemish arts center De Brakke Grond now stands. The guild chamber and dissecting room were located there from 1578 to 1619 and from 1639 to 1691, in the attic story of the meat market building. In the intervening period the surgeons' dissecting room moved to the Nieuwmarkt, above Amsterdam's weigh house. That is likely the place that inspired Rembrandt to paint *The Anatomy Lesson of Dr. Nicolaes Tulp* (1632), which now hangs in the Mauritshuis Museum in The Hague. The audience at public dissections consisted of a few hundred people who could afford the admission price of twenty cents. The heart, liver, and kidneys were passed around among the spectators. The justification for using bodies to train surgeons can still be read on the wall of what used to be the anatomy theater: "Those who did wrong during their life become useful after their death."

In 1656, Rembrandt records a crucial moment of such a dissection in his *Anatomy Lesson of Dr. Deijman* (fig. 35). The lecturer, Jan Deijman, stands behind the dissected cadaver of Joris Fonteijn, nicknamed "Black Jan," a Flemish tailor turned thief who was condemned to be hanged on January 27, 1656. He was executed that same month,

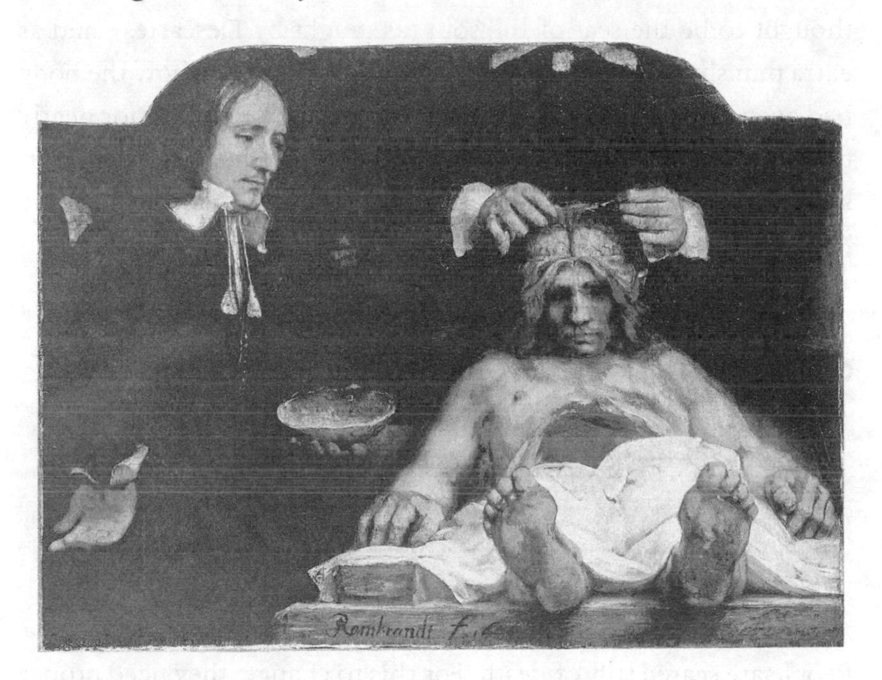

FIGURE 35. In 1656, Rembrandt recorded a crucial moment in the public dissection of a criminal in his *Anatomy Lesson of Dr. Deijman*. Surgeon Jan Deijman stands behind the dissected cadaver of "Black Jan," a Flemish tailor turned thief. Assistant surgeon Gijsbert Calcoen stands patiently holding the top of the skull in which the brain will be deposited. Meanwhile, Dr. Deijman is using a pair of forceps to hold up the falx cerebri, the crescent-shaped membrane that separates the left and right cerebral hemispheres, thus exposing the pineal gland. The protocol required surgeons to do this at the end of an autopsy, because in those days the pineal gland was thought to be the seat of the soul (as taught by Descartes), and as an extra punishment the criminal's soul was made to see how the body had been dissected. Amsterdam Museum.

probably on a temporary scaffold on Dam Square, in front of the old City Hall. His body was dissected at the surgeons guild in the former chapel of St. Margaret's cloister. Rembrandt's painting shows the assistant surgeon, Gijsbert Calcoen, patiently holding the top of the skull in which the brain will be deposited. Meanwhile, Dr. Deijman is using a pair of forceps to hold up the falx cerebri, the crescent-shaped membrane that separates the left and right cerebral hemispheres, thus exposing the pineal gland (fig. 2). The protocol required surgeons to do this, because in those days the pineal gland was thought to be the seat of the soul (as taught by Descartes), and as extra punishment the criminal's soul was made to see how the body had been dissected. Descartes lived in the Dutch Republic for nearly twenty years. He bought carcasses for his research at the cattle market near his lodgings, and his findings clearly left their mark on the Amsterdam of Rembrandt van Rijn.

CIVIC DISINTEGRATION COURSE: DEAD OBVIOUS

> I want to be the boss when it comes to decisions about my own life. I wasn't able to have any say in my conception and birth. But I demand to exercise this right in full when I get to the end of my life.

People are scared stiff of death. For this to change, they need proper information about the last stage of life, long before the time comes to die. In 2002 I responded to the fierce debate about introducing civic integration courses for foreigners—a discussion scarred by xenophobia—by urging the Health Council to make it compulsory for everyone in the Netherlands to do a civic *disintegration* course. At the request of the council's chair I drew up a debating paper on the issue, registered for posterity under the number 655-84. We had an animated discussion, but I wasn't at all surprised that the recommendation didn't become part of the formal advice given to the government by this respectable body.

Yet I would still very much like to see an "exit course" of this kind set up for the general public and, in an amended form, for trainee doctors. It would cover all of the problems associated with the end of life, like euthanasia, pain relief, palliative sedation, and terminal dehydration. (On this latter point: In their book *Way Out: A Dignified End to Life on Your Own Terms,* Boudewijn Chabot and Stella Braam write that terminal dehydration need not be a terrible form of self-euthanasia if the right preparations are made, the mouth is kept moist, and a doctor provides the necessary medication in the final stage.)

Assisted dying also needs to be discussed. The stance of the Dutch Voluntary Life Foundation (SVL) on this is, "We are entitled to choose our time of death and to obtain the means to achieve this in a humane manner." The Dutch Association for Voluntary Euthanasia distinguishes three groups whose needs aren't met by the current statutory procedures: people with dementia, chronic psychiatric patients, and old people who feel that their time has come. As far as the first two categories are concerned, the current euthanasia act has been applied in practice and is found to be usable (though doctors will only very exceptionally be prepared to countenance assisted dying in the case of psychiatric patients). However, for people who feel that they have reached the end of their lives but aren't terminally ill, the law needs to be changed. Yvonne van Baarle, former secretary of the Arts Council, set up a group to try to bring this about. We started an online signature campaign calling for this issue to be discussed in the House of Representatives. To our amazement, we collected the forty thousand required signatures within four days. Given the enormous positive interest, it's amazing how much abuse we received; our group was dubbed the "death squad."

Decisions by patients to refuse treatment (living wills) are also frequently problematic. Although doctors are obligated by law to abide by such decisions, they almost never do. Yet resuscitation can sometimes be unwise or even, as the Dutch doctor Bert Keizer memorably summed it up, "an extreme form of abuse." While still an intern, I once saved somebody's life in this way, and I regret it to this day. The patient suffered a heart attack as the nursing staff were wheeling his bed into the ward. I immediately put my newfound knowledge into practice and managed to resuscitate him. Shortly afterward we received his medical file. It turned out that he had a lung carcinoma that had grown into his heart. In the days that followed, I sat by his bed day and night as he struggled for breath, constantly clearing his blocked airways. I could have saved him all of that misery if I hadn't resuscitated him! Things are changing, though, and Ruud Koster, a cardiologist at Amsterdam's Academic Medical Center, has shown that the chances of survival after resuscitation have improved considerably. Indeed, the odds have more than doubled over the last decade, with people now surviving heart attacks in 20 percent of cases. The availability of increasingly effective automatic external defibrillators (AEDs) and better treatment prevent significant brain damage in over half of heart attack survivors. Cooling a patient's body temperature down right after resuscitation can prevent much of the brain damage caused by the release of toxins after a period of oxygen deficiency. And having a heart attack near an AED gives you an optimal chance of survival. There seems to be less and less reason for NVVE members to carry "do not resuscitate" medic alert emblems. On the other side of the coin, in nine cases out of ten, resuscitating a newborn baby for more than ten minutes without a heart response leads to severe brain damage and should therefore be abandoned. How many would-be parents are aware of that?

After death—which, like it or not, is inevitable—you can donate your body "to science," which means that it will be used for medical students to learn anatomy. There's nothing wrong with that, but if you really want to help science, you'd be better off donating your brain to the Netherlands Brain Bank, which has provided five hundred research groups around the world with brain tissue from over three thousand postmortems, resulting in hundreds of publications with new insights into neurological and psychiatric disorders (see later in this chapter). It can also be clinically important to consent to a postmortem, which is carried out to establish whether the diagnosis and treatment were correct. At present, such consent is requested at the moment when someone has just died and all the relatives are numb with grief. I was recently present on one such occasion, and it was obvious from the way the trainee internist put the request that he didn't expect the relatives to agree to a postmortem. Indeed, everything he said seemed to militate against it, perhaps because it would save him a lot of work. Doctors are insufficiently trained to discuss this topic, which is especially sensitive when someone has just died. Such a fraught time is of course the worst possible moment to approach the family on this matter, and it's no wonder that the number of postmortems has declined considerably.

Topics that need to be discussed well in advance include brain death, tissue transplants, and the transplant of organs and corneas. Of course the same applies to burial or cremation. Other issues that an exit course would need to cover are Alzheimer's cafés, active donor registration, preparing a body for burial, embalming, coma and related situations (see chapter 7), near-death experiences (chapter 16), advice on euthanasia, cultural differences affecting the issues surrounding the end of life, the legal aspect of these issues, the molecular biology of life and death (see earlier in this chapter), mummification, non-heart-beating donation, psychological problems at the end of life, and living wills.

So even if there isn't a life after death, death itself gives us a lot to think about. It's so much easier for everyone involved if they know your views on this issue and have plenty of time to discuss them with you. As for me, my brain is going to the Netherlands Brain Bank. If I have time, I will give my colleagues written instructions as to what they should especially look for, along with some technical suggestions on how to go about it, which will undoubtedly irritate them. The rest of my organs and tissue can be used for transplants—that's if anybody still wants them, because they've seen quite a lot of use already. And if doctors think that a postmortem is useful, they have my consent to perform one. I don't care what happens to the rest of me. That's a matter for my family to decide.

If you have any more suggestions for the exit course, I'd like to hear them. Are you looking forward to it? The course, I mean. For I wish you a healthy and enjoyable life that lasts as long as you want it to.

NETHERLANDS BRAIN BANK

The bank that takes your thoughts into account.

To find out the cause of brain disorders, you need to study the brain tissue of dead patients. In the late 1970s, however, it took me four years to obtain five clinically well-documented brains of Alzheimer's patients, even though there were one hundred thousand such patients in the Netherlands. This was because they didn't die in a teaching hospital but at home or in nursing homes. And it was impossible to obtain control material, because no one saw any point in carrying out postmortems of patients who didn't have brain disorders. But every piece of brain tissue from a patient with a brain disorder has to be compared with exactly the same piece of tissue from someone of the same age and sex who didn't die of a brain disease, who died at the same time of day, from whom the tissue was obtained at the same interval after death, and so on.

That's why in 1985 I took the initiative to set up the Netherlands Brain Bank (Nederlandse Hersenbank, NHB), which provides researchers with well-documented brain tissue. The neuropathologists of Amsterdam's VU University were involved right from the start. In just over twenty years, the NHB has provided five hundred research projects in twenty-five countries with many tens of thousands of pieces of brain tissue from over three thousand donors (www.brainbank.nl). In 1990 it won an award for providing a good alternative to animal research, and in 2008 Princess Máxima, the wife of the Dutch crown prince, honored it with a visit.

At present, the NHB has two thousand registered donors who have given consent for a postmortem of their brain and for their brain tissue and medical data to be used for research purposes. When a donor dies, his death is established by an independent doctor and the NHB is contacted immediately. The donor has to be transported as quickly as possible, usually within two to six hours, to the VU Hospital for a postmortem. Around seventy pieces of brain tissue are dissected, eight of which will be used to establish the diagnosis. The remaining pieces are frozen at −80°C, cultured, or prepared in some other way and sent to the research groups. What is unique to the NHB is that the tissue is made available so soon after death. This is only possible because donors and their families have completed all of the paperwork in advance and know exactly what will happen when the donor dies. The undertaker, too, knows that speed is of the essence. I was once called by the police, who had stopped a speeding undertaker and were reluctant to believe his story that the deceased needed to be taken to the hospital as quickly as possible. But when another undertaker got stuck in rush-hour traffic, a police officer on a motorcycle escorted him down the hard shoulder with flashing lights.

The donors show great commitment. A donor with MS once called me to say, "I want to see the enemy." So he came to the brain bank, where we mounted a microscope on his wheelchair table and the director showed him sections of brain tissue from MS patients. We get the strangest requests. Somebody once asked whether a relative could combine NHB donorship with donating organs for transplantation and their body to science. When I asked which relative of his he had in mind, he answered, "My mother-in-law." He seemed eager to make sure that nothing of her came back! We have also had our fair share of legal problems. In 1990, we launched a campaign to recruit MS donors, which resulted in our being sued by the husband of someone with MS. He thought that MS wasn't a brain disease but a muscular disease, arguing, "My wife's not mad!" We were also able to reassure a donor who asked if we would postpone the postmortem until her aura had dissipated.

It certainly isn't an easy decision to register as an NHB donor. Sometimes it helps when I tell people that I'm always very reassured to think that, whatever stupid things I say or do while I'm alive, at least my brain will be put to good use by the NHB after I'm dead.

HERBS FOR LONG LIFE AFTER DEATH

> That it will never come again
> Is what makes life so sweet.
>
> Emily Dickinson

Traditional Chinese medicine has countless remedies that are believed to prolong life. It's also often said of the many delicious dishes eaten in China that the food in question is good for your body or for a certain organ and that it guarantees longevity. When I say that I'm not so much interested in a long life as a good and interesting life, people tend to look a bit bewildered.

However, while in China I did witness the power of medicinal herbs to conserve bodies over a long period of time. My family and I had returned to Hefei, to the Medical University of Anhui, where I'm a visiting professor, when I heard for the first time of a region known as the Jiuhua Mountains. A monk by the name of Wuxia, who lived there at the time of the Ming Dynasty (1573–1619), succeeded, over a twenty-eight-year period, in copying out eighty-one parts of the Buddhist scriptures using blood from his tongue and gold dust. He's alleged to have died at the age of 126, and his body is said not to have decayed at all in the three years after his death. The other monks, who believed that he was the reincarnation of the living Buddha, gilded his body and preserved his mummy, known as

"Monk Longevity," in the Longevity Palace. Apparently, five-hundred-year-old mummies were also preserved and venerated in other monasteries in the Jiuhua Mountains. I didn't understand how that was possible, because the climate in that region is extremely damp. My first Chinese PhD student, Zhou Jiang-Ning, who had meanwhile become a professor at Hefei, suggested that if I was doubtful, I should go and take a look. My wife and daughter decided to join me. The university lent us a car and driver, and we were accompanied by a Chinese doctor, Dr. Bao Ai-Min, who interpreted for us.

After a six-hour drive in the darkness we reached the mountains at such a late hour that the monasteries and the many temples were shut, so we spent the night in the little town of Jiuhua. The next morning we returned to the monasteries, where Buddhist monks were praying around a glass case. In it we could indeed see a mummy covered in gold paint and sitting in a prayer position. The living monks praying in front of it were instructed by the head monk to make way for us so that we could inspect the mummy. The structure of the body was perfectly intact; it could have been used for an anatomy demonstration. The individual muscles were clearly visible through the dry, thin skin. Every monastery in the Jiuhua Mountains had one or more of these "flesh bodies"—the rather blunt name given to the mummies. With the help of our Chinese interpreter, I asked the head monk how it was possible that the body of this particular monk had remained intact so long after his death. "Because he is holy," came the enlightening answer. In a jocular mood, I called Zhou in Hefei to tell him that we had found the solution to the puzzle: "He is holy." According to Zhou, monks who felt they were nearing the end of their lives stopped eating normal food. Instead they ate special herbs, sitting in a vat in which they were submerged up to the neck in a mixture of herbal solution, carbon, and lime. In that way they could sometimes dry and preserve their own bodies before they actually died. Those who did so were deemed holy. Meanwhile, my daughter had been invited to join the monks in prayer. They were extremely kind to her, explaining the mysteries of

Buddhist prayer. The striking combination of the small, shaven-headed Chinese monks and my tall daughter, with her long blond hair, joined in prayer made everyone cheerful. To what extent her participation in the prayer contributed to the mummies' further preservation is something that only time will tell. I'm afraid I haven't yet managed to get ahold of the recipe of the herbal solution.

20

Evolution

> It is notorious that man is constructed on the same general type or model as other mammals. . . . The brain, the most important of all the organs, follows the same law, as shown by Huxley and other anatomists. Bischoff, who is a hostile witness, admits that every chief fissure and fold in the brain of man has its analogy in that of the orang[utan]; but he adds that at no period of development do their brains perfectly agree; nor could perfect agreement be expected, for otherwise their mental powers would have been the same.
>
> Charles Darwin, *The Descent of Man*

NEGOTIATION AND INCREASE IN BRAIN SIZE

> Bigamy is having one husband too many. Monogamy is the same.
>
> Oscar Wilde

Over the course of evolution, our brain size and intelligence have increased enormously. Intelligence entails problem-solving ability, speed of thought, capacity to act purposefully, rational thought, and the ability to deal effectively with one's surroundings. There are many different kinds of intelligence—linguistic, logical, mathematical, spatial, musical, motor, and social—so IQ is rather a limited way of testing it. The link between brain size and intelligence has nothing to do with the absolute size of the brain. The human brain, weighing in at three pounds, is of course by no means the largest: That record belongs to the sperm whale, with its nearly twenty-pound brain, while the brains of elephants weigh ten and a half pounds on average. In fact, an elephant named Alice who lived in Luna Park, Coney Island, had a brain that weighed thirteen pounds. But whales and elephants are by no means as intelligent as humans. The *relative* size of the brain compared to the animal's body, however, does have a clear correlation with the quality of the brain as an information-processing machine, as Darwin established back in 1871 and the Dutch neuroscientist Michel Hofman calculated a century later.

A better measure of the level of evolutionary brain development is the encephalization quotient (EQ), a relative measure of an animal's brain weight on top of what is needed to regulate body functions. Humans indeed score by far the best using this measurement. EQ is largely determined by the development of the cerebral cortex. The increase in our brain size during evolution was caused by an increase in the number of building blocks (neurons) and their connections. So the number of neurons in the cerebral cortex is a good measure of intelligence. These are grouped in functional units called columns. Although the cerebral cortex grew enormously over the course of evolution, the cross sections of the columns remained almost identical, around half a millimeter. It was an increase in the number of columns that caused our brains to grow bigger and the cortex to become convoluted in the process. Despite all these changes, the blueprint for the brain remained the same, so the difference between the brains of humans and those of other primates is largely one of size. This evolutionary increase in brain size greatly increased our information-processing ability and went hand in hand with longer pregnancy, a longer period of development and learning, longer life expectancy, and fewer offspring. During the course of human evolution, skull content has more than tripled and life span has doubled in a "mere" three million years.

Various hypotheses have been put forward to account for the evolutionary pressure that led to larger brains. An initial theory was that primates' brains provided an evolutionary advantage through the ability to use tools, which increased food supply. It was then suggested (the Machiavellian intelligence theory) that larger brains were a response to the demands of a socially complex existence, causing individuals to invest in social strategies that promoted long-term survival of the group. A clear correlation has indeed been found between the size of primates' cerebral cortex and the size and complexity of the social group. Primates started to live in social groups around 52 million years ago, when they abandoned their nocturnal existence and it became safer to band together. The complexity of life in groups is strongly determined by pair formation and monogamy, both of which place considerable demands on the brain. They require optimal selection of a fertile partner as well as complex negotiations between partners. The intricacy and intensity of such relationships—an issue familiar to us all—appears to have placed strong evolutionary pressure on the brain to grow. The mechanism of monogamous partner choice in humans is thought to have developed as far back as 3.5 million years ago. It has proved its evolutionary advantage in terms of protecting the family, but it continues to place an enormous burden on our brains.

THE EVOLUTION OF THE BRAIN

> We are here because one odd group of fishes had a peculiar fin anatomy that could transform into legs for terrestrial creatures; because the earth never froze entirely during an ice age; because a small and tenuous species, arising in Africa a quarter of a million years ago, has managed, so far, to survive by hook and by crook. We may yearn for a "higher" answer—but none exists.
>
> Stephen Jay Gould (1941–2002)

Humans are characterized by an amazing brain that weighs three pounds and is made up of cells known as neurons. We each have around 100 billion of them—fifteen times the number of people on earth. Each brain cell makes contact with around ten thousand other brain cells through specialized connections called synapses. Our brains contain over sixty thousand miles of nerve fibers. Yet the fundamental characteristics of the neuron, like the ability to receive, conduct, process, and transmit impulses, aren't inherently specific to nervous tissue. These functions (along with rudimentary forms of memory and attention) are also found in many other types of tissue in all living creatures, even single-celled organisms. But, as Cornelius Ariëns Kappers (who back in 1930 became the first director of what is now the Netherlands Institute for Neuroscience) observed, the nervous system has become vastly better at these functions as a result of evolutionary specialization. Whereas impulse speed in tissue other than the nervous system rarely exceeds 0.1 cm per second, the simplest neuron can transmit impulses at 0.1 to 0.5 meters per second. In fact, as Kappers calculated, our neurons can even reach conductivity speeds of 100 meters per second. And that's only one of the specialized characteristics of the neuron that provided a huge evolutionary advantage.

Sponges, the most primitive creatures, have only a few types of cells and lack both specialized organs and a true nervous system. But they do possess the precursors to neurons, and their DNA does have almost all of the genes it needs to build the proteins that are located in the postsynaptic membrane, the site of the receptor molecules between neurons. This shows how only a few small evolutionary adaptations are needed to create an entirely new system for the transfer of chemical messengers.

Primitive neurons developed as far back as the Precambrian era, between 650 and 543 million years ago. By then, coelenterates (aquatic organisms) already possessed a diffuse neural network with true neurons and synapses. We can trace the gradual molecular evolution of the chemical messengers used by these neurons to those found in our brains today. One of the most studied organisms in this context is the tiny polyp *Hydra,* which possesses only a hundred thousand cells. Its neural network is concentrated in its head and foot: a first evolutionary step toward developing a brain and spinal cord. *Hydra*'s nervous system contains a chemical messenger—a minuscule protein—that resembles two of our own: vasopressin and oxytocin. A protein of this kind is called a neuropeptide. In vertebrates, the gene for this particular neuropeptide first doubled and then mutated in two places, creating the two closely related but specialized neuropeptides vasopressin and oxytocin, which have recently become the focus of interest, partly because of their important role as messengers in our social brains (see chapter 9). Depending on their place of production, release, and reception, these two messengers can also be involved in kidney function (chapter 5), childbirth and milk secretion (chapter 1), day and night rhythms (later in this chapter), stress (chapter 5), love (chapter 4), erection (chapter 4), trust, pain, and obesity (chapter 5). By 2001, the Hydra Peptide Project had already isolated and chemically identified 823 peptides. These included neuropeptides that were subsequently found for the first time in vertebrates, like *Hydra*'s "head-activating peptide," which is also present in humans in the hypothalamus, the placenta, and brain tumors.

The chemical relationship between species is extremely close. An evolutionary basis for a rudimentary brain can be found in flatworms, in the form of a clump of neurons known as the head ganglion. The small, gradual structural and molecular changes that take place during the evolution of the brain show that the unique place often claimed for man in the animal kingdom needs to be put in perspective. As Darwin said in *The Descent of Man and Selection in Relation to Sex* (1871): "No one, I presume, doubts that the large proportion which the size of man's brain bears to his body, compared to the same proportion in the gorilla or orang[utan], is closely connected with his mental powers." And he hit the nail on the head there: The size of our brain is an extremely important factor in determining intelligence, but it isn't the only one. Tiny molecular differences have also had a huge impact.

MOLECULAR EVOLUTION

> How could it be that a not particularly bright young son of the English gentry managed to come up with the most important idea in the whole of human history?
>
> Midas Dekkers on Charles Darwin, *De Volkskrant,* January 2, 2010

In recent years, adherents of the Intelligent Design movement—some of them Dutch—have made desperate and completely futile attempts to undermine Darwin's theory of evolution. Of course, denying evolution isn't against the law, but to publicly deny the truth of scientific findings, as this movement does, is evidence of double standards: Blasphemy is still a crime in the Netherlands, but blaspheming Darwin isn't. Some Intelligent Design campaigners have sought to deny the considerable contribution of molecular biology to our understanding of evolution. In Cees Dekker's book on Intelligent Design (2005), the physicist Arie van den Beukel claims, "It's often said that the findings of molecular biology over the past few decades provide conclusive support for Darwin's theory. Nothing could be further from the case." I shall give a few examples to show just how nonsensical this sweeping assertion is.

It's scarcely credible that in 1859 Darwin was able, without any of the advanced molecular knowledge we have today, to theorize that life originated from a single primeval ancestor. Darwin could not have known that all living tissue is so *chemically* similar. Molecular biologists have fairly recently been able to provide this visionary notion with a firm foundation. Evolution can, for instance, be traced in DNA through gradual molecular changes in the genes that code for proteins, through the doubling of genes and the consequent formation of new functional genes, through the disappearance of genes and, lastly, through evolutionary mutations in those parts of RNA that don't code for proteins but that importantly regulate cellular functions. Molecular research is constantly generating new knowledge and theories about the course of evolution and its mechanisms. The genes of the nervous system are a case in point. Given the close molecular similarities in their nervous systems, worms, insects, and vertebrates—from fish to humans—must have had a single common ancestor who lived 600 million years ago. Take the tiny ragworm *Platynereis dumerilii,* which is considered to be a living fossil. Its embryonic development has been shown to proceed along the same molecular lines as mammals'.

Darwin would no doubt have greatly appreciated the molecular research done on the mitochondrial DNA of the famous finches that he discovered on the Galapagos Islands during the voyage of the *Beagle*. It showed that the thirteen species indeed had a common ancestor, as he suspected. That ancestor must have migrated from the South American continent to the Galapagos Islands around 2.3 million years ago. Molecular evidence has also been found for Darwin's belief that the ancestors of man originated in Africa: Both the "maternal" mitochondrial DNA and the "paternal" Y chromosomal DNA have been traced back to that continent. Darwin's theory of human migration from Africa through Europe and Asia has also been proved correct. It's now known that there were two waves of human migration out of Africa. The first, of *Homo erectus,* between 2 million and 1.6 million years ago, and the second, of *Homo sapiens* (modern man) around 50,000 to 60,000 years ago. The lack of genetic variation between populations outside Africa shows that only a few dozen *Homo sapiens* individuals originally migrated from Africa. Sexual intercourse between the two species of hominids in various regions of the world caused *Homo erectus* to be assimilated into *Homo sapiens.*

A recent field of study focuses on the molecular-genetic mutations that resulted in the emergence of humankind in the three hundred thousand generations following the split from chimpanzees. It's often said that the human and chimpanzee genomes differ by only 35 million or so DNA building blocks, or a mere 1 percent. (That figure has become something of a popular myth; the difference is more like 6 percent.) However, this considerable similarity is misleading, since only a few genes would have been needed for the tripling of our brain weight since we separated from the chimpanzees—as various findings now indicate. One of the characteristic differences between human and chimpanzee brains is that in our case, the genes involved in metabolic activity in the brain are much more strongly expressed—a difference for which only a couple of genes (transcription factors) are responsible. Efforts to identify the instrumental factors in humankind's development are lending weight to the "few genes" argument. These studies involve looking for the genes whose mutation can lead to undersized brains (microcephaly) and mental retardation in humans. The brains of people born with primary microcephaly, an inheritable condition, are just as small as those of the great apes, while their general structure remains intact. Such individuals have a normal appearance and show no neurological deviations. This developmental disorder can be localized in at least six different places in the DNA. All of the genes that have been identified are involved in cell division, making their contribution to increased brain size over the course of evolution plausible. One is the ASPM gene, the mutation of whose DNA building blocks accelerated after the split between humans and chimpanzees, around 5.5 million years ago. The theory has also been put forward that the human brain is still evolving, on the grounds that a genetic variant of ASPM is thought to have originated only 5,800 years ago and then spread rapidly through the population. A genetic variant of the microcephalin gene (D allele of MCPH1), which regulates brain size, is thought to have only entered the DNA of *Homo sapiens* during the last ice age, around 37,000 years ago—yet 70 percent of the current world population carries this variant. A rapid increase of this kind is only possible if a variant confers a clear evolutionary advantage.

Genes whose mutations are associated with human language have also been found. Mutations of the FOXP2 gene cause language and speech disorders that run in families. And ASPM and microcephalin also appear to have a linguistic connection.

In the course of evolution, new functional genes have also come into being. The best example is the gene that allows primates to see three colors. As a result of the doubling of the gene-produced pigment (opsin) that is sensitive to green, followed by mutation and selection, primates developed the red opsin. This conveyed the evolutionary advantage of being able to distinguish ripe fruit from unripe fruit. Humans are still programmed to find red exciting, whereas the dominant color in nature, green, has a calming effect, even in the case of placebos (see chapter 16). (That's why operating theaters are painted green.)

Genes have also been lost over time. Mice possess 1,200 olfactory receptor genes, while humans have only 350 left. The loss of a particular gene, MYH16, may have indirectly affected human brain size. This gene was responsible for the massive jaw muscles of our ancestors. Its loss is thought to have allowed skull size to increase in order to accommodate our larger brains.

Another strategy for identifying the genes that have crucially influenced the development of the brain entails the mapping of the entire genomes of various precursors in man's evolution. At the Max Planck Institute for Evolutionary Anthropology in Leipzig, the Swedish biologist Svante Pääbo is currently sequencing all the base pairs in the genome of Neanderthals, who died out 30,000 years ago. He has extracted DNA from three fossil bones of Neanderthal women who lived 38,000 to 44,000 years ago, inventing techniques to distinguish between the greatly fragmented Neanderthal DNA and contaminations caused by bacteria and modern humans. Within a few years he hopes to be able to compare Neanderthal DNA in its entirety to that of modern man and thus to identify the genetic mutations that enabled us to make such rapid strides in our evolution. Now that 60 percent of Neanderthal DNA has been mapped, the first surprising findings have already emerged. Europeans, Chinese, and Papuans bear traces of sexual contact with Neanderthals that must have taken place in the Middle East between 50,000 and 80,000 years ago. Between 1 and 4 percent of this group's DNA derives from Neanderthals. (Africans, by contrast, share no genetic material with Neanderthals.) This link makes one wonder what characteristics we have inherited from our Neanderthal ancestors. Until now, fifty-one genes that developed very rapidly after the split between *Homo sapiens* and Neanderthals have been found. Many differences have also been found in the parts of DNA that code for RNA and have regulatory functions (see below), and seventy-eight genes that are identical in all modern humans but differ in Neanderthals have been found. The differences affect a relatively large number of genes that relate to the brain and could therefore provide future insights about the emergence of the unique characteristics of modern humans.

As far as the 6 percent difference between human and chimpanzee DNA is concerned, it's important to remember that extremely small changes in genes, known as polymorphisms, can completely alter the structure of a protein and thus its function. A single gene can also produce many different proteins. A member of my research team, Tatjana Ishunina, discovered that our brains contain over forty variants of the estrogen receptor alpha, one of the proteins that receives the estrogen message. The production of these variants is influenced by age, brain area, cell type, and pathology. It has also recently been shown that that there is no need to focus so strongly on the genes that code for proteins when charting the evolution of the brain, because 98 percent of the genome doesn't code for proteins but only for RNA, and micro-RNA is thought to have been especially influential in the expansion of the human brain. Pieces of RNA regulate many cellular processes, and there are often great differences between humans and chimpanzees in this respect. As of now, the main difference has been found in the HAR1 (human accelerated region 1), a segment of a recently discovered RNA gene. The RNA that is expressed in early development (HAR1F) is specific to the reelin-producing Cajal-Retzius cells in the brain. HAR1F comes to expression together with reelin in the seventeenth to nineteenth weeks of fetal development, a crucial stage in the formation of the six-layered cerebral cortex. The mutations in this human gene are probably over a million years old and could have played a crucial role in the emergence of modern humankind.

Throughout our evolution, an enormous amount of junk and repetition has piled up in our DNA. These scars of our evolutionary history contain important information about our genesis but can hardly be seen as an argument for an Intelligent Designer and even less as evidence for regarding DNA as "God's language." Nothing has changed since Darwin concluded in 1871 that the key principle of evolution was undeniable, at least if you didn't look at natural phenomena with the eyes of a savage. Over 130 years later, adherents of Intelligent Design occupy a lonely place among the few remaining "savages" who deny evolution.

WHY A WEEK?

> Did we get the week from the Bible, or does the biblical week derive from our biological rhythm?

This book is based on a series of columns on the brain that I was asked to write for the Dutch newspaper *NRC Handelsblad* in response to readers' questions. One of the questions was: Why are societies all over the world structured around the week?

According to the Bible, God created the earth in six days and rested on the seventh. Your first thought might be that it wouldn't have hurt to devote an extra day to the creation of humans. But it also makes you wonder whether we have seven-day weeks because, according to the Bible, the creation took seven days or whether it's the other way around, and the creation story owes its seven-day structure to our having a seven-day biological rhythm.

All living things, from unicellular organisms to humans, have biological rhythms instilled over millions of years of evolution, enabling them to cope with the regular changes that affect our planet. The biological clock in our hypothalamus has a rhythm of approximately twenty-four hours; it warns us that it will soon grow dark and that it's time to return to the safety of the cave. As night ends, this clock prepares our bodies again for the activities that will start a few hours later by increasing the levels of the stress hormone cortisol. The day-night rhythm of our biological clock reflects the revolution of the earth. It also has an annual rhythm based on the Earth's rotation around the sun. The annual rhythm helped us estimate when we needed to sow, harvest, or prepare for winter. We have also internalized the rhythm of the moon, as shown by the female menstrual cycle.

Might we have a biological clock with a weekly rhythm? Weekly rhythms can indeed be seen in concentrations of substances in our blood and urine. Blood pressure, the incidence of heart attacks, brain infarcts, suicides, and births also fluctuate according to a weekly pattern. But these weekly fluctuations might reflect the rhythm of our social activities rather than an internal weekly rhythm. A scientist who measured hormone levels in his own urine for fifteen (!) years has plausibly demonstrated a biological weekly rhythm. Over a three-year period, the fluctuations showed patterns of around a week but weren't synchronous with the working week. A "free-running" rhythm of this kind points to the existence of a biological clock that measures time in roughly weekly installments. An experiment in which people spent one hundred days in a cave without time markers showed that they kept to a seven-day cycle. Studies of an insect, *Folsomia candida*, revealed that it stuck to its weekly egg-laying patterns even when kept in constant darkness. The last two examples seem to undermine the theory that our weekly biological fluctuations are caused by social factors. However, the strongest argument for a biological weekly rhythm is provided by fossils of human precursors discovered in East Africa. Two types of microscopic growth lines were found on the dental enamel of skull number 1500, "Turkana Boy": one with a twenty-four-hour periodicity and one with a periodicity of around seven days. In other words, the dental enamel of these hominids developed at a fixed rate for a six-day period and at a different rate on the seventh day. The same has been shown to apply to other primates. It's been suggested that this weekly periodicity might be accounted for by a fluctuation in the sun's radiation, but eminent astronomers oppose this theory. It's much more likely that our weekly rhythm dates back to the evolutionary period in which organisms made the transition from sea to land and started looking for food on beaches. The weekly fluctuations in tides caused by the gravitational pull of the sun and the moon would have had huge consequences for coastal foragers in terms of both food availability and type. Whatever the basis of that rhythm, a biological week existed 3.6 to 3.8 million years before the Bible was written and millions of years before our social week came into being. This biological rhythm may also have given rise to the idea that God completed his task of creation in a week and not, say, eight or nine days, let alone 4.5 billion years.

21

Conclusions

Altering the course of rivers and moving mountains is easy.
Changing someone's character is impossible.

Chinese saying

We come into the world with brains that are already unique, shaped by a combination of our genetic background and programming in the womb, and in which our characteristics, talents, and limitations are to a large extent already determined. This applies not just to things like IQ, being an early bird or a night owl, the extent to which we're spiritual, neurotic, psychotic, aggressive, antisocial, or nonconformist, but also to the risk we run of developing brain disorders like schizophrenia, autism, depression, and addiction. By adulthood there is very little about our brains that can be modified, and our characteristics have been established. This structuring of our brains determines their function: *We are our brains*.

Our genetic background and a host of factors that shaped our early brain development limit us in many ways. We aren't free to decide to change our gender identity, sexual orientation, level of aggression, character, religion, or mother tongue. That we lack free will isn't a new concept, and I find myself in the best of company here. Spinoza already demonstrated this with a few examples (*Ethics III*, proposition 2): "An infant believes that of its own free will it desires milk, a hot-headed youth believes he freely desires vengeance, a coward believes he freely desires to run away; a delirious man, a garrulous woman, a child . . . believe that they speak from the free decision of their mind, when they are in reality unable to restrain their impulse to talk." Certain qualities simply can't be changed. Charles Darwin came to the same conclusion in his autobiography, theorizing that "education and environment produce only a small effect on the mind of any one, and that most of our qualities are innate."

However, this view is at odds with the belief in social engineering that held sway in the 1960s and 1970s. Gender-based behavioral differences were blamed on the domineering patriarchal society, and the fact that women were twice as likely to become depressed was thought to be because their lives were harder. The thinking was that since society was causing these problems, something could be done to solve them. But the belief in progress and the importance attached to the social environment had its darker side. When things went wrong, they were attributed to upbringing and especially to the mother's role. A dominant mother would make her son homosexual, a cold mother would have an autistic child, and a mother who sent out mixed messages resulted in a schizophrenic child that needed to be "rescued from the claws of its damaging family." Transsexuals were considered psychotic, criminality resulted from having the wrong friends, stick-thin models were thought to cause epidemics of anorexia, and abuse or abandonment was said to result in borderline personality disorder. Nowadays very little is left of these credos.

That our characteristics, potential, and limitations are to a great extent determined in the womb doesn't, of course, mean that our brains are "complete" when we're born. A baby's brain continues to develop if provided with affectionate, secure, and stimulating surroundings. It's shaped by a constant process of learning as well as by native language acquisition and the imprint of religious beliefs. And, just as in the womb, the issue isn't about brain versus environment but about the strong interaction between the two. Crucially, however, the earlier that environmental influences come into play, the stronger and more lasting their effects will be. And the further a child's development progresses, the less scope there is for characteristics still to be programmed in the brain. Our character—that is, our innate qualities—emerges more and more strongly in the course of early development. Of course, what we learn is stored in our memory systems, which provide some ongoing plasticity. Moreover, after early development, society can influence our behavior, but our character can't be changed. And though behavioral changes—often achieved only with considerable difficulty by clinical psychologists and psychiatrists—can help people with personality disorders cope better with the character-related problems that emerged during early development, they don't eliminate such problems. It's not for nothing that the word *character* comes from a Greek word meaning "engraved," like the symbol stamped on a coin.

CONGENITAL ISN'T THE SAME AS HEREDITARY

Congenital isn't the same as hereditary. The latter term refers to inherited characteristics: From the moment that our father's and mother's genes were shuffled, we were permanently dealt a significant part of our character, IQ, and likelihood of brain disorders. But from the moment of conception, the uterine environment crucially influences brain development. The combination of inherited characteristics and intrauterine effects on brain development gives rise to congenital properties.

It's only very rarely that we can do anything about the genetic burden handed down to a child. In the case of Down syndrome and other chromosomal abnormalities, there's the option of prenatal diagnosis and, if a pathological gene has been found, abortion. Occasionally in vitro fertilization is a possibility, allowing the selection of an embryo that doesn't have the defect in question, which can then be replaced in the uterus. This can, for instance, be done in the case of early-onset or familial Alzheimer's. Heel pricks are used to test newborn babies for various genetic diseases, treatment of which can prevent damage to the developing brain. One such condition is congenital adrenal hyperplasia (CAH), which prevents the adrenal gland from producing cortisol, leading to overproduction of testosterone. This has the potential not only to disrupt the sexual differentiation of the brain but also to make the child acutely ill. Other disorders scanned for include congenital hyperthyroidism, a condition in which thyroid hormone deficiency impairs brain development, and phenylketonuria, a metabolic disease that can damage the brain but can be treated through a special diet. It isn't yet possible to repair genetic disorders of the brain using molecular technology.

Environmental factors are crucial to brain development, but contrary to what was thought in the 1960s and 1970s, it isn't so much the social environment after birth as the chemical environment before birth that's most influential—and the earlier the stage of development, the greater its impact. Very considerable health gains can be achieved during early development that will have lasting and significant effects on the rest of the child's life. To this end, pregnant women should take medication as sparingly as possible and avoid exposure to other chemical substances that can affect the development of their child's brain. They also need proper nutrition and sufficient iodine to ensure that both their own and their baby's thyroid hormones function properly. We know from experience of the Dutch famine ("Hunger Winter") of 1944–1945 that intrauterine malnutrition increases the risk of schizophrenia, depression, antisocial personality disorders, addiction, and obesity. Since my own intrauterine residence coincided with that same famine, I was fortunate that my pregnant mother's friends and acquaintances gave her some extra food. Heaven only knows where they got ahold of it. In fact, my mother produced enough milk even to feed another baby—a Jewish infant being secretly sheltered somewhere—whose identity we never discovered. The consignments of milk were delivered by a chain of female couriers. In today's world there are still two hundred million children who aren't as fortunate and who are caught in a vicious cycle as a result. Malnutrition in the womb disrupts the functioning of their brains, preventing them, when they reach adulthood, from providing the next generation with sufficient food and an optimal start in life. The only way this cycle can be broken is by improving the distribution of the world's food supply. There are also still many regions of the world with an iodine shortage, which leads to fetal thyroid malfunction, resulting in impaired brain development and mental disability. In principle, this problem can easily be solved with iodized salt, if a structural supply could be ensured for these regions.

FUNCTIONAL TERATOLOGY

The lasting effects of chemical substances on early brain development are thought to contribute substantially to psychological and psychiatric problems later in life. These disorders don't emerge until demands are made on the brain systems that were altered during early development. A newborn baby can appear healthy but later prove to have learning disorders because it was exposed to alcohol, cocaine, lead, marijuana, DDT, or antiepileptic drugs before birth. Children run a greater risk of depression, phobias, and other psychiatric problems if their mothers took DES or smoked during pregnancy and a greater risk of transsexuality if their mothers took drugs like phenobarbital or diphantoine. Chemical substances are also thought to contribute to developmental disorders caused by a mix of factors, like schizophrenia, autism, SIDS, and ADHD.

Sometimes achieving health gains seems potentially easy. Twenty-five percent of pregnant women still have the occasional glass of alcohol, and 8 percent smoke. If all pregnant women in the Netherlands were to stop smoking, very premature births would be cut by a third, the number of babies with low birth weight would be greatly reduced, and health costs would be cut by many millions. Add to that the gains resulting from the reduced incidence of

ADHD, which is linked to impulsive and violent behavior and juvenile delinquency, and you wonder why pregnant women still smoke. But theory is one thing and practice another. Achieving behavioral changes is extremely difficult, especially when addictive substances are involved. And nicotine patches aren't without risks to the unborn child either.

A lot of unnecessary medication is taken during pregnancy. Sometimes doctors prescribe drugs under pressure from the patient; sometimes they are passed on by a friend or neighbor. Even over-the-counter drugs like aspirin and acetaminophen can affect the fetus. The thalidomide tragedy raised awareness of the potentially damaging effect of chemical substances on unborn children, but it also left doctors inclined to think that the risk is confined to the first three months of pregnancy. This isn't the case; such substances continue to affect brain development right up to the moment of birth. Take the established practice of administering large quantities of adrenal gland hormones over a prolonged period to promote lung development both in premature babies and babies at risk of being born prematurely. In my inaugural speech thirty years ago I warned against this, having found from animal studies that these substances do help mature the lungs but at the same time impede brain development. Indeed, learning and behavioral disorders, smaller brains, a lower IQ, motor impairments, and an increased risk of ADHD have been found among children who were exposed to large amounts of these hormones. These days they are administered much more sparingly.

Sometimes, though, pregnant woman need treatment for conditions like epilepsy or depression. It's essential that doctors address these problems early on, so that if a patient plans to become pregnant, the safest drug can be prescribed. It would also be beneficial, if at all possible, to consider alternative therapies for pregnant women with depression, like light therapy, transcranial magnetic stimulation, or even placebos—especially since there are justifiable doubts about the effectiveness of antidepressants, while placebos are turning out to be surprisingly effective.

SEXUAL DIFFERENTIATION OF THE BRAIN

There can be little doubt that our gender identity and sexual orientation are programmed for the rest of our lives while we're still in the womb. Our sex organs differentiate in the first months of pregnancy, while sexual differentiation of the brain takes place in the second half of pregnancy. Since these processes are separate, it's impossible to determine, in those rare cases when children are born with indeterminate gender, whether the brain has developed along male or female lines. In the past, doctors have often been far too quick to operate. They would "make" the baby a girl, to establish clarity for the parents and child. We now know from patient associations that imposed gender identity frequently causes problems in later life. In cases of doubt regarding the sexual differentiation of the brain, it's better to assign a provisional sex until the child's behavior makes its gender identity clear. Some operations can be carried out in such a way as to be reversible.

Since our gender identity is determined so early on in development, it isn't necessary to defer a sex change until an advanced stage of adulthood to be sure that an individual really wants such a change. On the contrary, an early sex change has many advantages. First, it's much better for someone to get used to their new gender before they have finished their education and settled into a career and a relationship. It's of course also easier to turn a man into a convincing woman before he has grown into a hulking six-footer with broad shoulders and a deep voice.

The long-established idea that we're completely free to choose our sexual orientation not only is wrong but has also caused a great deal of misery. It has led to people being punished for homosexuality, which all religions castigate as a sin—that is, as a wrong choice. There's nothing optional about our sexuality, which is programmed in the womb. So it's ridiculous to persist in attempts to "convert" homosexuals into heterosexuals, a practice that still goes on in countries like the United States and Britain.

Pedophilia is also programmed into the brain at an early stage of development. The risk of pedophiles harming children can be reduced by teaching them how to control their impulses better. In addition to cognitive therapy, anti-androgens can sometimes also help. Providing sex offenders with a social network after they have been released from prison strongly reduces the risk of reoffending. Conversely, hounding and isolating pedophiles can have disastrous consequences.

THE FETAL BRAIN AND BIRTH

The signal for the onset of labor comes from the fetus, when it registers that the mother can no longer provide its growing body with sufficient food. A number of psychiatric disorders in adulthood turn out to be heralded by problems around the time of birth. It was long thought that schizophrenia resulted from brain damage caused by prolonged and difficult labor. We now know it to be an early developmental brain disorder that's largely genetic in origin. Intense interaction between the brains of both mother and child is necessary for labor to proceed well. So the difficult labor associated with babies that go on to develop schizophrenia could be seen as an early sign of the disorder. It's a counterintuitive conclusion, given that the typical symptoms of schizophrenia usually only manifest themselves in adolescence. (Though if one asks parents when they noticed that their child was different from other children, they often say, "Oh, he's always been different.") The same connection applies between a difficult birth and the subsequent emergence of autism, another early developmental brain disorder. A link has also recently been established between problems at birth and anorexia or bulimia. Again, these problems could be seen as the first symptom of a malfunction of the hypothalamus that later takes the form of an eating disorder. Consequently, a close eye should be kept on children whose birth was difficult to see if they develop a psychiatric disorder in early adulthood. Therapy in the early stages of schizophrenia has proved beneficial in preventing brain damage, and the same might apply to other developmental brain disorders.

THE IMPORTANCE OF GOOD POSTNATAL DEVELOPMENT

Stimulation during the earliest phase of life in surroundings that the child perceives as safe and familiar is crucial for optimum brain development. Neglect or abuse of young children can permanently impair such development and increase stress axis activity. It then only takes a relatively small problem in the child's surroundings for its stress axis to become hyperactive and cause depression. It's crucial that youth services act promptly to help children growing up in such circumstances, and they need to be much better organized than at present. A child's ability to bond also passes through a critical stage in which oxytocin, the "bonding hormone," plays an important role. Oxytocin levels are lowered for a long period, possibly even permanently, when children are deprived of parents or foster parents for too long. So the transfer from a children's home to a foster home should take place as early as possible to ensure that a child can still bond properly with its foster parents. A child's brain needs a stimulating and enriched environment to develop optimally after birth, and the linguistic environment plays a key role. Language stimulates many brain areas in a way that is culturally specific. As a result, Chinese or Japanese brains are configured differently than Western brains, seemingly without genetic factors being involved. Stimulation is perhaps even more important in the case of children with mental disability, as the effect on their developing brains can make a difference for the rest of their lives.

Young children have no religion; religious beliefs are imprinted in them by their parents at an early stage of development in which they accept without question anything they are told. In this way, beliefs are passed down from generation to generation and become fixed in our brains. Children should be taught not what to think but how to think critically and how to make their own ideological choices in adult life. Segregating young children in belief-based schools is pernicious—it not only prevents them from learning how to think critically but it also fosters an intolerant attitude toward other beliefs.

THE UNPRODUCTIVE: THEIR OWN FAULT?

> The variation that drives our evolution is still proving fatal to some.

Politicians have never lost faith in the misconception that social engineering can reprogram the brain. On the contrary, as of the 1980s, alarmed by recession and the top-heavy welfare state, they began to stress individual responsibility for prosperity and welfare. People were told that their fates lay in their own hands—a claim at odds with the many studies showing that people's capacities are largely determined by genetic factors and environmental influences in early development. An educational disadvantage is hard to eliminate. An innate lack of capacity can't be entirely compensated for. And more and more people are unable to meet the increasingly high demands placed on them by today's rat race society. Those whose capacities are lacking or who have mental health problems are now being unjustly blamed for their own failure. Young people with mild learning disabilities (IQ 50 to 85) are an especially vulnerable group. They make up 16 percent of the Dutch population but are strongly overrepresented among those tried for criminal offenses (accounting for 50 percent of them). They are held responsible for their own choices despite their inability to compete in the employment market. In the current repressive climate, misdemeanors that would previously have incurred a strong warning have now been criminalized. This group is easily influenced, a characteristic that often gets them into trouble. It's also the characteristic that should make it easy to help them to stay out of trouble. Unfortunately, the structures and facilities that would make this possible have fallen victim to budget cuts.

Children of poorly educated, poorly paid parents are more likely to drop out of school and end up in low-paying jobs; they also run a higher risk of poor health, criminality, addiction, gambling, and unemployment. There is a life expectancy gap of six years between the lowest- and the highest-educated. This discrepancy is largely due to lifestyle. Making tobacco and alcohol more expensive to encourage the former group to change their habits has proved ineffectual. Negative lifestyle factors are most pronounced when people live in deprived areas.

Deprivation often appears to be inherited, contagious even, since these problems accumulate in certain neighborhoods. According to the Netherlands Institute for Social Research, almost one hundred thousand Dutch children, 4 percent of the total, are socially excluded. They aren't members of a sports club, are almost never taken on outings, don't go on vacation, and hardly ever visit the homes of friends. If this definition is applied slightly less strictly, 11 percent of Dutch children can be considered to be socially excluded. This is largely because of their parents' financial situation. But their parents, too, are marginalized, and such families live in neighborhoods devoid of proper playgrounds and plagued by vandalism and antisocial behavior. These aren't circumstances in which a child's brain can develop optimally.

There's no quick fix for this problem. We can do our utmost to foster optimal development and to prevent damaging influences. But we must also accept that in a complex process like brain development things occasionally go wrong. As a result there will always be a small proportion of people who aren't properly equipped for life, who suffer from mental disability or have neurological or psychiatric problems. This can happen to any child in any family, and society must shoulder its responsibility by providing sheltered jobs, social benefits, and good, practical supervision. As things stand, there is a lot of room for improvement. Better education and public information are also needed to prevent society from laying the blame for failure at the door of those who, through no fault of their own, incurred a developmental brain disorder.

There's a similar tendency to assign blame to people struck by brain diseases in adulthood. They are often told that a positive attitude toward, say, MS, will promote healing. It's an appealing notion. However, not only is there absolutely no evidence for it, the result is that if the disease worsens, the poor patient will be blamed for not doing his best to overcome it! It's high time we abandoned the "own fault" myth.

THE BRAIN AND CRIME

Aggression levels are determined by gender (little boys are more aggressive than little girls), genetic background (tiny variations in DNA), fetal nourishment, and pre-birth exposure to smoking, drinking, or medication. The likelihood of boys displaying uninhibited, antisocial, aggressive, or delinquent behavior increases in puberty, as their testosterone levels rise. The level of violence shown by adult male criminals is also testosterone-related. So there are many factors beyond one's control that determine whether someone gets into trouble with the police or ends up in court. That doesn't mean that criminals shouldn't be punished, but criminal law should take these neurological factors into account. The development of the prefrontal cortex is a slow process, continuing at least until the age of twenty-five. It's only at that age that an individual is fully equipped to control their impulses and make moral judgments. On the basis of neurobiology, therefore, the age at which offenders are tried under adult criminal law shouldn't be reduced to, say, sixteen, as some politicians are urging in an effort to woo voters, but should rather be raised to an age at which the brain structures are mature, at around twenty-three to twenty-five. People who commit a minor offense as teenagers shouldn't have that held against them later on in life.

Some children are markedly more aggressive than others. A strikingly high incidence of psychiatric disorders is found among delinquent youths imprisoned for violent crimes—as high as 90 percent of the total group in the case of adolescent males. Genetic factors are also influential, as studies of twins have shown. The application of criminal law should be confined to individuals with healthy brains. But our criminal law system continually sins against this "M'Naghten rules" principle. Can a pedophile be held morally responsible for his sexual orientation, which he owes to his genetic background and atypical brain development? Can a child be deemed culpable for the combination of his genes and his mother's smoking during pregnancy, causing him to develop ADHD and to get into trouble with the police? We know, too, that malnourishment in the womb increases the likelihood of delinquency. And can an adolescent whose brain has just been completely reconfigured by sex hormones be considered fully responsible for committing a crime?

Moral responsibility is a tricky concept in these situations, free will an illusion. This doesn't mean that criminals shouldn't be punished, simply that punishments should be effective. In the field of medicine we have learned that to reach truly scientific conclusions, the effects of a drug or other therapy need to be ascertained in a properly controlled study. However, criminal law clings to the notion that punishment must be determined on an individual basis, rather than thinking in terms of well-defined groups. In point of fact, the criminal justice authorities are using the same arguments that psychoanalysts did in the days before it was systematically established whether a particular form of psychiatric therapy worked or not. As a result, it's impossible to determine the effectiveness of a punishment. Moreover, the criminal justice system is increasingly being pressured by politicians to devise ever more penalties, from community sentences to boot camps for young offenders, without establishing how they measure up against traditional forms of punishment. In the absence of a proper study involving a control group, the efficacy of measures will always be contentious. However, politicians have little interest in scientific methods. They only think in the short-term—particularly of the coming elections.

THE END OF LIFE

> Do you, like us, believe more in life *before* death?
>
> Dutch Humanist Association

Although all efforts in the medical field are focused on delaying the end of life, death always wins in the end. If at all possible, we want our brains to be healthy as death approaches, so that we can continue to make our own decisions in this final stage of our lives, right up to the last moment. People in the Netherlands who have cancer or some other serious illness are fortunate to have the option of euthanasia, which Dutch law considers acceptable as a means of preventing pointless suffering. If someone suffers from a brain disorder, however, assisted dying is a problem. Patients who are in a coma or similar state can no longer communicate their wishes, and patients with dementia or psychiatric disorders present doctors with the difficult task of assessing their mental competence. Views on euthanasia or assisted dying in these two latter categories are progressing, although the decision to perform euthanasia, which remains extremely difficult for doctors treating patients with chronic psychiatric disorders, is rarely made. Yet it can prevent horrible suicides. In general, Alzheimer's progresses so slowly that you have time to seek out and consult with a physician who can help you choose the right moment to end your life. But there are other forms of dementia, like vascular dementia, which can overwhelm you before you have had time to organize matters. So it's important that those close to you know your feelings on the subject, and that you make sure that your doctor will give you the assistance you want in the final stage of your life and that over time you can both prepare for this moment.

There is, however, a large and rapidly growing category of people for whom assisted dying still falls outside the scope of the legislation on euthanasia: elderly people who simply feel that their time has come. We're trying to address this problem by means of a civil society initiative whose aim is "to legalize assisted dying for elderly people who believe their time has come, under the proviso that this takes place at their express request, with all due care and in a manner that's open to scrutiny." The initiative, which was set up by Yvonne van Baarle, is called Uit Vrije Wil (By Free Will). Not wishing to complicate things unnecessarily, I refrained from mentioning to the group my view that free will is illusory until we had secured the forty thousand signatures that were needed to ensure discussion of this question by the Dutch parliament. In fact that only took four days. It would seem that people in the Netherlands are very concerned about this issue. We will see how politicians respond and how long it takes for the law to be amended.

NEW DEVELOPMENTS

When I started work as a student doctor in 1966, brain research was the domain of a few mavericks who were regarded with considerable suspicion by society at large. These days, the great social significance and huge potential of this field appear to be universally recognized, and neuroscience has become a top priority at universities and research institutes all over the world, where hundreds of thousands of scientists explore a wide range of technologies. The highly complex research techniques call for specialized scientists, who must work in multidisciplinary teams to achieve new insights. Networks are becoming ever larger and more international, as can be seen from the growing number of authors and affiliations cited at the top of publications.

In the years to come, insights into the molecular biology of brain disorders will produce new objectives for therapeutic strategies. Stimulation electrodes implanted at precise sites in the brain are being used to treat not just Parkinson's disease but also obsessive-compulsive disorders. Their effect is also being studied in such areas as minimally conscious states, obesity, addiction, and depression. As with all effective therapies, there are side effects. And these can be considerable in the case of Parkinson's patients undergoing stimulation of the subthalamic nucleus, ranging from obesity to changes in character, impulsive behavior, and even suicide. Psychosis, lack of sexual inhibition, and compulsive gambling have also been reported. Researchers are looking at the effect of transcranial magnetic stimulation on depression and tinnitus. This technique is also used to prevent hallucinations among schizophrenic patients. It's still too new for its side effects to be known. Neuroprosthetics—devices that can replace sensory systems—are becoming ever more sophisticated. One paraplegic had a plate with electrodes implanted into his cerebral cortex that allowed him to control a computer mouse and a prosthetic arm with his mind. Visual prostheses are being developed for the blind. Attempts are being made to carry out repairs in the brain and spinal cord by implanting fragments of fetal brain tissue or stem cells and by initiating gene therapy.

New discoveries are constantly being made, thanks to the huge growth in neuroscience and the technical advances made in the field. And they are vitally important, because 27 percent of Europeans suffer from one or more brain disorders. In the Netherlands, over 30 percent of the health care budget is spent on patients with brain diseases. You would expect that at least a proportionate amount of research funding would be earmarked for brain research, but in Europe only 8 percent of this budget is allocated to neuroscience. When will governments finally develop that much-needed long-term vision to ensure healthier brains for generations to come?

Acknowledgments

This book came about after the Dutch newspaper *NRC Handelsblad* asked me, in 2008, to write a column in which I answered readers' questions. I'm grateful for the help I received from Jannetje and Rinskje Koelewijn during this process. Portions of these chapters appeared in the *NRC*. I could never have written this book without being immersed in the international network of brain researchers and benefiting, in my own research group, from an amazing amount of new data and feedback supplied by a great many excellent, critical, and talented students, analysts, PhD students, postdocs, and staff members. Patty Swaab corrected all the chapters before I dared show them to anyone else and had her work cut out in doing so. I'm also indebted to the following people for suggestions and corrections: Bao Ai-Min, Els Boelens, Martijn Boelens, Kees Boer, Ruud Buijs, Wouter Buikhuisen, Hans van Dam, Marcel van Dam, Gert van Dijk, Cisca Dresselhuys, Frank van Eerdenburg, Tini Eikelboom, Michel Ferrari, Eric Fliers, Rolf Fronczek, Anton Grootegoed, Michel Hofman, Jan van Hooff, Witte Hoogendijk, Inge Huitinga, René Kahn, Bert Keizer, Felix Kreier, Jenneke Kruisbrink, Paul Lucassen, Martijn Meeter, Joris van der Post, Liesbeth Reneman, Carla Rus, Erik Scherder, Reinier Schlingemann, Eus van Someren, Roderick Swaab, Martijn Tannemaat, Unga Unmehopa, Joost Verhaagen, Wilma Verweij, Ronald Verwer, Geert de Vries, Linda de Vries, Frans de Waal, Katja Wolffenbuttel, Zhou Jiang-Ning, and many others. I enjoyed working with the staff of Uitgeverij Contact on this book, particularly Mizzi van der Pluijm, Bertram Mourits, Cindy Eijspaart, Kirsten van Ierland, Bieke van Aggelen, and Jennifer Boomkamp. I would moreover like to thank Maartje Kunen for her excellent drawings. Finally, I am extremely grateful to Jane Hedley-Prôle for her excellent English translation, the improvements to the text, and the most pleasant collaboration.

Index

Page numbers in italics refer to figures and illustrations.

ABOUT THE AUTHOR

D. F. SWAAB has been professor of neurobiology at the University of Amsterdam since 1979. From 1978 to 2005 he was director of the Netherlands Institute for Brain Research, now the Netherlands Institute for Neuroscience. In addition to heading the institute's research group on neuropsychiatric disorders, he holds three guest professorships in China and the United States. In 1985 he founded the Netherlands Brain Bank, whose tissue samples are used by researchers all over the world to increase understanding of the brain and to develop therapies for neurological and psychiatric diseases. In 2008 he was awarded the Medal of the Royal Netherlands Academy of Arts and Sciences for his contribution to neuroscience.

ABOUT THE TRANSLATOR

JANE HEDLEY-PRÔLE works at the Dutch foreign ministry and as a freelance literary translator. Her book translations include *Diaghilev: A Life* by Sjeng Scheijen (with S. J. Leinbach) and *The Fetish Room* by Rudi Rotthier. She is currently translating a book on the nature of identity.

ABOUT THE TYPE

This book was set in Dante, a typeface designed by Giovanni Mardersteig (1892–1977). Conceived as a private type for the Officina Bodoni in Verona, Italy, Dante was originally cut only for hand composition by Charles Malin, the famous Parisian punch cutter, between 1946 and 1952. Its first use was in an edition of Boccaccio's *Trattatello in laude di Dante* that appeared in 1954. The Monotype Corporation's version of Dante followed in 1957. Though modeled on the Aldine type used for Pietro Cardinal Bembo's treatise *De Aetna* in 1495, Dante is a thoroughly modern interpretation of that venerable face.